Mohamed Elzagheid
Chemical Laboratory

Also of Interest

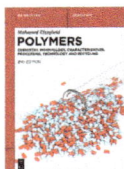

Polymers.
Chemistry, Morphology, Characterization, Processing, Technology and Recycling
Mohamed Elzagheid, 2025
ISBN 978-3-11-158565-9, e-ISBN (PDF) 978-3-11-158573-4

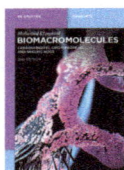

Biomacromolecules.
Carbohydrates, Lipids, Proteins and Nucleic Acids
Mohamed Elzagheid, 2025
ISBN 978-3-11-158298-6, e-ISBN (PDF) 978-3-11-158327-3

Organic Chemistry: 25 Must-Know Classes of Organic Compounds
Mohamed Elzagheid, 2024
ISBN 978-3-11-138199-2, e-ISBN 978-3-11-138275-3

Water Chemistry, Analysis and Treatment.
Pollutants, Microbial Contaminants, Water and Wastewater Treatment
Mohamed Elzagheid, 2024
ISBN 978-3-11-133242-0, e-ISBN 978-3-11-133246-8

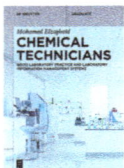

Chemical Technicians.
Good Laboratory Practice and Laboratory Information Management Systems
Mohamed Elzagheid, 2023
ISBN 978-3-11-119110-2, e-ISBN 978-3-11-119149-2

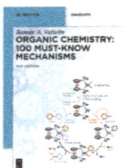

Organic Chemistry: 100 Must-Know Mechanisms
Roman Valiulin, 2023
ISBN 978-3-11-078682-8, e-ISBN 978-3-11-078683-5

Mohamed Elzagheid

Chemical Laboratory

Hazards, Safety, Good Lab Practice, Separation
and Analytical Techniques

2nd, expanded edition

DE GRUYTER

Author
Prof. Dr. Mohamed Elzagheid
Royal Commission for Jubail and Yanbu
Jubail Industrial College
Jubail Industrial City, Saudi Arabia
and
Center for Research and Strategic Studies
Libyan Authority for Scientific Research
Tripoli, Libya
melzagheid@gmail.com

ISBN 978-3-11-914383-7
e-ISBN (PDF) 978-3-11-221810-5
e-ISBN (EPUB) 978-3-11-221829-7

Library of Congress Cataloging-in-Publication Data
A CIP catalog record for this book has been applied for at the Library of Congress.

Bibliographic information published by the Deutsche Nationalbibliothek
The Deutsche Nationalbibliothek lists this publication in the Deutsche Nationalbibliografie;
detailed bibliographic data are available on the Internet at http://dnb.dnb.de.

© 2026 Walter de Gruyter GmbH, Berlin/Boston, Genthiner Straße 13, 10785 Berlin
Cover image: nikom1234/iStock/Getty Images Plus
Typesetting: Integra Software Services Pvt. Ltd.

www.degruyterbrill.com
Questions about General Product Safety Regulation:
productsafety@degruyterbrill.com

This work is dedicated to the discipline of chemical laboratory safety and techniques, where vigilance meets science, and every technique mastered becomes a step toward excellence in research and education. May this book serve as a beacon for safer, smarter, and more inspired scientific practice.

Preface

The chemical laboratory is the crucible of scientific discovery where theory meets practice, and ideas are tested, refined, or reimagined through meticulous experimentation. While the laboratory remains a place of innovation and learning, it is also a domain where risk and responsibility coexist. Understanding and managing these risks are fundamental to both scientific progress and human safety.

This book is designed to provide a comprehensive foundation for students, educators, laboratory personnel, and researchers. It offers a well-structured overview of essential safety principles, common laboratory hazards, and the best practices required to create and maintain a culture of safety and excellence. In addition, the book covers the principles and practical applications of core separation and analytical techniques that are central to chemical investigations.

The first chapters of the book focus on the identification of hazards, the use of personal protective equipment, chemical storage and waste management, emergency preparedness, and the implementation of institutional safety protocols. These chapters aim to cultivate a mindset of vigilance and responsibility among laboratory users.

Subsequent chapters introduce the reader to the fundamentals of good laboratory practice, emphasizing precision, reproducibility, and ethical conduct. The book then transitions into practical techniques, including both classical and modern methods of separation and analysis, ranging from distillation and chromatography to spectroscopy and titrimetry.

Throughout the text, real-world examples, diagrams, and step-by-step instructions are used to enhance understanding and encourage hands-on competence. Whether used as a teaching tool, a laboratory manual, or a professional reference, this book seeks to bridge the gap between theoretical knowledge and safe, effective laboratory practice.

It is my hope that this work serves not only as an educational guide but also as a contribution to the cultivation of a safety-first culture in scientific environments. By empowering readers with the knowledge and tools necessary for safe and effective experimentation, we ensure that the pursuit of discovery continues responsibly, ethically, and sustainably.

Mohamed Ibrahim Elzagheid, Chemistry Professor
Waterloo, Ontario, Canada
2026

https://doi.org/10.1515/9783112218105-202

Acknowledgments

First and foremost, I extend my deepest gratitude to my family for their unwavering support, inspiration, and love throughout my academic journey. Their constant belief in me, even during the most challenging times, has been a source of strength and determination. I am forever grateful for their patience and understanding, which have made it possible for me to achieve this milestone.

I would also like to express my sincere appreciation to my colleagues at Jubail Industrial College and the Libyan Authority for Scientific Research. Their encouragement and support have been immensely valuable. I am particularly proud to have contributed to the scientific community by helping to enrich our library with a substantial collection of textbooks, an enduring resource for students and scholars alike.

I extend my deepest gratitude to the dedicated publishing team at De Gruyter for their professionalism, commitment, and invaluable guidance throughout the publication process. My sincere appreciation also goes to Ute Skambraks, Helene Chavaroche, and Chandhini Magesh for their exceptional support and assistance.

https://doi.org/10.1515/9783112218105-203

Contents

Chapter 17
Heating, Cooling, and Reaction Techniques in the Chemical Laboratories —— 186

Chapter 18
Analytical Techniques in Chemistry Laboratories —— 197

Chapter 19
Separation Techniques in Chemical Laboratories —— 222

The Author

Professor Mohamed Elzagheid is a chemist, scientist, author, and teacher. In addition to his chemistry research, he specializes in chemical education, academic advising, curriculum innovation, vocational education, and higher education administration.

Throughout his 30-year career at Turku University in Finland, McGill University, SynPrep Inc. in Montreal, Canada, and Jubail Industrial College (JIC) in Saudi Arabia, he has been directly and indirectly involved in the education of laboratory technicians and chemists, as well as supervising numerous undergraduate and graduate students.

He has made significant contributions to various short-term and long-term training programs for Saudi enterprises, in addition to teaching a diverse variety of chemistry and chemistry-related courses at all levels. These courses include basic and advanced organic chemistry, polymer chemistry, macromolecular chemistry, biochemistry, laboratory techniques, safety in chemical laboratories, technician responsibilities, geology and petroleum geology, and water–wastewater analysis and treatment.

Dr. Elzagheid delivered training on laboratory techniques and chemical laboratory safety to industry trainees from leading companies such as ARAMCO, SADARA, SATORP, and MAADEN. He has also developed extensive expertise in Occupational Health & Safety (OHS) and Control of Substances Hazardous to Health (COSHH) management and risk assessment through over 25 years of experience in academic and industrial settings, complemented by various online certification courses.

Dr. Elzagheid is the author of nine textbooks, including *Introductory Organic Chemistry*; *Thoughts on Organic Chemistry*; *Macromolecular Chemistry: Natural and Synthetic Polymers*; *Chemical Laboratory Safety, and Techniques*; *Chemical Technicians: Good Laboratory Practice and Laboratory Information Management Systems*; *Water Chemistry, Analysis, and Treatment: Pollutants, Microbial Contaminants, Water and Wastewater Treatment*; *Organic Chemistry: 25 Must-Know Classes of Organic Compounds*; *Biomacromolecules: Carbohydrates, Lipids, Proteins, and Nucleic Acids*; and *Polymers: Chemistry, Morphology, Characterization, Processing, Technology, and Recycling*.

His work at Turku University in Finland, McGill University in Canada, and JIC in the Kingdom of Saudi Arabia has helped him build a solid reputation in both chemistry and chemical education, as demonstrated by his research papers and publications.

https://doi.org/10.1515/9783112218105-205

Chapter 1
Introduction to Chemical Laboratory Hazards, Safety, and Techniques

1.1 Overview

The chemical laboratory is not only a place of discovery and learning, but also one with inherent risks. Understanding laboratory hazards, practicing safety, and mastering foundational techniques are critical to effective and responsible scientific work. This chapter introduces key concepts and best practices for laboratory safety, good laboratory conduct, and fundamental analytical and separation techniques.

1.2 Key Definitions

The following definitions provide foundational terms referenced throughout this chapter:
- Safety. Defined as protection or being away from danger, risk, or injury. Laboratory safety is not only the prevention of accidents at work, but also a developed system to prevent accidents by proactively identifying possible causes.
- Hazard. The term "hazard" refers to anything that can cause harm. This could include illness, injury, or physical harm of any kind.
- Risk. A likelihood that the hazard will cause actual harm. The level of risk depends on the way the chemicals are used or managed in the laboratory.
- Incident. A work-related occurrence by or during which injury, illness, or fatality happened; or injury, illness, or fatality could have happened. This unplanned event could have the potential to lead to an accident.
- Accident. An incident arising from carrying out work that results in personal injury.
- Dangerous Occurrence. An occurrence arising from work activities in a chemical laboratory that results in a hazardous situation. It could be a chemical spill, a fire involving any substance, or an unintentional explosion.
- Chemical Laboratory Hazards. Potential risks are present in the laboratory environment due to the use of hazardous chemicals, equipment, and procedures. These hazards may be physical (e.g., fire and explosion), chemical (e.g., toxic or corrosive substances), or biological (e.g., pathogens).
- Laboratories Safety. Procedures that are followed to prevent or lessen exposure to hazardous chemicals that might harm people, property, or the environment. This covers waste management, labeling, storage, safe chemical handling, personal protective equipment (PPE) usage, and appropriate training.

https://doi.org/10.1515/9783112218105-001

- GLP stands for Good Laboratory Practice. A quality protocol that concentrates on the organizational processes that control the design, execution, monitoring, recording, and reporting of laboratory studies.
- Dangerous Substances. Substances that endanger the environment, human health, or safety because they are corrosive, reactive, flammable, or dangerous.
- Safety Data Sheet. A document that offers comprehensive details on the characteristics, risks, safe handling, and emergency protocols of chemicals.
- Risk Assessment. A laboratory method of detecting possible hazards, determining the likelihood and severity of injury, and adopting risk-reduction measures.
- Personal Protective Equipment. Items used by laboratory personnel to protect against chemical, physical, or biological hazards. PPE may include gloves, laboratory coats, goggles, face shields, and respirators.
- Separation Techniques. Methods used to separate mixtures into their components, typically based on differences in physical or chemical properties. These include distillation, chromatography, extraction, and filtration.
- Filtration. A separation process that removes solid particles from liquids or gases using a porous medium.
- Chromatography. An analytical technique for isolating, identifying, and quantifying components in a mixture using their mobility across a stationary phase with a mobile phase.
- Distillation. A method for identifying liquid substances according to their boiling points. This method is frequently applied to separate volatile chemicals and purify liquids.
- Titration. An analytical technique that involves reacting a solute with a standard solution of known concentration to determine the solute's concentration in a solution.
- Gravimetric Analysis. An analytical method that uses the mass of a chemical to determine how much of an analyte is in a sample.
- Laboratory Information Management System. A software-based system used to manage samples, laboratory workflows, and associated data to improve efficiency, traceability, and compliance.
- Standard Operating Procedure. A precise, written instruction intended to ensure that a certain laboratory operation is performed consistently.
- Emergency Response Plan. A set of procedures to be followed in the event of an accident, chemical spill, fire, or other emergencies to ensure safety and minimize damage.

1.3 Laboratory Hazards Safety and Risk Assessment

1.3.1 Safety Culture

Safety culture is a behavior-based approach that promotes, regulates, and rewards safety. A safety culture maturity ladder (Figure 1.1A) represents an overall approach to managing safety. The ladder consists of five levels. Starting at level 1, where production is emphasized, people disregard the rules and managers are not visible, level 2 managers set standards and supervisors follow up to make sure that everyone fully complies with the set rules.

Level 3, where everyone gets involved, including managers and supervisors, in the level 4 environment, health, and safety becomes an integral part of everyday business, and reaching level 5, where everyone demonstrates excellent behaviors. In this approach, the negative, directive, engaging, embedding, and demonstrated terms are acceptable and can form a customized or adapted framework for describing levels of health and safety culture maturity, particularly in educational or organizational contexts. These terms can be readily mapped to widely recognized models, such as the Hudson or Keil Centre maturity ladders (Table 1.1).

Figure 1.1A: Health and safety culture maturity ladder.

Table 1.1: The newly adapted terms can be mapped to widely recognized models such as the Hudson or Keil Centre safety culture maturity ladders.

New terms	Hudson model	Keil Centre maturity model	Description
Negative	Pathological	Emerging	Safety is ignored or only addressed after accidents.
Directive	Reactive	Managing	Rules are enforced from the top; compliance is the focus.
Engaging	Calculative	Involving	Employees are increasingly involved and aware.
Embedding	Proactive	Cooperating	Safety is integrated into systems and practices.
Demonstrated	Generative	Continually improving	Safety is a core value, consistently visible in behavior and leadership.

The five levels of the health and safety culture maturity ladder can also be described using an alternative framework, such as the Hudson model (Figure 1.1B):
- Pathological. Safety is not a top concern; the corporation does not care unless compelled to.
- Reactive. Safety is vital, but only after an accident occurs.
- Calculative. Systems are in place, and data is used, but safety is more of a checklist.
- Proactive. The organization actively seeks to avoid accidents and values safety feedback.
- Generative or Progressive. Safety is deeply integrated into the culture.

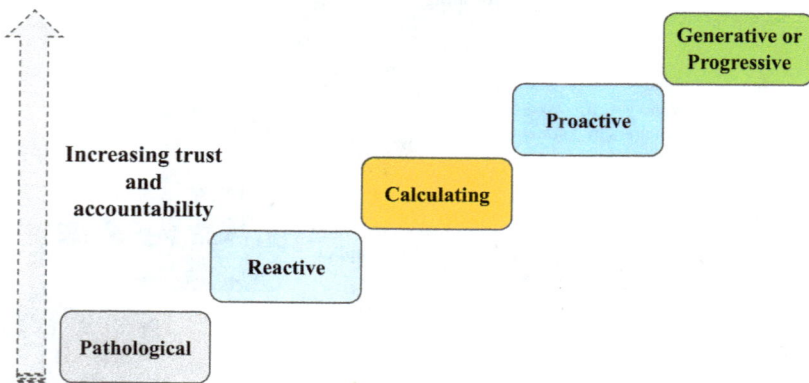

Figure 1.1B: Hudson safety culture maturity model.

Another valuable alternative approach is provided by the Keil Centre's safety culture maturity model (Figure 1.1C), which typically comprises five progressive levels:

- Emerging. Safety is viewed as a box to check or a legal requirement. Reaction to events is frequently blame-oriented and lacks proactive safety behavior.
- Managing. There are basic safety procedures and systems in place. Although they prioritize regulations and compliance, management demonstrates a commitment to safety.
- Involving. Workers begin to participate in safety programs. It is acknowledged that input from all areas of the organization is necessary for safety improvement.
- Cooperating. All share responsibilities for safety, strong teamwork, and communication between management and employees.
- Continually Improving. One of the fundamental values ingrained in the culture is safety. Performance improvement, innovation, and ongoing learning are essential.

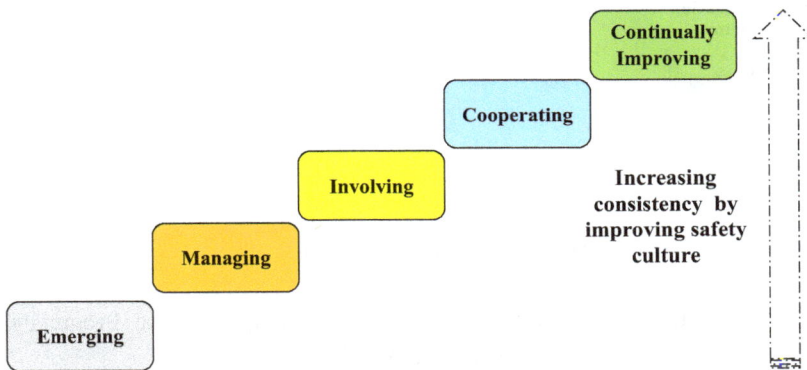

Figure 1.1C: The Keil Center's safety culture maturity model.

Safety culture isn't a collection of rules or processes. It is a mindset, lifestyle, and habit. It is also a practice that is introduced, maintained, supported, and followed by everyone. A successful safety culture should contain the following features (Figure 1.2):
- Provide high-quality, standardized tools and equipment for all jobs and procedures.
- Established and consistently adhered to clear and simple operational procedures.
- Implemented company-wide communication systems to gather, analyze, and share safety-related information and incident data. In addition, there is a framework in place to promote safety improvement proposals.
- Comprehensive training program with monthly refresher training.

1.3.2 Safety cycle

In institutions without a safety culture, incidents tend to rise and fall in a safety cycle. An increase in incidents results in management intervention, usually in the form of a greater emphasis on safety and training. This, in turn, may lead to a decrease in incidents. When

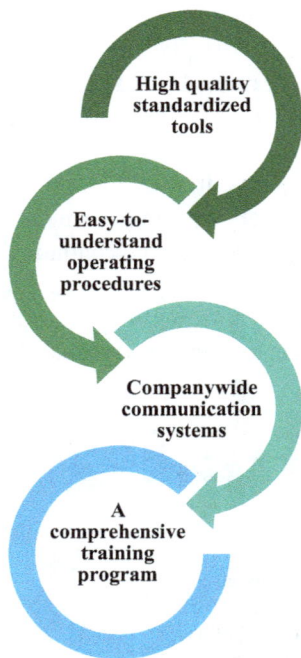

Figure 1.2: Safety culture elements.

the incident rate goes down, the emphasis on safety and training is reduced. Eventually, incidents increase, and the cycle begins again (Figure 1.3).

1.3.3 Risk Levels

A low-risk level is one in which there is a very minimal likelihood of damage, and the injury is modest if it occurs. Medium, if there is a risk of harm, which might be serious. High, if there is a significant risk of harm, which might be severe or extremely severe.

1.3.4 Risk Evaluation Matrix

The risk matrix for hazards can range from negligible to severe as shown in Table 1.2. Each severity rating has a clear definition as follows:
- *Severe.* If the risk results in several fatalities and/or the destruction of equipment and resources.
- *Significant.* If the risk results in a significant loss in safety margins, bodily suffering, or an increase in effort. This will lead to incorrect completion of any given task.
- *Moderate.* If the risk results in a significant loss of safety margins and the capacity to deal with adverse operating circumstances.

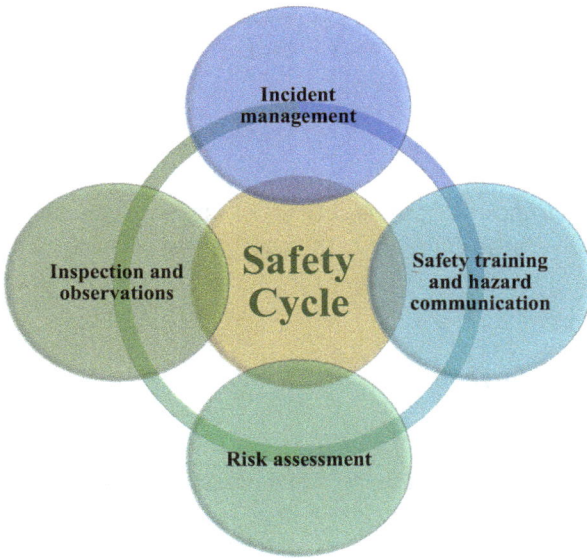

Figure 1.3: Safety cycle.

- *Minor.* If the risk causes a little cost and slight disturbance.
- *Negligible.* When the risk has only a few consequences but no general effect.

Table 1.2: Risk matrix for hazards.

Risk probability	Risk severity				
	5 Severe	**4 Significant**	**3 Moderate**	**2 Minor**	**1 Negligible**
5 Very likely	Extreme risk 25	Extreme risk 20	High risk 15	Moderate risk 10	Low risk 5
4 Likely	Extreme risk 20	High risk 16	High risk 12	Moderate risk 8	Low risk 4
3 Possible	Significant risk 15	Significant risk 12	Moderate risk 9	Moderate risk 6	Low risk 3
2 Unlikely	Moderate risk 10	Moderate risk 8	Moderate risk 6	Low risk 4	Low risk 2
1 Very unlikely	Low risk 5	Low risk 4	Low risk 3	Low risk 2	Low risk 1

Based on Table 1.2, the risk can be assessed as follows:
- *1–5 Low Risk.* Review existing measures.
- *6–10 Moderate Risk.* Improve control measures.

- *12–16 Significant to High Risk.* Consider stopping activity.
- *20–25 Extreme Risk.* Do not proceed and stop everything.

1.3.5 Hazardous Substances

A hazardous substance is any substance that can be corrosive, explosive, flammable, toxic, or easily oxidized or can develop one or more of these properties when contacts water or air.

1.4 Routes of Chemical Exposure

Routes of chemical exposure are:
- Inhalation. Breathing in toxins like formaldehyde or ammonia, which are prevalent in cleansers and disinfectants.
- Absorption. Chemicals can enter the eye or mucous membranes of the nose and mouth, as well as the skin through open wounds or regular skin contact with cleaning chemicals and disinfectants.
- Ingestion. Pesticides, cleaning and sanitizing solutions, or hazardous metals like mercury (used in thermometers) can enter the body through contaminated food or hands. Poor personal hygiene or housekeeping might lead to this type of exposure.
- Inoculation. Chemical exposure to antineoplastic (cancer) medications may occur during preparation, delivery, or disposal if the skin is punctured by a sharp item such as a needle.

1.5 Contributing Factors

There are many factors that contribute to an accident or incident. Among these factors are operating without authority at an unsafe speed, or carrying out unsafe loading, placing, mixing, combining, or failing to use PPE. It can also be due to improper PPE, improper or defective equipment, improper ventilation or lighting, or unsafe dress. In addition, personal factors such as fatigue, defective hearing, or eyesight, or physical or mental impairment can be among the contributing factors.

1.6 Nine Principles of Prevention

The principles of prevention need to be applied by employers to avoid hazards. If hazards cannot be avoided, then a risk assessment must be done in order to reduce them by using measures that protect all workers or by using safer work procedures. Where

risk prevention or reduction cannot be achieved, then employers must put in place appropriate measures for the protection of workers, such as training and signage. The nine principles of prevention are shown in (Figure 1.4) and described as follows:
- Avoidance. The first option to consider when dealing with risk is to avoid the hazard completely and so eliminate the risk. If the dangerous item is removed, then the hazard no longer exists.
- Evaluation. When risk cannot be removed, it should be at least evaluated. This allows the analysis of the situation and helps to come up with solutions.
- Risk Reduction. After evaluation, an act is required to reduce the level of risk. This might mean, for example, making sure that there is somebody holding the bottom of a ladder when in use.
- Adaptation. Getting acquainted with work in a new workplace can be very challenging. Writing down the way of carrying out each task with the least level of risk and having safety procedures and controls that are indicated in the safety statement, and keeping them close to the work area, will be very helpful.
- Replacement of Dangerous Items. Dangerous items should be replaced with nondangerous or less dangerous items.
- Appropriate Instructions to Everyone. A clear and understandable instructions should be given to all employees.
- Policy Development. A clear policy should be developed and well-enforced. This can greatly reduce the incident rates.
- Training. Training should also be introduced on safe practices. First aid and manual handling training are two areas of training often required.
- Protective Equipment. PPE should be the final option considered and implemented alongside training. When there are no adequate means to eliminate the risks involved in a task, ensure that the staff members or students wear suitable PPE. Employees are obliged to wear PPE when instructed to do so by their employer.

1.7 Safety Protocols and Emergency Preparedness

Safety in the chemical laboratory is not optional but an essential component of scientific practice. Safety protocols are structured procedures and guidelines designed to prevent accidents, protect personnel and equipment, and ensure compliance with legal and institutional safety standards. Every laboratory worker must be trained in these protocols before engaging in experimental work.

General safety protocols include the consistent use of PPE such as laboratory coats, gloves, and safety goggles; proper labeling and storage of chemicals; use of fume hoods when handling volatile or toxic substances; and regular maintenance of laboratory equipment. Good housekeeping, such as keeping workspaces clean and uncluttered, is also vital for minimizing hazards and ensuring quick access to exits and safety equipment.

Figure 1.4: The nine principles of prevention.

Equally important is emergency preparedness, which refers to the laboratory's readiness to respond effectively to incidents such as chemical spills, fires, injuries, or equipment failures. All personnel should be familiar with the location and operation of fire extinguishers, eye wash stations, safety showers, first aid kits, and emergency exits. Clear signage and accessible safety equipment are critical.

Laboratories should also include an emergency response plan that specifies protocols for various sorts of occurrences. This comprises evacuation procedures, communication chains, incident reporting systems, and designated safety officers. Regular safety exercises and training sessions ensure that everyone in the laboratory understands how to respond quickly and efficiently in an emergency. By incorporating safety standards and emergency planning into daily laboratory operations, institutions build a safe culture that safeguards both people and research results.

1.8 Separation Techniques

Separation methods are useful tools in the chemical laboratory for separating, purifying, and identifying molecules in mixtures, as well as removing contaminants and isolating valuable products. Their strategies are based on variations in physical or chemical properties such as boiling point, solubility, particle size, density, or polarity. Table 1.3 summarizes common laboratory separation procedures, including concepts and applications. The kind of component characteristics, necessary purity level, and availability of laboratory equipment all have an impact on the optimal technique and separation method to adopt. Several methods are frequently employed to achieve total separation.

Table 1.3: Common laboratory separation techniques.

Technique	Principle	Application
Filtration	Particle size	Separating solids from liquids
Distillation	Boiling point	Separating liquid mixtures
Chromatography	Polarity or molecular interaction	Identifying or isolating components in a mixture
Centrifugation	Density	Separating blood components or cell debris
Decantation	Density and immiscibility	Separating oil from water
Extraction	Differential solubility	Isolating compounds using solvents

1.9 Analytical Techniques

Analytical procedures (Table 1.4) are crucial tools in chemistry for determining the composition, structure, and quantity of compounds in a sample. These techniques enable scientists to identify unknown substances, verify product purity, monitor processes, and comply with safety and environmental regulations. They also guarantee precision and repeatability in research and industry. These techniques, whether used to verify medicine purity or test for pollutants in food or water, give quantitative and qualitative evidence for decision-making. They are essentially divided into two categories: qualitative analysis, which establishes which compounds are present, and quantitative analysis, which determines the amount of a material present.

Table 1.4: Common analytical techniques.

Technique	Purpose	Example application
Titration	Quantitative analysis by neutralization	Determining acid concentration in samples
Spectroscopy (UV-Vis and IR)	Molecular identification based on light absorption	Identifying functional groups or analyte concentration
Gravimetric analysis	Measurement based on mass	Determining sulfate ons by precipitating barium sulfate
Chromatography (TLC, HPLC, and GC)	Separation and identification of mixture components	Detecting food addit ves or pollutants
Electrochemical analysis	Measures electrical properties	Monitoring heavy metals in water

1.10 Questions and Answers

Questions	Answers
1. What is the difference between hazard and risk?	Hazard refers to anything that can cause harm, while risk is a likelihood that the hazard will cause actual harm.
2. How many levels are there in the health and safety culture maturity ladder?	There are typically five levels in the health and safety culture maturity ladder. These levels represent the progression of an organization's attitude and behavior toward health and safety.
3. List the effective safety culture elements.	High-quality, standardized tools and equipment. Clear, easy-to-understand operating procedures. Company-wide communication systems. A comprehensive training program.
4. What are the levels or types of risk severity?	Severe, significant, moderate, minor, and negligible.
5. What could be a hazardous substance?	Any substance that can be corrosive, explosive, flammable, toxic, or easily oxidized or can develop one or more of these properties when contacts water or air.
6. What are the routes of chemical exposure?	Inhalation, absorption, ingestion, and inoculation.
7. What are the factors that may contribute to an accident or incident?	Operating without authority at an unsafe speed, or carrying out unsafe loading, placing, mixing, combining, or failing to use PPE.
8. List down the nine principles of prevention.	Avoidance, evaluation, risk reduction, adaptation, replacement of dangerous items, appropriate instructions to everyone, policy development, training, and protective equipment.
9. What does PPE stand for?	Personal protective equipment.
10. What are the most common laboratory separation techniques?	Filtration, distillation, chromatography, centrifugation, decantation, and extraction.

Chapter 2
General Laboratory Safety

Teachers, technicians, chemists, researchers, and, to some extent, students are exposed to different chemicals, and among those are hazardous chemicals. This depends on the route of contact, which can be through inhalation, ingestion, injection, and/or absorption. This can happen anytime and anywhere and sometimes cannot be fully eliminated; therefore, the main goal of the laboratory safety procedures and protocols is to minimize this possible contact with chemicals in the laboratory.

2.1 General Laboratory Safety Rules

A standard set of fundamental laboratory safety guidelines is provided below, which must be observed in all laboratories that utilize hazardous products or operations. These fundamental standards give cleanliness and behavior safety advice to help prevent laboratory mishaps. Additional laboratory-specific safety guidelines may be necessary for certain operations and material handling:

- Locate laboratory safety equipment (e.g., safety showers, eyewash stations, and fire extinguishers) and know how to use them in an emergency.
- Know emergency exits.
- Avoid needless chemical exposures and presume all unknown toxins are very dangerous.
- When visiting a chemistry or chemical industrial laboratory, understand how to interpret warning signs for unexpected risks, dangerous products, and equipment.
- Display warning signs in laboratories for dangerous compounds, equipment, and specific conditions.
- Prioritize safety and chemical hygiene.
- Place emphasis on safety and chemical hygiene at all times.
- Wear laboratory safety glasses when handling or storing chemicals to prevent splashes or particles from entering the eyes.
- Avoid skin and eye contact with chemicals at all times.
- Wash chemicals exposed areas of the skin prior to leaving the laboratory.
- Use only approved garbage containers as directed. Avoid disposing of chemicals in drains or sewers.
- Use equipment and glassware only for their intended use.
- Do not use or dispose of unlabeled chemicals.
- Identify any dangers and take necessary safety procedures before starting any laboratory activity.
- Add reagents in the correct sequence, e.g., acid to water, not vice versa.
- Do not add solids to hot liquids.

https://doi.org/10.1515/9783112218105-002

- Do not leave chemical containers open after usage.
- Ensure that all containers have suitable labeling.
- Avoid purposely sniffing chemicals.
- Never use mouth suction when pipetting.
- Avoid using fume hoods for evaporation and disposal of volatile solvents.
- When dealing with hazardous chemicals, utilize a properly functioning fume hood to minimize exposures.
- Authorized staff should only have access to laboratories and support spaces, including stockrooms, chemical waste rooms, and specialist laboratories.
- Avoid consuming and storing food or beverages in places with harmful substances.
- Never wear contact lenses around hazardous chemicals, even when wearing safety glasses.
- Wear closed-toe shoes at all times in the laboratory. Perforated shoes or sandals are not appropriate.
- Avoid distracting or startling any persons working in the laboratory.
- Keep all sink traps filled with water by running water down the drain at least monthly.
- Avoid working alone in a building. Do not work alone in a laboratory if the procedures being conducted are hazardous.
- As a laboratory employee, request to have access to the chemical inventory list, applicable safety data sheets (SDSs), departmental laboratory safety manual, and relevant standard operating procedures when working with hazardous chemicals.
- Maintain equipment according to the manufacturer's requirements and records of certification and maintenance.
- Have designated and well-marked waste storage locations in the laboratory.
- Avoid using cell phones or earbuds in the active portion of the laboratories or during experimental operations.
- Store laboratory coats in offices or break rooms, and don't wear them in the coffee rooms or meeting rooms, as this spreads contamination to other areas.
- Do not leave your gloves on when using a computer, as inconsistent glove use around keyboards is a source of potential contamination.
- Avoid wearing jewelry in the laboratory as this can pose multiple safety hazards.

2.2 Laboratory Housekeeping

Keeping the laboratory clean and well-organized is a good way of having a safe hazard-free environment. This can be achieved by storing untouched equipment in a designated area, keeping equipment and glassware always clean after each experiment, returning chemicals to storage after use, keeping the laboratory always tidy, keeping all exits all the time unobstructed, cleaning up any spills immediately, allowing easy access to emergency equipment and utility controls, and storing any personal belongings outside the laboratory. These are all good practices that laboratory technicians, chemists, researchers, and students must do all the time.

2.2.1 Housekeeping Methodologies, Theories, and Principles

For maintaining proper housekeeping in laboratories and typically in any workplace, there are various methodologies for housekeeping theories that can be adapted to chemical laboratories. A comprehensive list of 10 housekeeping theories, including 5S, 6S, 7S, and seven additional organizational theories, is discussed in the following paragraphs, and each with its application in chemical laboratories is summarized in Table 2.1.

2.2.1.1 The 5S Methodology
The 5S methodology, Sort, Set in Order, Shine, Standardize, and Sustain, is a foundational workplace organization system developed in Japan to improve efficiency and safety. In chemical laboratories, this system is highly effective for promoting a clean, organized, and hazard-free environment. "Sort" involves removing unnecessary equipment, outdated chemicals, and broken tools to reduce clutter. "Set in Order" ensures that all items have designated, labeled storage locations for easy retrieval and return. "Shine" focuses on maintaining cleanliness through regular cleaning of benchtops, fume hoods, and common work areas. "Standardize" calls for the creation of consistent procedures and schedules for housekeeping tasks, while "Sustain" emphasizes the importance of discipline and regular audits to maintain the system over time.

2.2.1.2 The 6S Methodology
The 6S methodology builds upon the traditional 5S method by adding a critical sixth component: Safety. This addition places a deliberate focus on identifying, minimizing, and managing laboratory hazards in tandem with maintaining organization and cleanliness. In chemical laboratories, 6S encourages the integration of safety protocols, such as proper labeling of chemical containers, availability of **SDSs**, and correct storage of incompatible substances, into routine housekeeping tasks. The goal is to create a workspace that is not only efficient and orderly but also proactively managed to prevent accidents, chemical exposures, and other safety breaches.

2.2.1.3 The 7S Methodology

The 7S method further enhances the 6S framework by introducing "Spirit" or "Support," which emphasizes teamwork, morale, and a shared commitment to cleanliness and safety. In a chemical laboratory, the "Spirit" component involves cultivating a culture in which all laboratory members, from students to senior researchers, feel responsible for maintaining an organized and safe environment. This may include collective housekeeping duties, peer accountability, and recognition of good practices. The 7S methodology fosters a sense of community and pride in the laboratory, which can lead to greater adherence to safety protocols and continuous improvement in housekeeping standards.

2.2.1.4 Kaizen

Kaizen, a Japanese concept meaning "continuous improvement," focuses on making small, incremental changes to enhance safety, efficiency, and organization. Kaizen may be utilized in chemical laboratories by organizing regular reflection meetings when laboratory workers identify inefficiencies in storage, equipment organization, or workflow. Suggestions for improvement are welcomed from all team members and may result in changes to the layout of bench space, the rearranging of inventory systems, or the introduction of new cleaning standards. Over time, these minor improvements contribute to a more efficient and secure laboratory environment.

2.2.1.5 Visual Management

Important information is readily available in a self-explanatory workplace created via visual management using labels, color coding, and diagrams. In chemical laboratories, this technique works for improving situational awareness and reducing human mistakes. For instance, chemical storage rooms may be color-coded by compatibility groups, glassware and pipettes can be organized with labeled bins or shadow boards, and flammable cabinets can be marked with warning symbols. Spill kits, eyewash stations, and fire extinguishers are examples of emergency equipment that may be easily identifiable by special labels or color coding to facilitate a prompt response in case of an emergency.

2.2.1.6 Just-in-Time Inventory

The just-in-time (JIT) inventory concept originates from manufacturing and promotes the acquisition and use of materials only as they are needed. In chemical laboratories, this principle helps reduce waste, minimize clutter, and avoid the risks associated with storing large quantities of hazardous chemicals. Implementing JIT requires accurate tracking of reagent usage and demand, which can be supported by electronic inventory systems. This approach ensures that the laboratory remains clean and efficient while also reducing the potential for expired chemicals and associated safety hazards.

2.2.1.7 "A Place for Everything" Principle

This concept emphasizes that each item, whether a beaker, reagent, or safety goggles, should have a specific and properly labeled storage area. Adhering to this idea in chemical laboratories saves time searching for instruments and guarantees that key things are immediately available when needed. This strategy may include the use of drawer organizers, labeled shelves, glassware pegboards, and separate cabinets for flammables, acids, or personal protective equipment (PPE). A well-organized laboratory increases efficiency, avoids cross-contamination, and facilitates adherence to safety measures.

2.2.1.8 ABC Inventory Analysis

ABC analysis categorizes inventory depending on its value or relevance, with labels such as A (most crucial), B (moderately important), and C (least critical). In laboratory settings, this method can be utilized to prioritize housekeeping and stock management tasks. "A" items may contain extremely reactive or costly chemicals that need safe storage and continuous monitoring. "B" goods may be common solvents or glassware, whereas "C" items are low-use, low-risk supplies. This prioritizing enables laboratory workers to concentrate their housekeeping efforts on those areas that have the most immediate influence on safety and productivity.

2.2.1.9 Six Sigma

Six Sigma, as a quality management system, uses data to reduce mistakes and improve operations. Its approaches, which include the five whys (a root cause analysis method) and the DMAIC (define, measure, analyze, improve, and control) cycle, can be used for laboratory housekeeping and safety. For example, if chemical spills are widespread in a certain area, the five whys approach can help identify the underlying reason, which might be poor container design, insufficient location, or improper handling. Implementing the Six Sigma approach helps laboratories not only to resolve housekeeping issues, but also to prevent them from recurring via systematic improvements.

2.2.1.10 Risk-Based Housekeeping

The goal of risk-based housekeeping is to organize and clean the laboratory while also identifying and mitigating potential hazards. This method prioritizes housekeeping duties according to the amount of danger provided by certain regions, materials, or equipment. For example, storage spaces for caustic chemicals, high-temperature equipment, or flammable solvents may necessitate more regular cleaning and inspection than low-risk locations. This focused strategy helps to guarantee that the laboratory's most dangerous areas are kept to the highest safety standards and that resources are deployed efficiently to prevent accidents.

Table 2.1: A comprehensive list of 10 housekeeping methodologies, theories, and principles.

Theory	Description	Adaptation in chemical labs
5S	Japanese lean management: sort, set in order, shine, standardize, and sustain	Organize chemicals and equipment, label drawers, clean daily, and standard protocols
6S	5S + safety	Emphasizes safety alongside 5S
7S	6S + spirit (teamwork and morale)	Encourages culture and motivation
Kaizen	Continuous improvement (lean philosophy)	Promote gradual enhancements
Visual management	Using visuals to control and organize workspaces	Use hazard labels, color codes, floor markings, and signage for quick identification
Just-in-time (JIT)	Reduce inventory and waste	Maintain minimal chemical stocks to reduce clutter and risk
The "Place for Everything" principle	Classical organization principle	Clearly defined locations for all tools, chemicals, and PPE; use shadow boards
ABC inventory analysis	Inventory management tool classifying items by importance	A: items – critical chemicals; B: regular use; C: nonessential – manage accordingly
Six Sigma (five whys, DMAIC)	Process improvement and root-cause analysis	Apply to investigate accidents or inefficiencies in housekeeping
Risk-based housekeeping	Prioritize based on hazard and exposure	Extra housekeeping near corrosives, flammables, or reactive substances

2.2.2 Housekeeping Procedures

In addition to the above-mentioned housekeeping theories, there are also housekeeping procedures that are commonly implemented in well-run laboratories. These procedures are highly relevant for general laboratory safety and emphasize daily practices, accountability, and compliance. These include:

– Daily Housekeeping Checklist. Arrange for end-of-day checks for spills, clutter, chemical containers, fume hoods, and waste bins. A checklist can be printed or digital, signed off by responsible personnel.

– Clean-As-You-Go Policy. Encourage ongoing tidiness during experimental work and minimize buildup of clutter, accidental spills, and cross-contamination.

– Weekly Deep Cleaning Routines. Assign a schedule for bench wiping, fridge/freezer checks, cabinet rearrangement, fume hood cleaning, and checking labels/expiry dates.

- Spill Response Preparedness. Check and replenish spill kits regularly and have clear protocols and designated trained personnel for chemical spill response.
- Waste Segregation and Timely Disposal. Ensure correct disposal of organic, aqueous, sharps, glass, and biohazard waste and prevent overflow and unsafe conditions.
- Proper Storage Practices. Store chemicals by compatibility (e.g., acids separate from bases and organics). Heavy containers on lower shelves; no storage on benchtops or fume hood sashes.
- Broken Glassware Protocol. Clearly label broken glass containers and immediately remove broken items from work areas.
- Equipment Return Policy. Clean and return tools, glassware, and instrumentation to designated places and report damaged or malfunctioning items immediately.
- Housekeeping Responsibilities Chart. Assign tasks to specific laboratory members or groups and rotate responsibilities weekly or monthly.
- Housekeeping Audits. Do monthly internal audits (checklist-based) with scoring and feedback.
- Chemical Labeling and Inventory Review. Ensure all containers are labeled with chemical name, concentration, hazards, and date and conduct periodic inventory audits to remove expired or unlabeled materials.
- Emergency Access Maintenance. Keep fire extinguishers, eyewash stations, and exits completely clear. No equipment or boxes should obstruct emergency paths.

2.2.3 Housekeeping Tips

The next 11 tips are also absolutely essential for efficient workplace housekeeping:
- Prevent trips, falls, and tripping. The second most common cause of nonfatal occupational injuries or illnesses requiring days away from work was slips, trips, and falls.
- Get rid of fire risks. Employees are in charge of preventing unnecessary accumulations of combustible materials in the workspace. Combustible garbage needs to be "daily disposed of and stored in covered metal containers."
- Manage dust. A major explosive hazard exists when dust collection measures more than 1/32 of an inch, or 0.8 mm, and covers at least 5% of the surface of any given place.
- Do not trace materials. Work-area mats, which can be made of cotton or have a sticky top, should be kept tidy and in good condition. This helps stop the transfer of dangerous items to other workspaces.
- Avoid having objects fall. Toe boards, toe rails, and nets can help stop objects from falling and striking employees or equipment.
- Clear the clutter. Restricted movement by clutter in a crowded workplace, may develop ergonomic issues, and workers could experience injuries.

- Store items properly. Storage facilities shouldn't accumulate materials that could cause tripping risks, fires, or explosions.
- Make use of and examine tools and PPE. When doing housekeeping, it is advised to wear minimal PPE, such as closed-toe shoes and safety glasses.
- Establish the frequency. Every employee should help out with housekeeping, especially when it comes to keeping their workspaces organized, reporting any safety risks, and wiping up spills when necessary. Workers should check, tidy, and get rid of any unnecessary items from their workspaces before their shift ends. According to experts, this dedication can save on future cleaning time.
- Produce regulations in writing. The consensus among experts is that cleaning policies ought to be documented. Which cleansers, tools, and techniques to be utilized could be laid out in written procedures.
- Take a long view. Housekeeping should be ongoing through monitoring and auditing rather than being a one-time project. To support housekeeping, retain records, follow a regular walkthrough inspection schedule, flag hazards, and teach staff.

2.3 Laboratory Visitors

A laboratory visitor is anyone who is not assigned to be part of the laboratory and has no duty to handle in the laboratory. Therefore, he should only be allowed if permitted. Laboratory visitors should be made aware of hazards in the laboratory, emergency procedures in the event of a spill, fire, or alarm, and provided with appropriate PPE as necessary.

2.4 Running Experiments in the Laboratory

Due to the diverse nature of the chemical laboratory, experiments are diverse, and they are conducted by different chemists whose skills and backgrounds vary. The laboratory supervisor should put in place all information related to the level of experiment planning and appropriate documentation for each situation, including safety precautions and emergency procedures. Laboratory supervisor's prior approval is required when:

- Working with highly toxic chemicals
- Starting a new, unfamiliar experiment or procedure
- Running experiments unattended for a lengthy period or overnight
- Running experiments in which there has been an unexpected result or incident
- Working alone or after hours

2.5 Working Alone and After-Hours Work Practices

Staff working alone and/or working after hours in the laboratory may face risks that are not encountered during regular working hours because, in the event of an emergency, assistance may not be readily available; therefore:
- Keep the amount of laboratory work performed alone or after work hours to a minimum and choose low-risk work only if possible.
- Perform new or unfamiliar procedures during regular hours only.
- Ensure that the first aid kit, emergency shower, eyewash, and fire extinguisher are available in the laboratory.
- Have a communication system established so someone can provide aid in an emergency.

2.6 The Four Principles of Safety

The four principles of safety, abbreviated as RAMP (Recognize, Assess, Minimize, and Prepare), are shown in Figure 2.1:

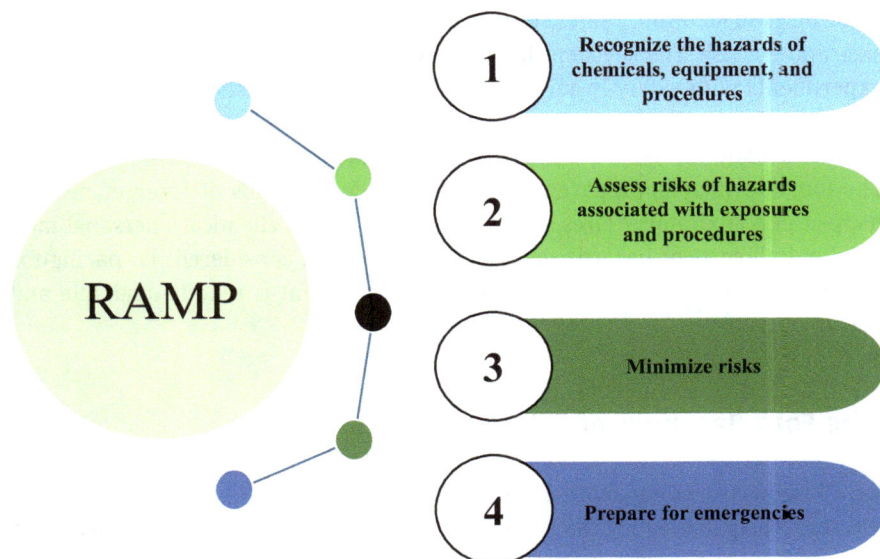

RAMP

1 — Recognize the hazards of chemicals, equipment, and procedures

2 — Assess risks of hazards associated with exposures and procedures

3 — Minimize risks

4 — Prepare for emergencies

Figure 2.1: The four principles of safety.

- The first principle, to recognize the hazards of chemicals, equipment, and procedures, requires that you know and recognize the hazards of the chemicals that you are using. There are millions of chemicals, of course, and knowing the hazards of all of them is not possible. To understand these hazards, you must understand the terminology and information that describes these various chemical properties. What does "flammable" mean? What is "toxic" or "corrosive"? And, how will you know if a chemical has any of these properties? Then, and more specifically, "getting to know your chemical" requires that you review and understand available information about its hazards, such as container labels, material safety data sheets, reference books, online hazard information, and talking with experienced people.

– The second principle, assess risks of hazards associated with exposures and procedures, is perhaps the most important of all the principles. This requires that you consider what kind of exposure to various chemicals could or will occur during a procedure or reaction, as well as the risk associated with the use of equipment. A few questions should be asked, such as is this reaction exothermic (releasing energy) in a way that might lead to a fire or explosion? Are there any flammable chemicals involved that might pose a fire hazard? What is the chance of some exposure to a toxic chemical?

– The third principle, minimize risks, requires careful attention to both the design and execution of an experiment. This requires that you take whatever reasonable steps are necessary to minimize, manage, or eliminate your exposure to a hazard by using good laboratory safety practices. This can only be done after careful consideration of risk. The key steps in minimizing risk are designing and performing experiments with safety in mind, using PPE (such as splash goggles) and other safety equipment (such as chemical hoods), and applying good housekeeping practices. Many accidents are caused by sloppy and cluttered work areas.

– The fourth principle, prepare for emergencies. What kinds of emergencies can happen in a laboratory? Fires, explosions, exposures to chemicals, personal injuries, and all sorts of hazards that have already been considered! Preparing for emergencies involves knowing what safety equipment is readily available and how to operate it.

2.7 The Four Baselines of Safety

The four baselines of safety abbreviated as 4Rs (recognize, react, report, and review), are selected to help in creating safety culture. These are briefly presented in Figure 2.2.

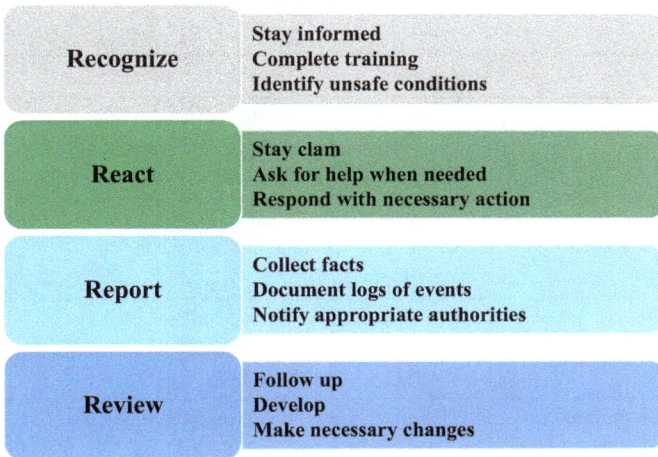

Recognize	Stay informed Complete training Identify unsafe conditions
React	Stay clam Ask for help when needed Respond with necessary action
Report	Collect facts Document logs of events Notify appropriate authorities
Review	Follow up Develop Make necessary changes

Figure 2.2: The four baselines of safety.

2.8 Questions and Answers

Questions	Answers
1. List five important general laboratory specific safety rules.	Use equipment, apparatus, and glassware only for its designated purpose. Never use or dispose of unlabeled chemicals. Avoid adding solids to hot liquids. Never leave chemical containers open after use. Make sure that all containers must have appropriate labels.
2. Why it is recommended not to leave your gloves used in the chemical laboratory on when using computers?	Because it can be a source of potential contamination.
3. Why it is advised to work with hazardous chemicals only in a properly working fumehood?	To reduce potential exposures to those chemicals.
4. List three names of housekeeping methodologies.	5S, 6S, and 7S methodologies.
5. How can the Kaizen methodology be applied in chemical laboratories?	Kaizen can be applied through regular reflection meetings where lab members identify inefficiencies in storage, equipment layout, or workflow.

(continued)

Questions		Answers
6.	What does the "A Place for Everything" principle stress on?	This principle stresses that each item, whether a beaker, reagent, or safety goggle, should have a designated and clearly labeled storage location.
7.	How can the visual management theory be adapted in the chemical laboratory?	It can be adapted by using hazard labels, color codes, floor markings, and signage for quick identification.
8.	What does the "Clean-As-You-Go Policy" mean?	It means encouraging ongoing tidiness during experimental work and minimizing buildup of clutter.
9.	What are the four principles of safety?	Recognize, assess, minimize, and prepare.
10.	List the four baselines of safety.	Recognize, react, report, and review.

Chapter 3
WHMIS in the Laboratory

WHMIS stands for Workplace Hazardous Materials Information System. It is some knowledge about the workplace risks associated with hazardous materials that everyone working in the laboratory should know. Many of the hazards of chemicals are not obvious or evident by smell, odor, appearance, or immediately detectable by the organs of the body.

Many hazardous materials (Table 3.1) lead to adverse physical or health effects in humans and also pose specific risks, such as skin irritation or burns, and long-term health problems that cause damage to humans.

Table 3.1: Common hazardous chemicals and their induced health effects.

Chemicals	Induced health effects
Acids and bases	Burn and corrode the tissues.
Alcohols	Strong CNS depressant.
Cyanide	Very dangerous, can cause collapse and death.
Hydrogen cyanide and carbon monoxide	May cause death (asphyxiation or asphyxia) by combining with the oxygen-carrying system in the blood.
Hydrogen sulfide (H_2S)	Extremely flammable, highly toxic poisonous gas with the odor of rotten eggs. Very hazardous, cause collapse and death if inhaled. Low concentration may cause irritation of mucous membrane, headache, and nausea.
Lead	Acute lead poisoning may lead to anorexia, vomiting, and permanent damage.
Mercury	In any form is toxic. Mercury may have toxic effects on lungs, skin and eyes.
Methyl alcohol	Dangerous to the optic nerve and may cause permanent damage and blindness.
Benzene	Extremely toxic and acts as an acute or chronic poison. Long-term exposure to high levels of benzene in the air can cause blood related cancer.
Carbon tetrachloride (CCl_4)	Acute exposure to CCl_4 may cause liver and kidney damage in humans.
Silver nitrate ($AgNO_3$)	Contact with skin or mucous membrane can be caustic and irritating.
Chloroform ($CHCl_3$)	Breathing or drinking large amount cause severe kidney and liver damage.

https://doi.org/10.1515/9783112218105-003

In addition to WHMIS, the Occupational Safety and Health Administration's (OSHA) Hazard Communication Standard (HCS) requires employees to be warned about potential risks and hazards connected with their usual job activities. This understanding is critical in lowering the danger of exposure to hazardous substances and the possibility of resulting injuries or health problems. Table 3.2 is an overview of the hazardous material classes found in the workplace.

Table 3.2: The workplace hazardous material classes.

Class	A	B	C	D1
Interpretation	Compressed gases	Flammable and combustible material	Oxidizing material	Materials causing immediate and serious toxic effects

Class	D2	D3	E	F
Interpretation	Materials causing other toxic effects	Biohazardous infectious material	Corrosive material	Dangerously reactive material

Although many hazardous chemicals or materials are easily recognizable, some are used routinely without workers being fully aware of their potentially harmful effects. Common examples of hazardous chemicals found in the workplace include:

- Disinfectants: Skin allergies and respiratory irritation can result from skin contact or inhalation.
- Glues: Direct skin contact might cause allergic reactions or irritation.
- Cleaning Agents: Volatile organic compounds, which are produced by a variety of cleaning products, can contribute to indoor pollution.
- Detergents: Extended exposure to detergents can result in dermatitis and other skin conditions.
- Paint: In addition to contributing to indoor and outdoor air pollution, paint exposure can be hazardous to one's health through inhalation or skin contact.
- Cosmetics: Certain cosmetics include potentially harmful substances that might endanger a person's health if not handled appropriately.

3.1 Worker Education and Training

Under WHMIS, employers at universities, colleges, companies, and other institutions are required to educate and train new employees who work with hazardous materials, most of which include chemicals commonly used in chemistry or laboratory settings. Faculty, staff, and students must receive this training before handling any of these substances, and regular refresher sessions are strongly recommended. Training should include instruction on WHMIS labeling, safety data sheet (SDS), and practical guidance on how to implement and manage WHMIS procedures at the departmental level. Designated individuals can then provide training within their respective departments. This should address the hazards associated with specific hazardous materials, as well as laboratory-specific protocols for safe handling, storage, spill response, and disposal.

3.2 Labeling

Labels are meant to notify users of chemicals and show them the precautions to take while handling chemicals or in case of an emergency. WHMIS recognizes two main types of labels, supplier labels and workplace labels (Figure 3.1). In addition, container labels and laboratory sample labels (Table 3.3) are also essential components of a robust chemical labeling system, especially in laboratory environments.

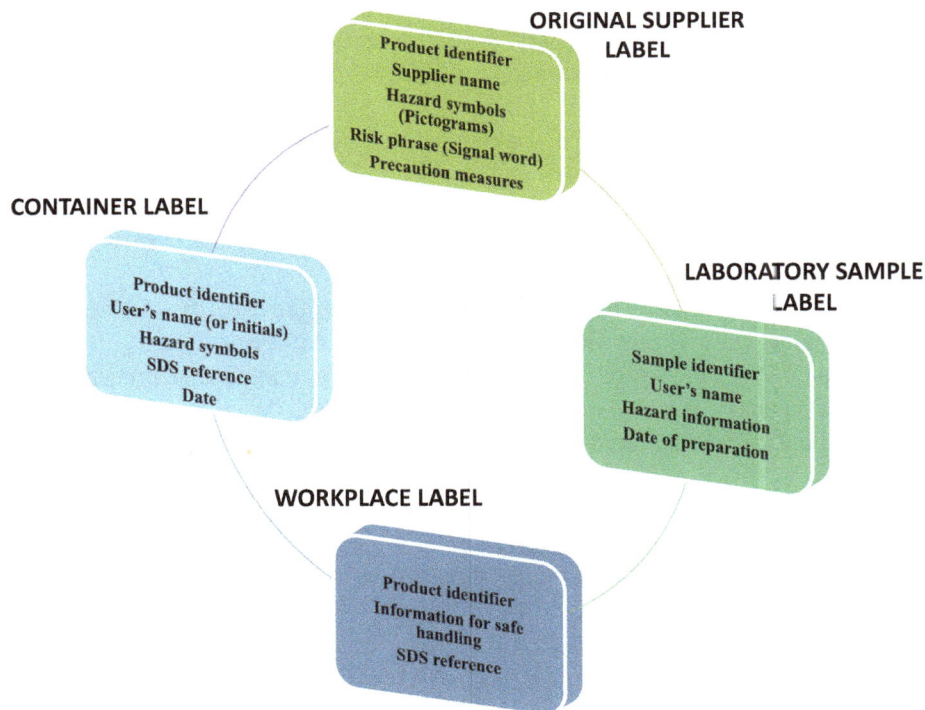

ORIGINAL SUPPLIER LABEL
Product identifier
Supplier name
Hazard symbols (Pictograms)
Risk phrase (Signal word)
Precaution measures

CONTAINER LABEL
Product identifier
User's name (or initials)
Hazard symbols
SDS reference
Date

LABORATORY SAMPLE LABEL
Sample identifier
User's name
Hazard information
Date of preparation

WORKPLACE LABEL
Product identifier
Information for safe handling
SDS reference

Figure 3.1: Chemical laboratory labeling.

Proper chemical labeling is a fundamental component of the WHMIS. Labels serve as the first line of communication to alert users about the hazards of a chemical product and provide essential safety instructions. Every hazardous chemical used in the workplace must have a legible and accurate label, and these labels must remain attached to the chemical containers.

When chemicals are transferred from original containers to the new ones in the workplace, new workplace labels must be used. If the chemical will be used only in the laboratory in which it was decanted or transferred, it only needs to have a chemical identifier. The same identifier labeling applies to chemicals produced in the laboratory, reaction vessels, mixtures undergoing testing or analysis, and hazardous waste. However, these simplified labels only apply as long as the chemical is in the original laboratory.

3.2.1 Supplier Labels

Supplier labels are affixed to chemicals, including hazardous materials, by the manufacturer, importer, or distributor before the product (chemical) is sold or supplied for use in the workplace. These labels are mandatory under WHMIS and must meet specific requirements to ensure consistency and clarity. A supplier label must include the following six elements:
- Product Identifier. The name of the chemical product as it appears on the SDS.
- Pictogram. Standardized symbol within a red diamond border that illustrates the type of hazard (e.g., flame for flammability, and skull and crossbones for toxicity).
- Signal Word. Either "Danger" (more severe hazard) or "Warning" (less severe hazard) is used to indicate the relative level of risk.
- Hazard Statement. Brief descriptions of the nature of the hazards associated with the product (e.g., "Causes severe skin burns and eye damage").
- Precautionary Statement. Recommended measures to minimize or prevent exposure, improper handling, or emergency response (e.g., "Wear protective gloves" and "Do not breathe fumes").

Supplier labels must be clearly legible, securely affixed, and in contrast with the background color. If a product is transferred from the original container to another, a workplace label must be applied (unless used immediately by the person who transferred it).

3.2.2 Workplace Labels

A workplace label is required when:
- A chemical product is transferred from its original container.
- The supplier label becomes illegible or is removed.
- A hazardous product is produced in the workplace for internal use.

Workplace labels must include:
- Product identifier (as it appears on the SDS)
- Safe handling information

3.2.3 Container Labels

Container labels (Figure 3.2) refer to labels placed on any container holding chemical hazardous materials, whether it's the original manufacturer's container or one used for internal handling and storage.

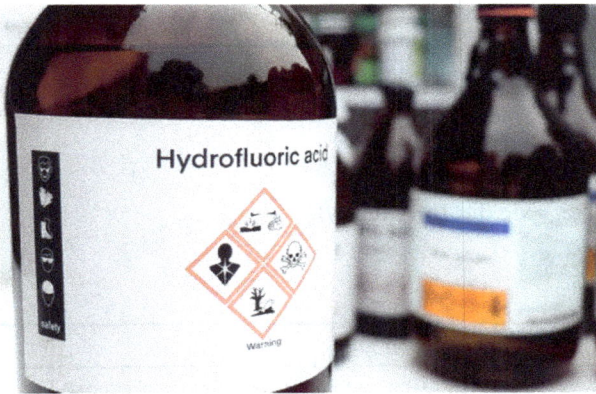

Figure 3.2: Container's label. Image credit: https://depositphotos.com/photos/hazardous-chemicals.html.

Proper labeling is crucial for secondary containers, reagent bottles, stock solutions, and waste containers. A container label is required when:
- A hazardous material is transferred from its original container into another container.
- A new solution or mixture is prepared in the laboratory.
- A waste chemical is collected in a dedicated container.

Information required on a container label:
- Product identifier (matching SDS)
- Hazard information (e.g., flammable, corrosive, and toxic)
- Date the material was prepared or transferred
- Initials or name of the person responsible for the contents
- Reference to the SDS

For containers used temporarily during a single shift and handled only by the person who transferred the chemical, labeling may not be mandatory, but it is strongly recommended for best practices.

3.2.4 Laboratory Sample Labels

Laboratory sample labels are used for research samples, experimental compounds, or small-scale preparations that may not yet be fully classified but are known or suspected to be hazardous. These labels ensure safe handling of research materials, especially when shared among laboratory members or stored for future use. A laboratory sample label should include at a minimum:
- Sample identifier or name
- Hazards information (known or suspected hazards based on chemical structure, origin, or synthesis route)
- Date of preparation
- Responsible person's name (researcher or student)
- Reference number or logbook code (for traceability)

Table 3.3: Label types and key elements.

Label type	Used for	Key information
Supplier label	Original commercial chemical containers	5 Standard elements (product identifier, supplier name, hazard symbols (pictograms), risk phrase (signal word), and precaution measures
Workplace label	Repackaged, decanted, or in-house products	Product identifier, information for safe handling, SDS reference
Container label	Storage and transfer containers	Product identifier, user's name (or initials), hazard information, SDS reference, and date
Laboratory sample label	Research or experimental samples	Sample identifier, user's name, hazard information, and date

3.3 Safety Data Sheet (SDS)

A standardized document that offers detailed information on a chemical molecule or mixture is called a safety data sheet (SDS). Before being superseded by the Globally Harmonized System (GHS), which was authorized by WHMIS 2015, it was known as a material safety data sheet. Its primary objective is to promote safe chemical handling, use, storage, and disposal. SDSs are essential parts of the WHMIS and the GHS framework. Each SDS contains crucial information about the chemical, including its general

characteristics, specific hazards, control strategies, and emergency protocols. It describes how chemicals should be used and handled to protect laboratory personnel and the environment during disposal. The purpose of SDS is:
- To inform users about the hazards associated with a chemical
- To provide instructions for safe handling, emergency measures, protective equipment, and spill or leak procedures
- To support risk assessment and ensure proper workplace controls are in place

And it can be used by:
- Employers and supervisors to develop safety procedures
- Workers, including laboratory personnel, to understand how to safely handle chemicals
- Emergency responders to know how to act during spills, fires, or exposures

The minimum required information that must appear on every SDS is illustrated in Figure 3.3.

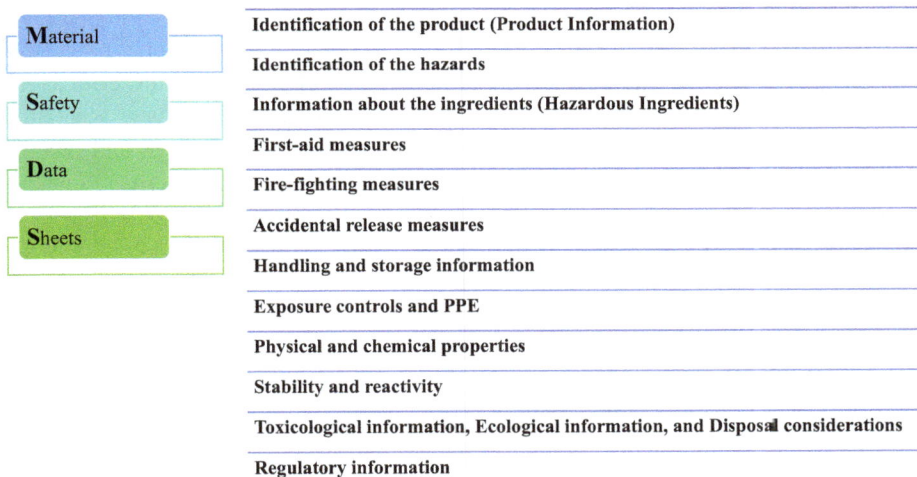

Material	Identification of the product (Product Information)
	Identification of the hazards
Safety	Information about the ingredients (Hazardous Ingredients)
	First-aid measures
Data	Fire-fighting measures
	Accidental release measures
Sheets	Handling and storage information
	Exposure controls and PPE
	Physical and chemical properties
	Stability and reactivity
	Toxicological information, Ecological information, and Disposal considerations
	Regulatory information

Figure 3.3: Material safety data sheet or safety data sheet.

Each hazardous material regulated under WHMIS must have an SDS available, except for the chemicals from a laboratory supply house that are labeled with all information required on an SDS or when the controlled products are produced and kept in the same laboratory. SDS must be frequently updated, and the current hazard information must always be available; the continuation or discontinuation of hazardous materials laboratory use should be updated. SDS should also be made available and readily accessible. They can be available in the following forms:

- Paper copies
- Computer-accessed SDS

If the SDS is only computer-accessible, the following must be ensured:
- The computer must be easily accessible at all times to laboratory personnel.
- Laboratory personnel should know how to access and retrieve the SDS information.
- Internet links to several SDS sources should be available, especially when the primary sources' server is not available.
- If a password and login are required, they must be set up in advance.
- If common laboratory reagents are ordered from different suppliers, a single SDS from one of these suppliers is acceptable when the reagent ingredients are the same, the product identifier on the label resembles the one on the SDS, and no variations in the hazard information.

3.4 Hazard Classification

Chemical hazard classification is an important aspect of laboratory safety and chemical management. Understanding the possible hazards of chemicals and effectively conveying these risks through standardized identification helps to prevent accidents, encourage responsible handling, and guarantees compliance with legal and institutional rules. Hazard classification is the process of determining a substance's or mixture's potential for damage, whereas identification is the communication of these dangers using words, symbols, and defined forms.

3.4.1 Purpose of Hazard Classification

The main objectives of hazard classification include:
- Protecting human health by identifying chemical dangers and minimizing exposure
- Promoting safe handling and storage of hazardous substances
- Providing clear communication of risks to laboratory workers, students, emergency responders, and the public
- Ensuring global consistency through standardized classification and labeling systems

3.4.2 The Globally Harmonized System (GHS)

The GHS for classification of chemicals is an internationally agreed-upon system developed by the United Nations. It was designed to replace the different classification standards used in various countries with a consistent, universal approach. The key components of GHS are:
- Hazard classes and categories
- Pictograms
- Signal words
- Hazard statements
- Precautionary statements
- Product identifier
- Supplier information

3.4.3 GHS Hazard Classification

The GHS classifies chemicals (hazards) into three broad categories. Examples are listed in Figure 3.4.

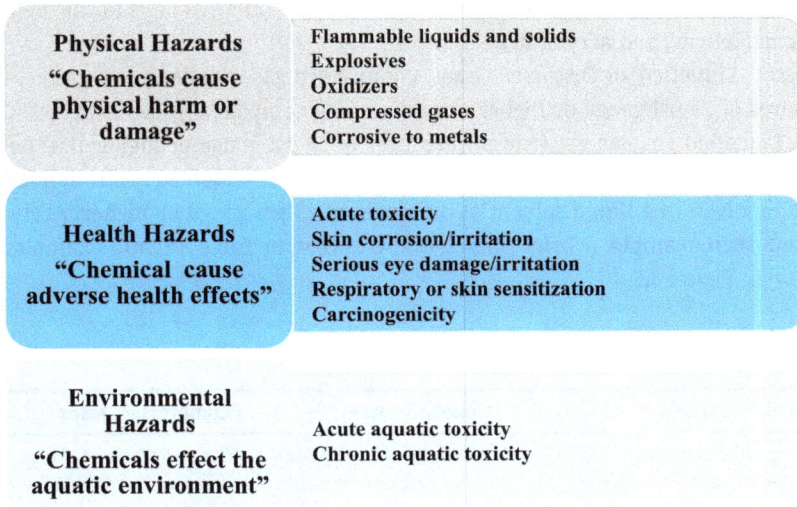

Physical Hazards **"Chemicals cause physical harm or damage"**	**Flammable liquids and solids** **Explosives** **Oxidizers** **Compressed gases** **Corrosive to metals**
Health Hazards **"Chemical cause adverse health effects"**	**Acute toxicity** **Skin corrosion/irritation** **Serious eye damage/irritation** **Respiratory or skin sensitization** **Carcinogenicity**
Environmental Hazards **"Chemicals effect the aquatic environment"**	**Acute aquatic toxicity** **Chronic aquatic toxicity**

Figure 3.4: Classes of chemical hazards.

3.4.4 GHS Label Elements

Each chemical container must include a label with the following GHS elements (Table 3.4).

Table 3.4: GHS label elements and their description.

Element	Description
Pictograms	Graphical symbols representing specific hazards
Signal word	Danger (more severe) or warning (less severe)
Hazard statements	Standardized phrases describing the nature and degree of hazard
Precautionary statements	Instructions to reduce or prevent adverse effects
Product identifier	Chemical name or code
Supplier identification	Name, address, and phone number of the responsible party

3.4.5 Definitions of Chemical Hazard Classes

– Carcinogens. Chemicals or physical substances that induce cancer or tumor growth, usually after repeated or prolonged exposure. Their effects may be delayed and have no immediate negative consequences. Examples are methylene chloride, benzene, formaldehyde, and acrylonitrile.
– Compressed, Liquefied, or Dissolved Gases. Compressed gas is defined as a gas kept at pressures of 29 psi gauge or higher and remaining completely gaseous at –50 °C (–58 °F). Liquefied gas is a gas kept at pressures of 29 psi gauge or higher that becomes partly liquid at temperatures above –50 °C (–58 °F). Dissolved gas is defined as a gas dissolved in a liquid solvent at pressures of 29 psi gauge or higher. Acetylene is one such example. A brief comparison is shown in Table 3.5, and examples are shown in Figure 3.5.

Table 3.5: A comparison of dissolved, liquefied, and compressed gases.

Feature	Dissolved gases	Liquefied gases	Compressed gases
Definition	Gases dispersed uniformly in a solvent (usually a liquid) at molecular level	Gases stored in liquid form under pressure or reduced temperature	Gases stored under high pressure but remain in gaseous state
Physical state in container	Dissolved in a liquid (invisible gas phase)	Liquid (may have some gas above the liquid)	Gas only

Table 3.5 (continued)

Feature	Dissolved gases	Liquefied gases	Compressed gases
Storage conditions	Ambient pressure or mild pressure; influenced by solubility and temperature	Moderate to high pressure or low temperature	High pressure
Examples	O_2 in water, CO_2 in soda, and NH_3 in wastewater	Propane, butane, liquefied CO_2, and anhydrous ammonia	Oxygen, nitrogen, hydrogen, and helium in gas cylinders
Common containers	Bottles, cans (carbonated drinks); aquaria; water treatment tanks	Steel or aluminum cylinders, tanks, and refrigerant bottles	Gas cylinders (steel or composite) and scuba tanks
Key hazards	Over-saturation → degassing, toxic exposure (e.g., NH_3 and H_2S)	Expansion on vaporization, flammability, and cold burns	High-pressure release, explosion risk, and asphyxiation
Industrial uses	Beverage industry, water treatment, and aquatic life support	Fuel (LPG), refrigeration, and fire extinguishers	Welding, medical gases, and laboratory gases
Phase change on release	No phase change and just diffusion	Liquid → gas (rapid expansion and cooling)	Gas → same gas (may expand rapidly)

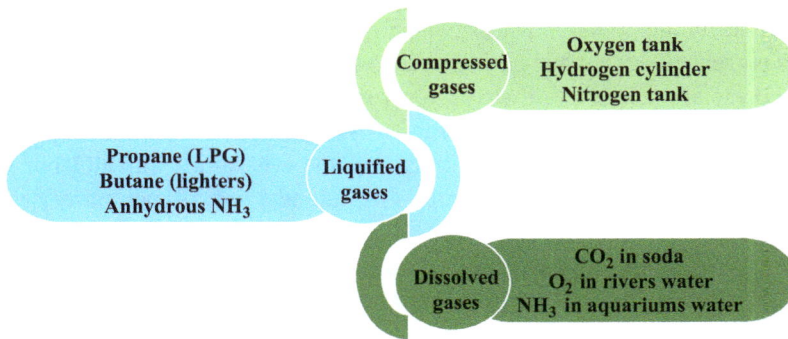

Figure 3.5: Examples of compressed, liquefied, and dissolved gases.

– Corrosive Materials. They cause irreparable damage to living tissue by chemical activity at the site of contact. Corrosive substances can be liquids, solids, or gases, and their most common impacts are on the skin, eyes, and respiratory tract. Corrosive materials include liquids such as acids and bases, gases such as chlorine and ammonia, and solids such as phosphorus and phenol.

- Cryogenic Liquids. Are materials with extremely low boiling points (i.e., less than −150 °F/−101 °C). One special property of cryogenic liquids is that they undergo substantial volume expansion upon evaporation or sublimation, which can potentially lead to an oxygen-deficient atmosphere where ventilation is limited. Some cryogenic liquids can also pose additional hazards, including toxicity and flammability (i.e., liquid carbon monoxide). Examples of cryogenic liquids include liquid nitrogen, liquid helium, and liquid argon.
- Flammable and Combustible Liquids. Flammable liquids have a flash point (*The flash point of a substance is the temperature at which enough substance will vaporize to form an ignitable mixture with air*) of less than 199.4 °F (93 °C), and combustible liquids have a flash point above 199.4 °F (93 °C). Examples of flammable and combustible liquids include acetone, methyl alcohol, acetic acid, toluene, and glycerol.
- Chemicals Having Significant Acute Toxicity. Substances with a high level of acute toxicity are classed based on their LC50 or LD50 values. They can have adverse effects after a single exposure/dose or many exposures/doses within 24 h. Many of these compounds can also be categorized as toxic gases, selective agent poisons, corrosives, irritants, or sensitizers. Acrolein, arsine, and nitrogen dioxide are all exceedingly toxic chemicals.
- Highly Reactive/Unstable Materials. They are those that have the potential to vigorously polymerize, decompose, condense, or become self-reactive under conditions of shock, pressure, temperature, light, or contact with another material. Examples of such substances are explosives (lead azide and nitroglycerin), peroxide formers (isopropyl ether, sodium amide, diethyl ether, and styrene), water-reactives (alkali metals such as lithium, sodium, potassium, and calcium carbide), and pyrophorics (Grignard reagent "RMgX," metal powders such as Al, Co, Fe, Mg, Mn, and Pd).
- Irritants. These are substances that cause reversible effects (e.g., swelling or inflammation) on the skin or eyes at the site of contact. A wide variety of organic and inorganic compounds are irritants; thus, skin and eye contact with all laboratory chemicals should be avoided. Examples include tear gas, pepper spray, solvents, and acids.
- Reproductive Toxins. These are chemicals that have an unfavorable effect on reproductive capability. Possible consequences include chromosomal damage (mutations), effects on fetuses (teratogenesis), negative effects on sexual function and fertility in adult males and females, and poor effects on child development. Many reproductive poisons induce harm after many low-level exposures. Long latency durations result in noticeable effects. Chemicals that are harmful to reproduction or development include chloroform, benzene, arsenic, and mercury compounds.
- Sensitizers. These are chemicals triggering allergic reactions in normal tissue when exposed to them repeatedly. The response might be as small as a rash (contact dermatitis) or as severe as anaphylactic shock. Compounds such as diazomethane, different isocyanates, formaldehyde, and benzylic and allylic halides might produce sensitivity in certain people.

3.5 Identification of Chemical Hazards

One of the main workplace issues is the inability to identify or recognize hazards associated with chemicals. In order to achieve proper hazards identification, the knowledge of hazard pictograms and hazards statements should be gained through proper training, and information should be put in place. Hazard pictograms give a quick indication of the hazards associated with the chemical and hazard statements provide details on the nature of the hazards associated with a hazardous product.

3.5.1 Hazard Pictograms

Hazard pictograms (Figure 3.6) are graphical symbols used on labels and SDS to quickly communicate the type of hazard a chemical product presents.

Figure 3.6: Hazard pictograms. Image credit: https://www.conceptdraw.com/How-To-Guide/osha-haz com-pictograms.

3.5.1.1 WHIMS Pictograms

The WHMIS is the national hazard communication standard for Canada. To guarantee global compliance, WHMIS 2015 is completely integrated with the GHS system. Most

WHMIS pictograms (Table 3.6) employ the GHS diamond form with a black hazards sign, a white backdrop, and a red border to indicate risks including flammability, corrosivity, and acute toxicity. For biohazardous infectious items, WHMIS still uses a different circular logo that shows a black biohazard sign inside a black circle on a white background.

Table 3.6: List of WHMIS pictograms, their symbols, hazard classes, and examples.

Pictogram	Symbol	Hazard classes	Examples
Flame		Flammable gases, liquids, and solids; self-reactive substances; and pyrophoric materials	Gases such as hydrogen and propane; and solvents such as acetone and ethanol
Flame over circle		Oxidizing gases, liquids, and solids	Gases such as chlorine, nitrous oxide, and oxygen; liquids such as hydrogen peroxide; and solids such as sodium hypochlorite
Gas cylinder		Gases under pressure	Compressed nitrogen and carbon dioxide
Exploding bomb		Explosives and self-reactive substances	Organic peroxides and TNT
Corrosion		Corrosive to metals, skin corrosion, and eye damage	Hydrochloric acid, sodium hydroxide, and sulfuric acid
Skull and crossbones		Acute toxicity (fatal or toxic)	Cyanide and arsine
Health hazard		Carcinogens, respiratory sensitizers, and reproductive toxins	Benzene, asbestos, and busulfan

Table 3.6 (continued)

Pictogram	Symbol	Hazard classes	Examples
Exclamation mark		Irritation, skin sensitization, acute toxicity (less severe), and narcotic effects	Acetone and ammon a (in low concentrations)
Environment		Aquatic toxicity	Pesticides and heavy metals
Biohazardous infectious materials		Biological toxins, viruses, and bacteria	Blood samples, bacteria, and viruses

3.5.1.2 CLP/GHS Pictograms

The GHS of Classification and Labeling of Chemicals uses standardized pictograms to visually communicate chemical hazards in a consistent way worldwide. GHS pictograms are red diamonds with a white background and a black hazard symbol in the center. These pictograms are designed to be easily recognizable and are required to appear on labels and SDS for chemicals classified as hazardous. By harmonizing hazard communication, GHS pictograms ensure that workers, users, and transporters can readily identify risks and take appropriate safety measures regardless of country or language.

The Classification, Labelling, and Packaging (CLP) regulation is the European Union's implementation of the GHS system. CLP pictograms are identical to GHS pictograms: red diamonds with a white background and black hazard symbols. They cover the same hazard categories as GHS, including physical, health, and environmental hazards. However, CLP labels also include specific regulatory details required by the European Union, such as multilingual hazard statements and precautionary statements, which ensure that chemicals are properly labeled across EU member states. CLP pictograms appear on chemical containers, transport packaging, and SDSs to inform users about chemical hazards in a consistent, recognizable format CLP/GHS hazardous symbols (pictograms) are shown in Table 3.7.

In conclusion, the pictograms used in the GHS and CLP systems are identical, consisting of nine red diamonds that provide universal chemical hazard communication. In contrast, WHMIS pictograms also use these nine diamond symbols but include an additional circular biohazard symbol unique to Canada. A comparison of these three systems is presented in Table 3.8.

Table 3.7: CLP/GHS hazard pictograms.

Physical Hazard

| Explosive | Corrosive | Gases under pressure | Oxidizing | Flammable |

Health Hazard

| Serious health hazard | Toxic | Harmful/irritant |

Environmental Hazard

Hazardous or dangerous to the environment

Table 3.8: Comparison of GHS–CLP–WHMIS pictogram systems.

Feature	GHS	WHMIS (2015)	CLP
Shape	Diamond	Diamond (GHS-aligned) plus circle for biohazardous materials	Diamond
Border color	Red	Red (diamond); black (circle) for biohazardous infectious materials	Red
Background	White	White	White
Symbol color	Black	Black	Black
Number of pictograms	9	9 (GHS-aligned) + 1 unique biohazard circle	9
Biohazard symbol	Not included	Black biohazard symbol in black circle	Not included
Applicable region	Global standard (UN)	Canada	European Union

3.5.2 Hazard Statements

The hazard statements describe the nature of the risks connected with a hazardous product. In some circumstances, it also identifies the level of risk. The EU's CLP legislation, which replaced the Dangerous Substances Directive (DSD) and the Dangerous Preparations Directive (DPD), assigns a unique alphanumerical code to hazard statements consisting of one letter and three numbers (e.g., H200 – unstable explosives). The letter "H" for hazard statement; a digit denoting the kind of hazard, e.g., "2" for physical hazards; and two numbers relating to the sequential numbering of hazards, such as explosivity (codes 200 to 210) and flammability (codes 220 to 230). Additional instances are provided in Table 3.9.

Table 3.9: Selected CLP hazard codes.

Code	Hazard statement
H201	Explosive; mass explosion hazard
H202	Explosive; severe projection hazard
H203	Explosive; fire, blast, or projection hazard
H204	Fire or projection hazard
H220	Extremely flammable gas
H221	Flammable gas
H222	Extremely flammable aerosol
H223	Flammable aerosol
H224	Extremely flammable liquid and vapor
H225	Highly flammable liquid and vapor
H226	Flammable liquid and vapor
H228	Flammable solid
H300	Fatal if swallowed
H301	Toxic if swallowed
H302	Harmful if swallowed
H319	Causes serious eye irritation
H330	Fatal if inhaled
H331	Toxic if inhaled
H400	Very toxic to aquatic life
H410	Very toxic to aquatic life with long-lasting effects
H411	Toxic to aquatic life with long-lasting effects

Hazard statements carried through from the DSD and the DPD, and the ones that are not included in the GHS are codified as "EUH." Examples for EUH codes are shown in Table 3.10.

CLP serves as an identifying tool that ensures effective communication of chemical classification and labeling information from producers to end users across the supply chain. By doing this, the environment and human health are protected.

The same coding scheme is not required by the U.S. OSHA Hazard Communication Standard (HCS); instead, hazard statements must be marked as such and adhere to

Table 3.10: Selected EUH hazard codes.

Code	Hazard statement
EUH 014	Reacts violently with water
EUH 018	May form flammable/explosive vapor–air mixture
EUH 019	May form explosive peroxides
EUH 031	Contact with acids liberates toxic gas
EUH 032	Contact with acids liberates very toxic gas
EUH 202	Cyanoacrylate (danger): bonds skin and eyes in seconds
EUH 203	Contains chromium(VI); may produce an allergic reaction
EUH 204	Contains isocyanates; may produce an allergic reaction

the standards' phrasing. Chemical manufacturers and importers are required by HCS to categorize their products and identify whether they pose a health risk (toxic or poisonous) or a physical one (flammable or explosive). An illustration of a chemical hazard identification system is presented in Figure 3.7.

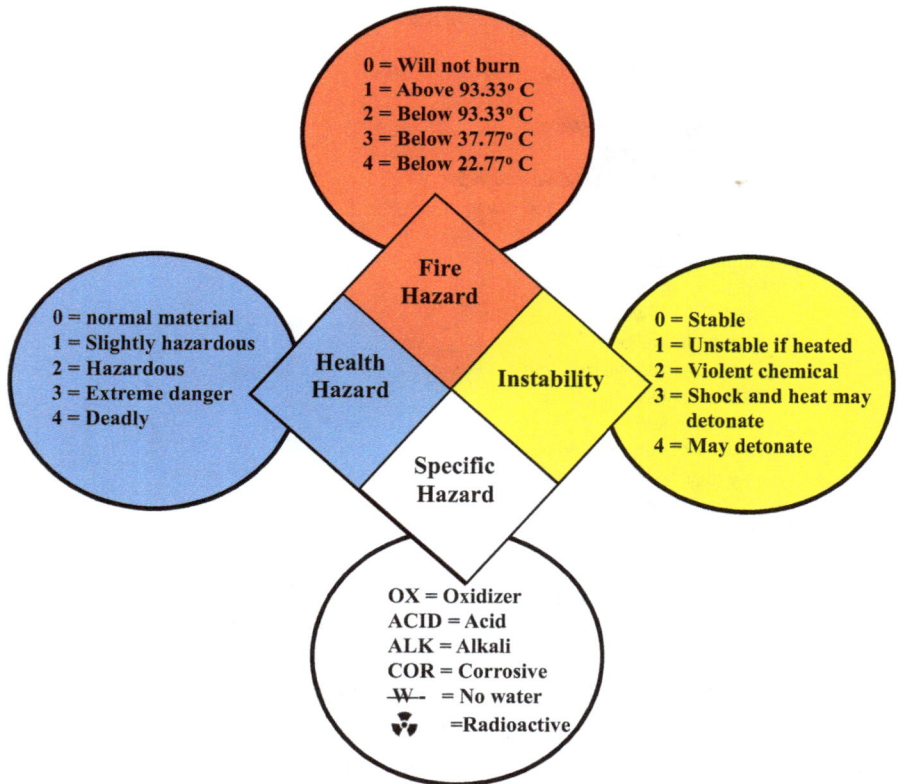

Figure 3.7: Chemical hazard identification system.

3.6 Questions and Answers

Questions	Answers
1. What is the primary goal of WHMIS?	To ensure workers receive information about hazardous materials used in the workplace.
2. What is the main purpose of GHS?	To standardize chemical hazard communication through consistent labeling, classification, and safety data sheets worldwide.
3. What does WHMIS stand for?	Workplace Hazardous Materials Information System.
4. Name any three WHMIS/GHS hazard pictograms.	Flame for flammables, self-heating substances. Skull and crossbones for acute toxicity (fatal or toxic). Corrosion for skin corrosion/burns, eye damage, or corrosive to metals.
5. What is a hazard statement?	A phrase describing the nature and severity of a chemical hazard (e.g., "Causes skin irritation").
6. What is a precautionary statement?	A phrase giving advice on preventing or minimizing adverse effects (e.g., "Wear protective gloves").
7. What type of chemical uses the exploding bomb pictogram?	Explosives and self-reactive substances.
8. How many pictograms are there in the GHS system?	Nine (eight are required under WHMIS 2015 in Canada; environmental is optional).
9. What is the difference between "danger" and "warning" in GHS labeling?	"Danger" indicates a more severe hazard than "Warning."
10. Where must WHMIs/GHS pictograms appear?	On chemical container labels and safety data sheets (SDS), in the hazard identification section.

Chapter 4
Working with Chemicals

Working with chemicals in the lab necessitates a thorough comprehension of their chemical and physical characteristics as well as the application of stringent safety procedures and related risks. Lab staff must be proficient in deciphering safety data sheets (SDSs), chemical labels, and hazard symbols in compliance with international harmonization systems since chemicals differ greatly in their reactivity, toxicity, volatility, and corrosiveness. Before starting any activity, it is essential to evaluate the risks by carrying out a comprehensive hazard analysis and determining appropriate control measures such as the use of glove boxes for air-sensitive chemicals or fume hoods for volatile substances. To reduce direct exposure, personal protective equipment (PPE) such as lab coats, gloves, safety goggles, and occasionally respirators must be used regularly.

Training in proper handling techniques, spill response, and first aid is also vital, especially when dealing with corrosives, toxins, or reagents that may pose delayed health effects. Cultivating a culture of accountability and vigilance in chemical handling not only protects individual researchers but also promotes a safer, more efficient working environment for the entire laboratory.

When handling chemicals, four basic concepts are required to ensure both safety and effective laboratory processes (Figure 4.1):

- Plan, Recognize, and Review the Hazards. Know the physical and chemical properties of the items you are handling. Examining labels, SDS, and hazard categories (such as flammability, toxicity, reactivity, and corrosiveness) are all part of this. Recognizing potential risks makes it simpler to plan the appropriate precautions.
- Minimize Exposure. Reduce the exposure risk by using control measures such as engineering controls (fume hoods or glove boxes), administrative controls (standard operating procedures (SOPs)), and PPE (like lab coats, gloves, and goggles).
- Assess Risks. Consider how the chemicals will be used, the quantities required, the processes to be followed, and the possibility of exposure or mishaps. Risk assessment helps to identify the likelihood and severity of harm, which guides the selection of appropriate management measures.
- Prepare for Emergencies. Prepare to respond to spills, exposures, fires, and other mishaps. Understand the location and usage of safety equipment such as eyewash stations, safety showers, fire extinguishers, and spill kits. Emergency protocols and first-aid techniques should be thoroughly known and practiced.

Working with chemicals is an essential aspect of many scientific and commercial operations, but it necessitates a thorough grasp of the distinct risks that each category poses. Highly toxic chemicals, flammable and combustible liquids, highly reactive and explosive compounds, corrosive agents, cryogenic liquids, and compressed gas cylin-

https://doi.org/10.1515/9783112218105-004

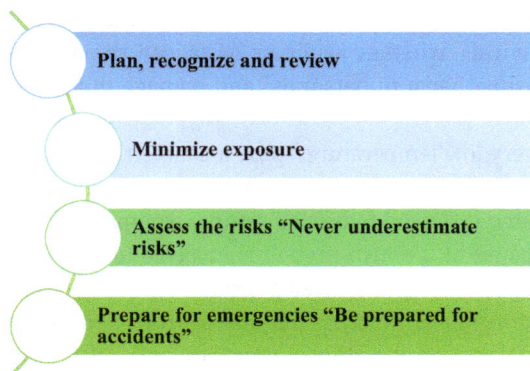

Plan, recognize and review

Minimize exposure

Assess the risks "Never underestimate risks"

Prepare for emergencies "Be prepared for accidents"

Figure 4.1: The four basic principles to consider when working with chemicals.

ders are among the most dangerous types of hazardous chemicals (Figure 4.2), each posing unique risks to health, safety, and the environment.

Figure 4.2: Classes of hazardous chemicals. Image credit: https://depositphotos.com/photos/hazardous-chemicals.html.

These substances, while indispensable in research, manufacturing, and medical applications, can cause severe injuries, fires, explosions, environmental damage, or even fatalities if not handled properly. Highly toxic chemicals can cause serious health effects even at very low exposure levels, affecting organs or systems with acute or chronic consequences. Flammable and combustible liquids present significant fire and explosion hazards, especially in poorly ventilated areas or in the presence of ignition sources. Similarly, highly reactive and explosive chemicals can undergo violent reactions under specific conditions, necessitating strict control measures to prevent accidental detonation. Corrosive chemicals, whether acidic or basic, can rapidly destroy tissue and materials, causing lasting harm to personnel and damage to laboratory infrastructure.

Cryogenic liquids, which exist at very low temperatures, pose a concern because of their quick vaporization and oxygen displacement, which can result in severe cold burns. Despite being widely used in laboratory and industrial settings, compressed gas cylinders require special handling because of the high pressures involved and the potential for leaks, fires, or explosions. These categories cover a broad spectrum of dangerous substances and require not just technical expertise but also a strong dedication to risk assessment, safety culture, safe handling and storage practices, and emergency readiness. It is also important to remember that improper transport of chemicals from storerooms to laboratories and between laboratories may lead to a chemical spill. To avoid this, a set of three basic protocols (Figure 4.3) is followed as a guide when chemicals are transported outside the laboratory.

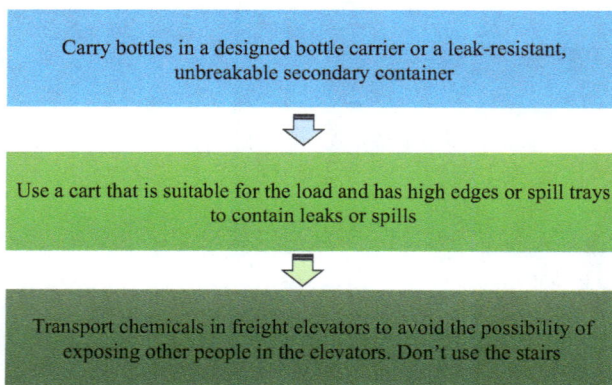

Carry bottles in a designed bottle carrier or a leak-resistant, unbreakable secondary container

Use a cart that is suitable for the load and has high edges or spill trays to contain leaks or spills

Transport chemicals in freight elevators to avoid the possibility of exposing other people in the elevators. Don't use the stairs

Figure 4.3: The three basic protocols for transporting chemicals.

4.1 Highly Toxic Chemicals

Chemicals that have long-term harmful consequences include carcinogens, poisons that affect reproduction or development, and mutagens. The SDS often contains information on possible mutagenicity, carcinogenicity, or reproductive toxicity. The lethal dosage 50% (LD50) or lethal concentration 50% (LC50) of a compound is used to evaluate its toxicity.

The dose at which half of the test animals (typically rats or mice) perished is known as the LD50. This usually happens 1–2 h after the animals are dosed via ingestion, injection, or skin exposure. The air concentration at which half of the test animals (typically rats or mice) perished, usually after a certain amount of time after being exposed by inhalation, is known as the LC50. Chemicals with high acute toxicity may be identified using the criteria presented in Table 4.1.

Table 4.1: Criteria for identifying chemicals with high acute toxicity.

Oral *LD_{50} (rats, mg/kg)	<50
Skin contact LD_{50} (rabbits, mg/kg)	<200
Inhalation **LC_{50} (rats, ppm for 1 h)	<200
Inhalation LC_{50} (rats, mg/m^3 for 1 h)	<2,000

Carcinogens are among the severe and very dangerous compounds that have been documented. These are chemicals that have the potential to cause cancer. Simply said, cancer is the uncontrolled proliferation of cells that can happen in any organ. Chemicals that cause cancer might be solid, liquid, or gaseous. One significant distinction between carcinogens and many other substances is that exposure to them does not always correlate with the development of cancer in a dose–response manner. As a result, every precaution must be made to prevent exposure to carcinogens. Keeping oneself clean, using engineering controls like fume hoods, biosafety cabinets, and glove boxes, wearing PPE like lab coats, gloves, and safety glasses, following the right procedures, and receiving training are all ways to protect oneself from carcinogens in the lab.

Increased fume hood speeds (average face velocity) of at least 100 fpm are the first step in handling carcinogens with the same alertness as highly dangerous compounds. Twenty-six known carcinogens are regulated by the Occupational Safety and Health Administration (OSHA):

- 1,2-Dibromo-3-chloropropane
- 1,3-Butadiene
- 2-Acetylaminofluorene
- 3,3'-Dichlorobenzidine (and its salts)
- 4-Aminodiphenyl
- 4-Dimethylaminoazobenzene
- 4-Nitrobiphenyl
- Acrylonitrile
- α-Naphthylamine
- Asbestos
- Benzene
- Benzidine
- β-Naphthylamine
- β-Propiolactone
- Bis-chloromethyl ether

- Cadmium
- Coke oven emissions
- Ethylene oxide
- Ethyleneimine
- Formaldehyde
- Inorganic arsenic
- Methyl chloromethyl ether
- Methylene chloride
- Methylenedianiline
- *N*-Nitrosodimethylamine
- Vinyl chloride

Before starting experiments with highly toxic chemicals, all work phases and steps should be checked. This includes storage and handling, experimental protocol, decontamination, disposal, and cleanup of spills. Each experiment should be evaluated individually, as the circumstances and amounts of the toxic chemical used will affect the types of precautions required. Experimental work should be carried out in a designated area of the laboratory, such as a fume hood or glove box.

All laboratory personnel are made aware of the nature of the toxic chemicals being used and the necessary precautions to take. Warning signs to alert others in the area should be posted, and they should clearly define boundaries. In addition to the abovementioned, the following protocols should be followed:
- Make sure that the fume hoods are working properly all the time while running the experiment.
- Handle glove boxes under negative pressure to prevent the escape of toxic vapor, specks of dust, or aerosols.
- Avoid releasing toxic dust, vapors, or aerosols to the atmosphere or into apparatus such as vacuum pumps and lines by having a high-efficiency particulate air filter or cold trap in the system.
- Choose equipment that can be easily decontaminated. For example, use vacuum pumps that can be decontaminated easier than vacuum lines. Any used equipment should be labeled and isolated from the general laboratory equipment and decontaminated before being removed from the designated work area.
- Wear long-sleeved clothing and appropriate PPE and select gloves that are suitable for the chemicals being handled.

4.2 Flammable and Combustible Liquids

Flammable (or Class I) liquids are those with a flashpoint below 37.78 °C. Combustible (Class II or III) liquids are those with a flashpoint between 37.78 and 93.33 °C (Figure 4.4). The great challenge of handling these liquids in the laboratory lies in their potential to cause fire or explosion. It is dangerous to heat flammable liquids with an open flame, and it is advisable to handle them in a free ignition area.

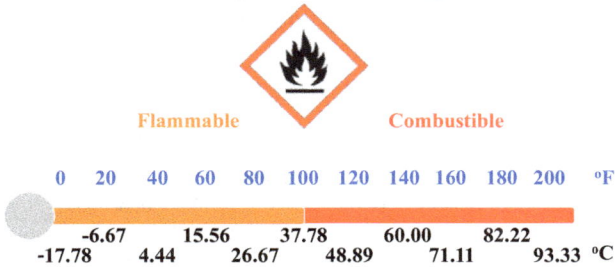

Figure 4.4: Flammable and combustible liquids' flash point range.

For liquids with very low flash points, such as diethyl ether and carbon disulfide, hot surfaces are avoided. Appropriate ventilation should be used to prevent the formation of flammable or explosive gas mixtures in the air. Keep containers of flammable liquids (Figure 4.5) always closed except during transfer of contents.

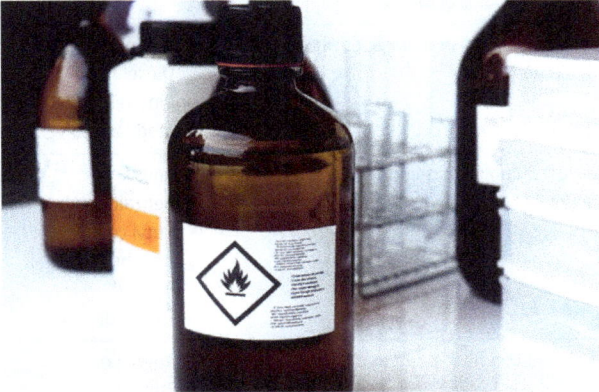

Figure 4.5: Flammable liquids. Image credit: https://depositphotos.com/photos/flammable-liquids.html?qview=624353884.

The National Fire Protection Association (NFPA) classifies flammable and combustible liquids into six classes: IA, IB, IC, II, IIIA, and IIIB (Table 4.2).

The OSHA has somewhat varied classifications for flammable liquids. OSHA no longer refers to any liquids as "combustible" following amendments to the U.S. Labor Law in 2012. Any liquid that has a closed-cup flash point lower than 200 °F (93.33 °C) is referred to as a "flammable liquid." Table 4.3 summarizes the four categories into which flammable liquids are divided.

Table 4.2: Flammable and combustible liquid classes with description and examples.

Class	Description	Examples
Class IA	Liquids have a flash point below 73 °F (22.77 °C) and a boiling point below 100 °F (37.78 °C).	Diethyl ether Pentane Ligroin Heptane Petroleum ether
Class IB	liquids have a flash point below 73 °F (22.77 °C) and a boiling point at or above 100 °F (37.78 °C).	Acetone Toluene Benzene Isopropyl alcohol Cyclohexane Methyl ethyl ketone Ethanol
Class IC	liquids have a flash point between 73 and 100 °F (22.77–37.78 °C). There is no specific boiling point requirement with this classification.	Xylene Naphtha Turpentine Styrene Butyl alcohol
Transition from flammable liquids to combustible liquids according to NFPA's definitions		
Class II	A Combustible liquid with a closed-cup flash point at or above 100 °F (37.78 °C) and below 140 °F (60.00 °C).	Camphor oil Diesel fuel Pine tar Stoddard solvent
Class IIIA	A combustible liquid with a closed-cup flash point at or above 140 °F (60.00 °C) and below 200 °F (93.33 °C).	Aniline Benzaldehyde Butyl cellosolve Nitrobenzene Pine oil Formaldehyde
Class IIIB	A combustible liquid with a closed-cup flash point at or above 200 °F (93.33 °C).	Animal oils Ethylene glycol Glycerin Transformer oils Triethanolamine Benzyl alcohol Hydraulic fluids Vegetable oils

Table 4.3: Flammable liquid categories from the Occupational Safety and Health Administration (OSHA) with descriptions and examples.

Class	Description	Examples
Category 1	Any liquid with a closed-cup flash point below 73.4 °F (23 °C) and with a boiling point below 95 °F (35 °C).	Diethyl ether Pentane Ligroin Heptane Petroleum ether
Category 2	Any liquid with a closed-cup flash point below 73.4 °F (23 °C) and with a boiling point above 95 °F (35 °C).	Acetone Benzene Cyclohexane Isopropyl alcohol Methyl ethyl ketone Toluene Ethanol
Category 3	Any liquid with a closed-cup flash point at or above 73.4 °F (23 °C) and below 140 °F (60 °C).	Xylene Naphtha Turpentine Camphor oil Diesel fuel Pine tar Stoddard solvent
Category 4	Any liquid with a closed-cup flash point at or above 140 °F (60 °C) and below 200 °F (93 °C).	Aniline Animal oils Benzaldehyde Benzyl alcohol Butyl cellosolve Ethylene glycol Formaldehyde Glycerin Hydraulic fluids Lubricating, quenching, and transformer oils Nitrobenzene Pine oil Triethanolamine Vegetable oils

Ground metal lines and containers are used to dispense flammable liquids to prevent the buildup of static electricity. In the flammable liquid storage area, drums should also be grounded during dispensing. Drums are grounded by connecting the container to an already grounded object that will conduct electricity.

4.3 Highly Reactive and Explosive Chemicals

Highly reactive and explosive chemicals are those that may be detonated by mechanical shock, elevated temperature, or chemical action to produce a violent release of energy and a large volume of gas, heat, and possibly toxic vapors. Even milligram quantities of some highly reactive substances can turn small fragments of glass or other material into potentially seriously injurious or lethal missiles. They have the potential to vigorously polymerize, undergo a vigorous condensation or oxidation–reduction reaction, or become self-reactive due to shock, pressure, temperature, light, or contact with another material. Examples of highly reactive chemicals are organic peroxides, pyrophorics, water-reactive chemicals, and explosives. When handling these chemicals, observe the following safety precautions:

– Handle reactive chemicals with great care. Store them separately and never mix even small quantities with other chemicals unless proper procedures are followed and appropriate PPE is used.
– Conduct chemical reactions at non-ambient temperatures or pressures in a way that minimizes hazards such as explosions or vigorous reactions. Provide reliable systems for temperature control and the safe dissipation of excess heat and pressure.
– Use shielding to protect personnel and facilities from hazards like over-pressurization or implosion. Minimize the amount of reactive chemicals used or produced, limiting it to the smallest necessary quantity.
– For highly reactive or explosive chemicals, restrict reaction vessel quantities to no more than 0.5 g of reactants and limit product synthesis to no more than 0.1 g per run. Any exceptions must be specifically approved by the laboratory supervisor, with appropriate written procedures, training, and mitigation controls in place.
– Use shields, barricades, and PPE, such as face shields with throat protectors and heavy gloves, whenever there is a risk of explosion, implosion, or vigorous reaction.

4.3.1 Organic Peroxides

Organic peroxides are among the most toxic compounds utilized in labs. They are extremely combustible and will react aggressively to heat, friction, light, and both oxidizing and reducing agents. When handling organic peroxides, use the following safety measures:

– Avoid Metal Contact. Do not use metal spatulas, stirring bars, or other metal equipment since they can contaminate peroxides and cause explosive decomposition.
– Minimize the Effect. Avoid friction, grinding, and other mechanical impacts, especially when using solid peroxides.

- Proper Storage. Do not keep organic peroxides in glass containers with screw tops or glass stoppers. Instead, use polyethylene containers with suitable lids or stoppers.
- Clean Thoroughly. After each use, clean the container's neck, cap, and threads with a towel before resealing.
- Spill Cleanup. Immediately clean up any peroxide spills to prevent accidents.

4.3.2 Pyrophoric Chemicals

Pyrophoric chemicals are substances that can ignite spontaneously in air within 5 min of exposure, even in small quantities, and this poses a significant fire and injury risks to personnel. Due to their unique fire hazards, they require special handling procedures to prevent exposure to air. Laboratories that use or store pyrophoric chemicals must develop written SOPs outlining safe practices for storage, use, and disposal. Additionally, all personnel who work with pyrophoric chemicals must receive specific training on safe handling procedures, and this training must be documented.

Examples of pyrophoric chemicals include:
- Grignard reagents: RMgX (R = alkyl and X = halogen)
- Metal alkyls and aryls: alkyl lithium compounds such as *n*-butyllithium (*n*-BuLi) and *tert*-butyllithium (*t*-BuLi)
- Metal carbonyls: nickel tetracarbonyl ($Ni(CO)_4$)
- Metal powders (finely divided): cobalt, iron, zinc, zirconium, and palladium on carbon
- Metal hydrides: sodium hydride
- Nonmetal hydrides: Diethylarsine and diethylphosphine
- Nonmetal alkyls: R_3B, R_3P, and R_3As; tetramethyl silane and tributyl phosphine
- White phosphorus
- Alkali metals: lithium, sodium, potassium, and other alkali metals
- Gases: silane (SiH_4), dichlorosilane (SiH_2Cl_2), diborane (B_2H_6), phosphine (PH_3), and arsine (AsH_3)

4.3.3 Water-Reactive Chemicals

Water-reactive chemicals are substances that react vigorously with moisture. Common examples include:
- Alkali metals such as sodium (Na), lithium (Li), and potassium (K)
- Alkali metal hydrides such as lithium hydride (LiH), calcium hydride (CaH_2), lithium aluminum hydride ($LiAlH_4$), and sodium borohydride ($NaBH_4$)
- Alkali metal amides like sodium amide ($NaNH_2$)
- Metal alkyls including lithium and aluminum alkyls

- Grignard reagents (RMgX)
- Nonmetal halides such as boron trichloride (BCl_3), boron trifluoride (BF_3), phosphorus trichloride (PCl_3), phosphorus pentachloride (PCl_5), silicon tetrachloride ($SiCl_4$), and disulfur dichloride (S_2Cl_2)
- Inorganic acid halides like phosphorus oxychloride ($POCl_3$), thionyl chloride ($SOCl_2$), and sulfuryl chloride (SO_2Cl_2)
- Anhydrous metal halides such as aluminum chloride ($AlCl_3$), titanium tetrachloride ($TiCl_4$), zirconium tetrachloride ($ZrCl_4$), and tin tetrachloride ($SnCl_4$)
- Phosphorus pentoxide (P_2O_5) and calcium carbide (CaC_2)

A selected list of these water-reactive chemicals and their corresponding reactions to water is summarized in Table 4.4.

Table 4.4: A selected list of water-reactive chemicals and their corresponding reactions to water.

Water-reactive chemical	Resulting reaction with water
Aluminum chloride	Violent decomposition forming HCl gas
Butyl lithium	Ignites on contact with water
Lithium hydride	Violent decomposition
Phosphorous pentachloride	Violent reaction
Phosphorous pentoxide	Violent exothermic reaction
Potassium hydride	Releases hydrogen gas
Sodium hydride	Reacts explosively with water
Titanium tetrachloride	Violent reaction that produces HCl gas
Zirconium tetrachloride	Violent reaction with water

4.3.4 Explosives (Explosive Chemicals)

According to the OSHA laboratory standard, an explosive is a substance that produces a quick, virtually immediate release of pressure, gas, and heat when exposed to a sudden shock, pressure, or high temperature. Fortunately, most laboratories do not employ many explosives; but certain compounds can become unstable and/or potentially explosive over time as a result of contamination with air, water, and other materials such as metals or drier conditions. Explosives can cause damage to adjacent items (such as hoods, glassware, windows, and persons), the release of hazardous gases, and flames. Whenever doing investigations with potentially explosive compounds:
- Use the smallest quantity of chemicals feasible.
- Conduct your studies in a fume hood with a properly approved safety shield.
- Remove any unnecessary equipment and chemicals from the work area, particularly those that are extremely harmful or flammable.

- Ensure that everyone in the lab is informed of the experiment including any risks and the time.
- Mixing, chopping, or scraping potentially explosive compounds using metal or wooden implements is not recommended. Instead, use non-sparking plastic devices.
- To lessen the possibility of a catastrophe, use additional safety measures such as high temperature controls and water overflow devices in tandem.
- Dispose of hazardous waste responsibly and indicate any special precautions required for potentially explosive compounds on the label.
- When working with potentially explosive substances, always wear proper PPE such as gloves, a lab coat or apron, safety goggles with a face shield, and explosion-proof shields.

Reactive or explosive materials requiring special attention. A list of selected examples is provided in Table 4.5; however, this list is not all-inclusive.

Table 4.5: A selected list of reactive or explosive materials.

Chemical	Types of reactions
Ammonia (NH_3)	Reacts with iodine to give nitrogen triiodide, which detonates on touch.
Azides such as sodium azide	Can displace halide from chlorinated hydrocarbons such as dichloromethane to form highly explosive organic polyazides.
Chromium trioxide–pyridine complex (CrO_3–C_5H_5N)	May explode if the CrO_3 concentration is too high.
Diazomethane (CH_2N_2)	They are very toxic, and the pure gases and liquids explode readily even from contact with sharp edges of glass.
Diethyl, diisopropyl, and other ethers including tetrahydrofuran and 1,4-dioxane	Sometimes it explodes during heating or refluxing because the presence of peroxides has developed from air oxidation.
Dimethyl sulfoxide (DMSO) and (CH_3)$_2$SO	Decomposes violently on contact with a wide variety of active halogen compounds such as acyl chlorides. Explosions from contact with active metal hydrides have been reported.
Ethylene oxide (C_2H_4O)	Explodes when heated in a closed vessel.
Chloroform ($CHCl_3$) and carbon tetrachloride (CCl_4)	Should not be dried with sodium, potassium, or other active metal; violent explosions usually result.
Lithium aluminum hydride ($LiAlH_4$)	The reaction of $LiAlH_4$ with carbon dioxide has reportedly generated explosive products.

Table 4.5 (continued)

Chemical	Types of reactions
Phosphorus trichloride (PCl_3)	Reacts with water to form phosphorous acid with HCl evolution; the phosphorous acid decomposes on heating to form phosphine, which may either ignite spontaneously or explode.
Potassium (K)	Can form explosive peroxides on contact with air.
Sodium amide ($NaNH_2$)	Can undergo oxidation on exposure to air to give sodium nitrite in a mixture that is unstable and may explode.

4.4 Corrosives

Corrosives (Figure 4.6) are chemicals that, through chemical action at the point of contact, can cause irreversible damage or destruction to living tissues (such as skin, eyes, and the respiratory tract) as well as to metals. Examples include strong acids (such as sulfuric acid and nitric acid), strong bases (such as sodium hydroxide), and certain strong oxidizers (like concentrated hydrogen peroxide).

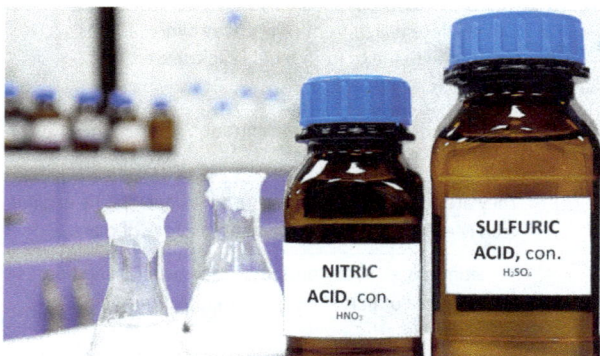

Figure 4.6: Corrosives. Image credit: https://postapplescientific.com/the-differences-between-nitric-acid-and-sulfuric-acid/?srsltid=AfmBOopB8jQ9KAsMrwCVQyTTRFqmGIWJmaJ-Vl8dN4nNHb6fWEwF7Fi5.

Before working with corrosive substances, it is crucial to be fully aware of their hazards and the potential risks associated with improper handling. Corrosives can cause serious burns to the skin and eyes, irritate the respiratory tract, and damage laboratory equipment and containers. Accidental spills or contact can lead to severe injuries or property damage. Therefore, it is vital to understand the nature of these chemicals

and to implement appropriate safety measures to maintain a safe working environment. Consider the following precautions:
– Avoid storing corrosive liquids above eye level.
– Always add acids or bases to water, never the opposite.
– Store acids and bases separately to prevent harmful reactions.
– To dissolve corrosive substances in water, gently add the solid and stir frequently. Additional cooling may be required.
– Use a fume hood for work that may cause substantial dust.
– To prevent explosions or extreme temperatures, utilize additional shielding like a glove box or lower the fume hood sash. Portable shields can also give further protection.
– Keep corrosive substances away from heat, fires, oxidizers, and water. Make sure that the containers are sealed and the manufacturer's labels and warnings are intact.

To handle potentially corrosive gases:
– Use a fume hood or wear appropriate respiratory protection.
– Protect exposed skin areas from corrosive or irritating gases and vapors.
– When not in use, close gas regulators and valves and clean them with dry air or an inert gas like nitrogen. When discharging corrosive gases into liquids, use a trap, check valve, or vacuum break device to avoid harmful backflow.

Personal protective equipment (PPE):
– Wear OSHA-compliant safety eyewear when working with caustic substances. If there is a possibility of flying particles (such as glass or plastic), use safety glasses with side shields. In addition to safety glasses, utilize chemical splash goggles or a face shield to protect against substantial splash dangers.
– Wear gloves while handling corrosives; nitrile gloves provide enough protection in most lab environments. Always study the SDS to ensure sufficient skin protection, especially for long-term or high-risk exposures.
– When dealing with caustic substances, it is suggested to wear a lab coat or apron. Open-toed shoes are not allowed.

4.5 Compressed Gases

Due to their high pressure, compressed gas cylinders (Figure 4.7) represent additional physical hazards. In the case of a leak, inert gases in the cylinders can create an oxygen-deficient atmosphere, while toxic gases can create toxic atmospheres, and flammable gases can result in fire. The following precautions are advised when handling the cylinders:

- Transport cylinders with a handcart equipped with a restraining strap.
- Never drag, roll, or slide cylinders.
- Keep the valve cap in place during transport and remove it only when the cylinder is securely strapped to the wall.
- Use only approved cylinder regulators as per the specifications.
- Do not lubricate oxygen regulators, as the cylinder contents may oxidize the oil or grease and cause an explosion.

Figure 4.7: Compressed gas cylinders. Image credit: https://hazchemnetwork.co.uk/2022/05/25/present ing-gas-cylinders-for-carriage/.

To minimize static ignition, fireback, and flashbacks, place a flash arrestor in the line and ground all cylinders, gas lines, and combustible gas equipment. Check your cylinder connections and gas lines for leaks on a regular basis. To clean all joints, use a leak detector or a solution of soap and water. In the case of a leak, bubbles will form around the leaking area. If a leak is detected, switch off the gas before making any repairs. If the leak remains after shutting off the cylinder valve, consider it an emergency uncontrolled release. Cylinders should not be left empty to prevent possibly dangerous flashbacks or backflow of air or other impurities.

When removing a cylinder from use:

- Close the main valve.
- Bleed the system.
- Shut off and remove the regulator, and replace the valve cap.
- Mark the cylinder "empty" or "MT" and return it to the appropriate storage area for pickup by the supplier.

4.6 Cryogenic Liquids

Cryogenic liquids (Figure 4.8) are liquefied gases that exist in their liquid state at very low temperatures. They are extremely cold. Examples of cryogens include nitrogen, helium, hydrogen, argon, methane, and carbon monoxide. When handling these chemicals, the following precautions should be followed:
- Wear full-coverage clothing with no cuffs or pockets, which could catch the liquid in the event of a spill.
- Use insulating gloves and wear chemical splash goggles or a face shield for protection in case of cryogenic liquid splash.
- Store and transport cryogens only in Dewar flasks designed for that purpose. Always fill Dewar flasks slowly to reduce temperature shock effects and minimize splashing.
- Keep cryogens always covered to prevent condensation of atmospheric moisture that may cause a plug to form in a narrow vessel neck, resulting in an over-pressurized vessel.
- When using cold traps, ensure that they do not become plugged with frozen material.

Figure 4.8: Cryogenic liquids. Image credit: https://www.thoughtco.com/cryogenics-definition-4142815.

4.7 The Control of Substances Hazardous to Health (COSHH)

The Control of Substances Hazardous to Health (COSHH) is a major UK regulation designed to ensure safe handling of hazardous chemicals in the workplace. It instructs employers to identify risks, carry out assessments, and put in place suitable measures to reduce or eliminate exposure. Continuous monitoring and review are also emphasized to maintain a safe working environment.

Employees who handle chemicals need to be aware of the potential dangers and be trained in proper safety practices. Under COSHH, employers have a legal duty not only to minimize chemical exposure and protect worker health but also to prevent misuse or unauthorized possession of hazardous substances. To support compliance, COSHH outlines an eight-step process, summarized in Table 4.6, that provides practical guidance for controlling hazardous substances in daily operations.

Table 4.6: The eight-step approach to the COSHH requirements.

Step 1	Assess any risks to health arising from hazardous substances used in, or created, by your workplace activities.
Step 2	Make necessary precautions and assess the risks first before allowing any work.
Step 3	Prevent or properly control employees' exposure to hazardous substances.
Step 4	Apply control measures and maintain them. Ensure that safety procedures are followed.
Step 5	Watch employees' hazardous substances exposure.
Step 6	Carry out appropriate inspection and control on any COSHH-specific requirements.
Step 7	Put in place emergency procedures to handle any accidents, incidents involving hazardous substances.
Step 8	Make sure that employees are properly informed, trained, and supervised.

4.8 Questions and Answers

Questions	Answers
1. What is a highly toxic chemical?	A substance that can cause serious health effects or death even in small quantities through inhalation, ingestion, or skin contact.
2. How should flammable liquids be stored?	In approved flammable storage cabinets away from ignition sources and incompatible materials.
3. What distinguishes a combustible liquid from a flammable one?	Combustible liquids have higher flash points (above 100 °F or 37.78 °C) than flammable ones.
4. What makes a chemical highly reactive?	Its tendency to undergo rapid or violent chemical changes under certain conditions.
5. How should corrosive chemicals be handled?	With proper PPE, in well-ventilated areas, using spill containment and resistant containers.
6. What are the dangers of cryogenic liquids?	They can cause severe cold burns and displace oxygen, leading to asphyxiation.

(continued)

Questions	Answers
7. How should cryogenic liquids be stored?	In specially designed, well-ventilated, insulated dewars, or cryogenic containers.
8. What are the risks of compressed gas cylinders?	They can become projectiles, leak toxic or flammable gases, or explode if damaged.
9. How should compressed gas cylinders be secured?	Upright with appropriate clamps or chains, away from heat and direct sunlight.
10. What PPE is essential when handling toxic chemicals?	Gloves, lab coat, eye protection, and sometimes respirators depending on the exposure risk.

Chapter 5
Storage of Chemicals

The proper storage of chemicals, particularly hazardous substances, is crucial to laboratory safety. Chemical storage is a complicated process; there is no one-size-fits-all solution, but regulations, workplace standards, and best practices may assist in managing it. The ultimate goal is to make sure that chemicals do not harm people, property, other chemicals, or the environment. Safe storage begins with a current inventory of chemicals and an awareness of the risks associated with each.

5.1 General Storage Requirements

Chemical storage must be done properly to ensure a safe and efficient laboratory environment. Inadequate storage can result in accidents, pollution, and compliance issues. Understanding and following general storage standards help avoid chemical reactions, deterioration, and exposure to dangerous situations. The instructions below define the essential principles for safe chemical storage in the laboratory:
- Store all chemicals in a secure area.
- Ensure that shelves are level, stable, and securely linked to walls or other sturdy structures.
- Store chemicals away from direct sunlight, heat sources, and exit routes.
- Hazardous substances should be stored below eye level.
- Avoid storing chemicals on the floor, window ledges, or balconies.
- Keep containers closed except for dispensing or adding chemicals.
- Use supplementary containment for liquids whenever practical. Dishpans and plastic trays both will work.
- Chemicals should not be kept in sinks or fume hoods.
- Label containers and ensure that they are compatible with the chemical.

5.2 Chemical Segregation

Chemicals should always be segregated based on their specific hazards and properties to prevent unintended or hazardous reactions. Generally, chemicals can be divided into nine different categories, as illustrated in Figure 5.1.

Chemicals should be stored apart from incompatible substances to avoid harmful reactions. Some essential conditions for segregation are listed below:
- Keep combustible substances in authorized safety containers.
- Separate acids from bases.

https://doi.org/10.1515/9783112218105-005

1. Pyrophorics

2. Water reactive chemicals

3. Flammables

4. Corrosives

5. Oxidizers

6. Toxic chemicals

7. Compressed gases

8. Explosives

9. Cryogens

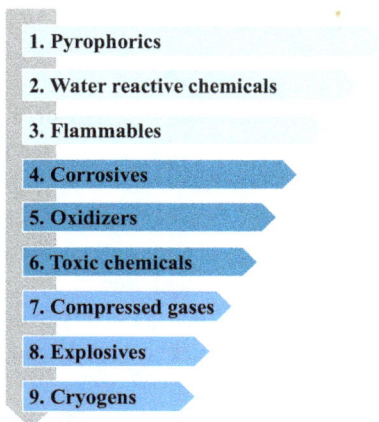

Figure 5.1: Chemical segregation categories.

- Keep oxidizers away from flammables and combustibles. Keep corrosives away from compounds that may react and produce corrosive, poisonous, or flammable vapors.
- Avoid storing chemicals alphabetically unless they are compatible. Many compounds belong to many chemical families or danger classes. Chemicals must often be analyzed individually. Ideally, rules for each category should be followed; however, this may not be achievable in every case.
- Prioritize the harmful effects of a certain chemical. A pyrophoric chemical, for example, may be a flammable liquid, but the pyrophoric feature should exceed the flammability when it comes to storage.

5.3 Chemicals and Compatibility Groups

Chemicals should be stored and separated according to their compatibility to minimize the risk of accidental reactions. Incompatible materials must be kept apart to prevent contact in the event of a spill or container failure. A good practice is to organize chemicals into groups based on similar properties, ensuring that these groups are physically separated with appropriate barriers or safe distances. When assigning a chemical to a compatibility group, keep in mind that some substances may fall into multiple categories. In such cases, determine the chemical's primary hazard and check for any specific incompatibilities that could affect its placement in a group. The most reliable way to ensure safe storage is by consulting the safety data sheet (SDS) for guidance on reactivity and compatibility. Examples of common compatibility groups are discussed in the following paragraphs but remember that this guidance is not comprehensive; it is intended as a helpful reference.

5.3.1 Pyrophorics

Pyrophoric chemicals are substances that can ignite spontaneously when exposed to air, often due to their high reactivity with oxygen or moisture. Examples include orga-nolithium reagents, alkylaluminum compounds, finely divided metals such as iron sulfide (FeS) and uranium hydride (UH_3), and metal alkoxides. When working with these materials, the following precautions should be observed:

- Store these chemicals under an inert atmosphere (e.g., nitrogen or argon) or in an appropriate liquid medium (e.g., mineral oil) to prevent contact with air.
- Store separately from water, oxidizers, and other incompatible materials (e.g., halogens and acids). Keep away from general lab traffic to avoid accidental exposure to air.
- Use airtight, moisture-free containers, typically sealed under an inert atmosphere (e.g., nitrogen or argon).
- Always ensure that containers are tightly sealed and in good condition (check regularly for corrosion, leaks, or damage).
- Store in dedicated, fire-resistant cabinets or approved safety cabinets rated for flammable and reactive materials. If cabinets are ventilated, they should be spark-free and designed to minimize oxygen ingress.
- Store in cool, dry, well-ventilated areas, away from ignition sources (open flames, sparks, and hot surfaces).
- Use sturdy, leak-proof trays or secondary containers to contain spills and prevent leaks.
- Make sure that the containment is compatible with the stored chemical (some pyrophorics can react with certain plastics).

5.3.2 Water-Reactive Chemicals

Water-reactive chemicals react violently or hazardously when they come into contact with water or even humidity in the air. Such reactions can generate flammable gases, intense heat, toxic vapors, or even cause explosions. When handling these substances, it is essential to follow these precautions:

- Store in a dry area away from any moisture.
- Store separately from aqueous solutions, acids, bases, and other chemicals that could generate moisture or vapors.
- Keep them isolated from water sources such as sinks, pipes, sprinklers, or steam lines.
- Use airtight, moisture-proof containers.
- Containers should be made of materials compatible with the stored chemical (some water-reactive chemicals can corrode certain plastics or metals).

- Seal containers firmly to avoid moisture entry.
- Store water-reactive products in a cold, dry, and well-ventilated environment, preferably in a separate cabinet.
- Ensure that the cabinet is properly labelled and placed away from emergency showers and sinks.
- Use waterproof, chemically resistant trays or bins.
- Store away from things that may leak water or aqueous solutions.

5.3.3 Flammable Liquids

Flammable liquids burn rapidly when there is an ignition source present. They often have low flash points, which allows them to produce flammable vapors at or near room temperature. The amount of flammable and combustible liquids that are allowed in the laboratory should be kept to a minimum. If the nature of the laboratory work requires more, one container of each flammable or combustible liquid will be sufficient. They should never be vented into the laboratory since their cabinets are designed to protect their contents from external fire. Vents must either be sealed or vented to the outdoors using materials or piping that provide fire protection equivalent to the cabinet itself. Refrigerators and freezers used for storing flammable or combustible liquids must be rated as "flammable material storage" or "explosion-proof" models.

5.3.4 Corrosives

When corrosive chemicals come into contact with living tissues, metals, or other materials, they can cause significant damage. Acids should be maintained apart from bases, flammable solvents, and oxidizing agents, whereas bases (alkalis) should be kept separated from acids and incompatible organic compounds. Inorganic acids (like nitric and sulfuric acid) and organic acids (like acetic and formic acid) should be kept apart. Organic acids can be stored alongside other suitable organic compounds but should be kept apart from incompatible compounds.

Concentrated nitric acid and hydrofluoric acid require special care; they should be stored separately from each other and all other chemicals. These acids can be housed in acid- or corrosion-resistant cabinets equipped with polyethylene or polypropylene compartments that help isolate them within the same cabinet. When handling volatile corrosive liquids, use fume hoods or other local exhaust ventilation. Ensure that containers are tightly closed and stored away from incompatible substances to prevent dangerous reactions.

5.3.5 Oxidizers

Oxidizers are substances that may rapidly release oxygen or other powerful oxidizing species, therefore initiating or speeding up the combustion of other materials. Nitrates, chlorates, and peroxides are only a few examples. To ensure safe operation, keep oxidizers apart from organic compounds, reducing agents, flammable or combustible materials, and other incompatible chemicals. Store containers well sealed in a cool, dry place away from heat sources and direct sunlight. Use marked containers to avoid unintended mixture or contact with incompatible goods. When dealing with oxidizers, always use appropriate personal protective equipment such as a lab coat, gloves, and goggles. When working with volatile or reactive oxidizing chemicals, use fume hoods as needed and provide enough ventilation.

5.3.6 Toxic Chemicals (Toxics)

Toxic substances can impair human health by inhalation, ingestion, or skin contact. These compounds can impact individual organs or systems and range in severity from irritants to very fatal poisons. When working with dangerous toxic compounds, the following precautions should be taken:
– Store toxic substances in well-ventilated areas, preferably in dedicated cabinets or containers that are clearly labeled and sealed.
– Use appropriate personal protective equipment, including gloves, lab coats, goggles, and, if necessary, respiratory protection.
– Handle toxic materials within a fume hood or other suitable containment to prevent inhalation exposure.
– Keep containers tightly closed and segregated from incompatible materials to prevent accidental reactions or exposures.
– Develop and practice spill response and emergency procedures specific to the toxic chemicals you work with.

5.3.7 Compressed Gases

Compressed gases are held at high pressures and can cause fires, explosions, asphyxiation, and chemical toxicity. When handling compressed gases, use the following precautions:
– To prevent tipping, secure gas cylinders upright using chains or straps. Store them in well-ventilated rooms away from heat sources and direct sunlight.
– Separate incompatible gases (e.g., combustible gases from oxidizers) to avoid hazardous reactions.

- Use proper regulators and check for leaks before use. Do not attempt to fix cylinders or valves yourself.
- Keep protective valve caps on cylinders when not in use and handle with caution to prevent damage.
- Develop and implement emergency protocols for compressed gas leaks, fires, and mishaps.
- Transport cylinders only in approved carts.

5.3.8 Explosives

Explosives are chemicals that can undergo rapid decomposition, producing heat, light, gas, and pressure, often with destructive force. Even small quantities of these materials can present serious hazards. When working with explosives, observe the following precautions:
- Store explosives in designated, approved storage facilities equipped with appropriate safety features, such as blast shields and temperature controls.
- Segregate explosives from flammable, oxidizing, and incompatible materials to prevent accidental initiation.
- Handle explosives in small quantities and only when necessary; avoid friction, shock, or impact that could trigger detonation.
- Use personal protective equipment including face shields, lab coats, and gloves rated for explosive work.
- Maintain proper labeling and documentation and regularly inspect stored materials for signs of degradation or instability.
- Develop and practice emergency response procedures for accidental detonation, spills, and fires involving explosives.

5.3.9 Cryogens

Cryogens are materials that exist as extremely cold liquids or gases. They can pose asphyxiation risks, pressure build-up, frostbite, and material embrittlement:
- Store in well-ventilated areas in specialized containers designed to withstand extremely low temperatures and pressure buildup.
- Use Dewar flasks, vacuum-insulated containers, or cryogenic storage vessels specifically designed to handle cryogenic temperatures.
- Containers must have pressure-relief valves or venting mechanisms to prevent pressure buildup and possible explosions.
- Ensure that container seals and valves are compatible with low temperatures and designed for safe venting.

- Store cryogens in well-ventilated areas to prevent the displacement of oxygen, which can cause asphyxiation.
- Never store cryogens in cold rooms, walk-in freezers, or other confined spaces without proper ventilation and oxygen monitoring.
- Store liquid oxygen separately from flammable materials, organic solvents, and combustible items because it can greatly enhance combustion.

5.4 Storage Limitations

It is a best practice to minimize the quantities of chemicals, and hazardous chemicals in specific, stored in any laboratory or work area whenever possible. Reducing the amount of chemicals on hand decreases the likelihood and potential severity of incidents such as fires, chemical spills, or exposures and facilitates safer emergency response and cleanup. Chemical storage limits are established by several regulatory bodies and codes to maintain safety and compliance, including:
- National Fire Protection Association (NFPA 45): Standard on Fire Protection for Laboratories Using Chemicals, which specifies maximum allowable quantities of flammable and combustible liquids per control area or laboratory space.
- Occupational Safety and Health Administration (OSHA) regulations: Address hazards and storage requirements for particular chemicals.
- Campus or institutional policies: Often provide further guidance tailored to local conditions and operational needs.

These limits may apply not only to individual laboratories but also to groups of labs, entire floors, or defined "fire control areas" within buildings. For example, the total volume of flammable solvents allowed on a floor may be restricted, regardless of how many labs share that space. Adhering to these limits protects personnel, property, and the environment by reducing risks and enhancing overall laboratory safety.

To further assist laboratories in managing chemical storage, below is a quick-reference checklist highlighting key actions (Figure 5.2), along with a table of typical chemical storage limits (Table 5.1). These tools can help ensure compliance with applicable regulations, support safe laboratory practices, and guide proper storage planning.

5.5 Storage Cabinets and Safety Cans

5.5.1 Flammable Storage Cabinets

The objective of flammable storage cabinets is to comply with particular rules outlined by the NFPA and OSHA, among others. They are designed primarily to confine flammable goods and prevent a fire from spreading throughout the cabinet. Venting

√ Minimize quantities of hazardous chemicals stored whenever possible.

√ Know the maximum allowable quantities (MAQs) for your chemicals according to NFPA, OSHA, and local fire codes.

√ Check if limits apply to individual labs or entire floors/fire control areas.

Chemical storage limitations "Quick checklist"

√ Label and segregate chemicals according to hazard class and compatibility.

√ Store flammable liquids in approved safety cabinets when quantities exceed threshold limits.

√ Consult Chemical Safety/EHS staff for an evaluation or questions about your specific space.

√ Keep documentation of chemical inventories and storage evaluations up to date.

Figure 5.2: Quick reference checklist for chemical storage limitations.

Table 5.1: Typical chemical storage limits.

Chemical type	Typical storage limit		Notes
	NFPA	**OSHA**	
Flammable liquids	No more than 60 gallons of flammable materials (flashpoint below 140 °F) or 120 gallons of combustible (flashpoint at or above 140 °C).	Not more than 60 gallons of Categories 1, 2, and/or 3 flammable liquids or 120 gallons of Category 4 flammable liquids shall be stored in any one storage cabinet.	Not more than 25 gallons of flammable or combustible liquids may be stored in a room outside of an approved storage cabinet. Higher volumes require storage in a flammable safety cabinet and not more than three storage cabinets may be present in a single storage area.
Flammable solids	Since no numeric limit is given by NFPA or OSHA, many institutions adopt a small-quantity approach to flammable solids. Store only what is needed for immediate work, typically less than a few pounds.		For flammable solids, OSHA and NFPA generally emphasize storage in approved, fire-resistant containers or rooms, focusing on preventing ignition sources and limiting the quantity of flammable materials in any one location.

Table 5.1 (continued)

Chemical type	Typical storage limit		Notes
	NFPA	OSHA	
Corrosive chemicals	NFPA and OSHA do not give hard limits for corrosives. Many institutions adopt a 10-gallon (38-L) guideline for open-use areas per laboratory.		Anything beyond that typically requires approved corrosive storage cabinets (polyethylene-lined or metal cabinets with corrosion-resistant coating), segregation from incompatible chemicals (e.g., separate acids from bases and oxidizers), and secondary containment trays.
Compressed gases	NFPA and OSHA do not set a strict numeric limit on compressed gas storage.		Many institutions adopt these best practices: Store only essential cylinders in the lab, often one cylinder in use and one in reserve per gas type. Store spare cylinders (especially flammables or toxics) in a dedicated gas cylinder storage room or gas cage outside the lab. Segregate incompatible gases (flammables and oxidizers). Use proper signage and secure cylinders with straps or chains to prevent tipping.
Reactive chemicals	Neither NFPA nor OSHA specifies a numeric storage limit for reactives.		Institutions typically: Store only small amounts needed for immediate use, often a few hundred grams or less depending on the chemical. Segregate reactives from incompatible materials. Use dedicated cabinets for particularly hazardous reactives (e.g., peroxides or pyrophorics) and label them accordingly. Maintain a reactive chemical inventory and rotate stock to prevent accumulation and degradation.

flammable storage cabinets is not needed and is usually not advised. Improper venting may render a cabinet's fire protection ineffective. They should be approved by Factory Mutual (FM Global), an insurance company that specializes in loss prevention and provides insurance and safety consulting services or listed with Underwriters Laboratories (UL), an international safety science organization that develops and tests products to ensure they meet safety standards.

UL approval or listing means that the cabinet is tested by Underwriters Laboratories for compliance with safety standards and indicates product meets recognized construction, fire resistance, and operational safety. FM approvals mean that the cabinet is tested by Factory Mutual for loss prevention and property protection. Indicates product meets rigorous safety, fire resistance, and risk mitigation standards.

5.5.2 Safety Cans

Safety cans are approved containers specifically designed for the safe storage, dispensing, and handling of flammable or combustible liquids like solvents, fuels, and certain chemicals. They typically range from 1 to 5 gallons (4–20 L) in capacity. They are constructed from metal (usually steel) or high-density polyethylene, with spring-loaded, self-closing lids and flame arrestors. The flame arrestor is a metal screen inside the spout that prevents an external ignition source (like a spark) from igniting the vapors inside the can. Many safety cans also have pressure-relief mechanisms to prevent over-pressurization. Safety cans prevent spills and leaks that could lead to fires or chemical exposures, minimize vapor release by using self-closing lids, reduce the risk of flashback ignition by using flame arrestors, and comply with OSHA (29 CFR 1910.106) and NFPA 30 requirements for safe storage and handling of flammable liquids.

5.5.3 Corrosive Storage Cabinets

A corrosive storage cabinet is a specialized cabinet designed to store corrosive chemicals, primarily acids and bases, that can degrade metals, cause burns, or react dangerously with other materials. Usually made from chemical-resistant materials like polyethylene, polypropylene, or epoxy-coated steel to resist corrosion. Equipped with a liquid-tight sump at the bottom to catch leaks or spills. Some models include ventilation ports for optional local exhaust or filters, though many labs keep them closed to avoid spreading fumes. Often segregated inside the cabinet (shelves and bins) to separate acids from bases, minimizing the risk of dangerous reactions. Clearly labeled with hazard warnings (e.g., "Corrosive-Acids" or "Corrosive-Bases"). Manual-close or self-closing, depending on lab policy and local fire code.

Different chemicals present varying hazards; some may be flammable, while others can be corrosive, toxic, or reactive. Selecting the appropriate type of storage cabinet (Figure 5.3) is essential for reducing the risk of fire or explosion, preventing chemical incompatibility (such as acids reacting with bases), containing potential leaks or spills, and ensuring compliance with OSHA, NFPA, and local fire codes. Table 5.2 provides a concise overview of the most common types of hazardous chemical cabinets and compares their key features.

Figure 5.3: Corrosives storage cabinet. Image credit: https://www.calpaclab.com/scimatco-sc1460-stak-a-cab-corrosive-cabinet/smc-sc1460.

Table 5.2: Common types of hazardous chemical cabinets.

Cabinet types	Purpose	Features
Flammable liquids cabinet	Stores flammable and combustible liquids such as solvents, alcohols, and fuels	Double-walled steel construction, self-closing doors, UL/FM approved, grounding connector, often yellow.
Corrosives cabinet	Stores acids and bases (corrosive liquids)	Polyethylene or epoxy-lined steel, resistant to chemical attack, separate from flammables. Often blue (bases) or white (acids).
Toxic/chemical storage cabinet	Stores general lab chemicals such as toxics or/and irritants.	May be similar to a general chemical cabinet, with good ventilation and spill containment.
Oxidizer cabinet	Stores oxidizers such as nitrates and peroxides.	Often requires segregation from flammables and reducing agents; may be labeled with an oxidizer symbol.

5.6 Chemical Inventory

A current and accurate chemical inventory is the foundation of every safe and efficient chemical storage program. It enables laboratories, workshops, and storage facilities to keep track of their chemical inventories, assuring regulatory compliance and encouraging safe handling and storage. Effective inventory management improves operating efficiency, facilitates emergency response, and eliminates chemical waste. A well-maintained chemical inventory not only helps to satisfy regulatory standards, but it also promotes a culture of safety and responsibility in the lab or storage area. Facilities that integrate inventory management with chemical storage techniques can reduce hazards, boost efficiency, and improve emergency preparedness.

The chemical inventory is an essential tool for ensuring proper storage, segregation, and comprehensive oversight of all chemicals on site. Its purpose is illustrated in Figure 5.4.

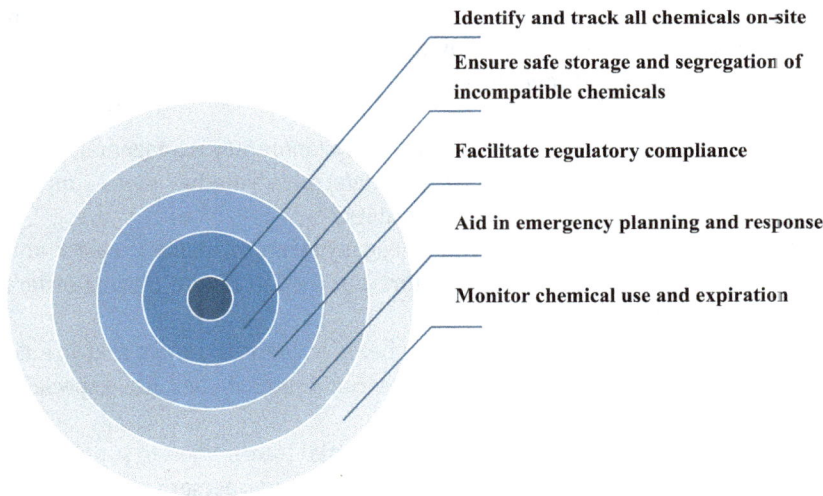

Identify and track all chemicals on-site

Ensure safe storage and segregation of incompatible chemicals

Facilitate regulatory compliance

Aid in emergency planning and response

Monitor chemical use and expiration

Figure 5.4: Chemical inventory purposes.

An effective chemical inventory should include the following information for each chemical:
- Chemical Name. Includes both the common name and synonyms.
- CAS Number. Provides a unique identifier for each chemical.
- Quantity. Indicates the amount on hand (weight, volume, or container count).
- Storage Location. Specifies building, room, cabinet, and shelf to ensure easy retrieval and proper segregation.
- Hazard Classification. Includes information such as flammability, corrosivity, toxicity, reactivity, and special storage considerations.

- Date of Acquisition. Helps with shelf-life management and identifying old or potentially degraded chemicals.
- Manufacturer/Supplier. Useful for reordering and ensuring correct product information.
- Expiration Date (if applicable). Particularly important for peroxide-forming chemicals and other time-sensitive reagents.

A good inventory system should be regularly updated to reflect current stock and usage. Updates should occur:
- When chemicals are received.
- When chemicals are used or disposed of.
- At least annually, during a formal chemical audit.

Electronic inventory systems are recommended, as they allow for easier searching, reporting, and updating. Many laboratories now use barcode or RFID systems to track chemicals and ensure efficient updates. However, smaller facilities may opt for paper-based systems, provided they are consistently maintained.

A well-maintained chemical inventory offers several important benefits, including those outlined below:
- Efficient Ordering. An accurate inventory will help laboratory management identify what things are available in the lab and determine whether more supplies are needed before placing duplicate/excess orders.
- Locating Reagents. Defining and tracking reagent locations make it easier and faster for researchers to locate reagents stored in the lab before placing orders for materials that are already available.
- Deliberate Disposal. An accurate inventory helps to identify things that are no longer in use as well as track commodities having a limited shelf life, especially those that may become harmful with time.
- Effective Reaction. Chemical inventories give essential information to supplement Emergency information posters, allowing fire and hazmat personnel to respond to campus crises safely and effectively.
- Meet Hazardous Material Restrictions. Fire and building rules limit the amount of hazardous material that can be stored in a certain place, depending on a variety of construction-related criteria. Accurate chemical inventory reporting (both subjectively and quantitatively) enables the identification and mitigation of the life safety risk posed by hazardous substance overages.
- Effective Hazard Communication. An accurate chemical inventory detects and conveys the risks associated with substances to potential consumers. The inventory also serves as a reference point, allowing users simple access to SDSs.
- Improved Storage Mechanisms. An accurate inventory can help to reduce some of the effort involved in finding locations where incompatible materials are stored

together as well as simplify planning for how to adopt more effective and safer chemical storage and separation procedures.

5.7 Questions and Answers

Questions	Answers
1. What are chemical compatibility groups?	Categories that group chemicals based on their reactivity and storage compatibility.
2. Can acids and bases be stored together?	No, they should be stored separately to avoid violent neutralization reactions.
3. Where should oxidizers be stored?	Away from flammables, organics, and reducing agents.
4. What happens if incompatible chemicals are mixed?	They can react violently, causing heat, fire, toxic gases, or explosions.
5. Are flammable liquids compatible with oxidizers?	No, they should be kept far apart to avoid fire hazards.
6. Why should water-reactive chemicals be stored away from sinks?	Because contact with water can cause fire, heat, or toxic gas release.
7. Can organic peroxides be stored with other flammable liquids?	No, they are unstable and may detonate or burn violently.
8. Should chemicals be stored alphabetically?	Only within compatible groups, to avoid storing incompatible substances together.
9. Should heavy chemical containers be stored on high shelves?	No, they should be stored on lower shelves to prevent injury from falls.
10. How should chemicals be grouped in storage areas?	By compatibility group, not alphabetically.

Chapter 6
Chemical and Hazardous Chemical Waste Management

6.1 Chemical Laboratory Waste

Chemical waste management (also known as hazardous waste management) is the process of collecting, transporting, processing, recycling, and disposing of waste materials. Chemical waste in laboratories refers to any compounds, by-products, or materials that are left over or created during experimental operations. This garbage can be classed as hazardous or nonhazardous, and it includes a wide range of objects such as organic and inorganic compounds, broken glassware, and sharps. Figure 6.1 offers a quick summary of the various waste categories.

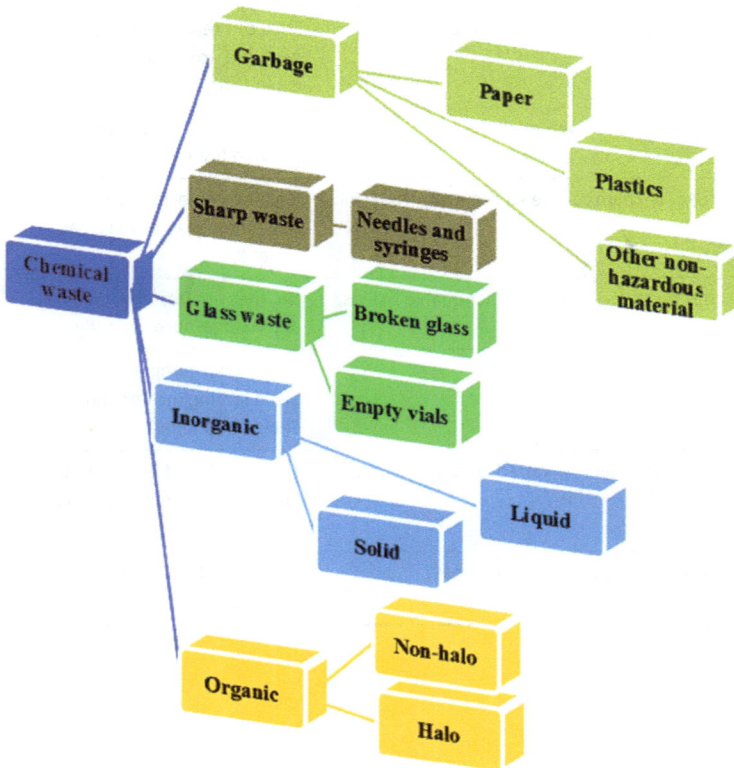

Figure 6.1: Different types of waste materials.

https://doi.org/10.1515/9783112218105-006

Hazardous chemical waste can be any solids, liquids, or gases containing or polluted with compounds that endanger human health, safety, or the environment. These may include flammable or combustible liquids, reactive compounds including oxidizers and water-reactive materials as well as corrosives, reducing agents, and acutely or chronically poisonous substances. Managing these hazardous items requires special care since their existence necessitates rigorous adherence to handling, storage, and disposal regulations in order to avoid accidents and environmental pollution. Understanding the nature and categorization of these wastes is critical for successful chemical waste management and meeting regulatory requirements.

6.2 Minimizing Chemical Hazardous Waste

Waste reduction is an important part of chemical and hazardous chemical waste management, with an emphasis on lowering the volume and toxicity of waste created in labs, industrial facilities, and other organizations. Effective waste reduction not only lowers disposal costs and regulatory hassles, but it also encourages sustainability and resource conservation. Strategies include:
- Source Reduction. Modifying procedures or replacing less hazardous materials to decrease waste at the source. For example, conducting microscale experiments in teaching laboratories or using less harmful chemicals in manufacturing processes.
- Process Optimization. Improving operational efficiency through higher reaction yields, more precise chemical measurement, and process monitoring to reduce by-products and off-spec materials.
- Inventory Management. Implementing strict inventory controls to avoid over-ordering and unnecessary stockpiling of chemicals that may degrade or expire. This includes first-in, first-out rotation systems and regular inventory audits.
- Reuse and Recycling. Identifying opportunities to reuse chemicals within a process or recover valuable components for other applications. For instance, distillation of solvents for reuse or recovering metals from waste streams.
- Product Substitution. Replacing hazardous chemicals with safer alternatives wherever feasible, thereby reducing the volume and hazard level of waste.

By integrating waste minimization strategies, organizations can significantly reduce the environmental impact of hazardous waste and support a safer, more sustainable laboratory or workplace environment. Unattended leftover chemicals pose significant hazards if mishandled by anyone, and they can also cause environmental harm. Therefore, it is essential for everyone working in a chemical laboratory to strive to minimize the quantity of chemicals used and, consequently, the volume of waste produced. To effectively manage chemical waste, it is also recommended to follow three key principles (Figure 6.2) that serve as the foundation of a comprehensive waste minimization:

- Reduce Chemicals. Keep only a minimum amount of chemicals in the laboratory. Bring to the laboratory only the amount of chemicals required for the experiment, if possible. Use the smallest possible amount for the experiment.
- Reuse Chemicals. Borrow and lend chemicals to fellow researchers, technicians, and chemists. This will reduce the accumulation of chemicals by purchasing the same ones from different individuals.
- Recycle Chemicals. Recycle chemical waste whenever possible.

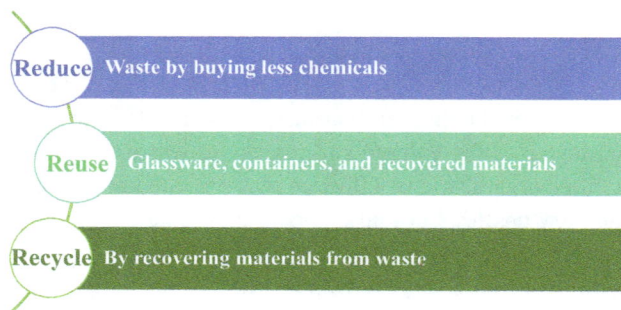

Reduce Waste by buying less chemicals

Reuse Glassware, containers, and recovered materials

Recycle By recovering materials from waste

Figure 6.2: Three key principles of chemical waste management.

6.3 Handling and Storage

Generally, all the precautions one has to follow when handling, storing, and using laboratory chemicals apply to hazardous laboratory waste. Safe handling and storage of hazardous waste generated at the worksite, through proper identification and worker education, is a must in any workplace. Laboratory leaders and/or supervisors have to put in place safe work procedures to deal with hazardous waste from receiving through storage until disposal. They must ensure that all laboratory personnel are trained on those procedures. A few points to keep in mind when storing hazardous laboratory waste:

- Keep containers free of chemical contamination.
- Segregate chemicals according to compatibility.
- Do not store incompatible chemicals together.
- Leave 10–20% air space in the chemical waste containers to allow for vapor expansion and to reduce the spills from occurring when moving overfilled containers.
- Never accumulate large volumes of hazardous waste and have it disposed of regularly.

Waste containers should be kept closed at all times, except when contents are being added. Do not leave filter funnels in the open necks of containers, even if the waste is in a fume hood. Fume hoods are not to be treated as a worry-free method of waste containment or disposal. Wastes should be separated as follows:
- Separate liquid and solid waste.
- Separate liquid organic waste from liquid aqueous waste.
- Separate strong acids and bases from other aqueous waste.
- Separate halogenated from non-halogenated waste.

6.4 Labeling Hazardous Waste

The person who creates the garbage is accountable for accurate waste labeling. All garbage containers must be appropriately labeled so that the contents may be identified. Containers should have the following information:
- Building name and lab number
- Name of principal investigator or researcher working in the laboratory and a contact phone number
- The main contents should be broken down by approximate percentage

Waste containers should not be labeled with broad, ambiguous terminology like "chemical waste," "inorganic waste," or "solvent waste." Use explicit identifiers to identify the contents; avoid abbreviations, acronyms, trademarked names, and chemical formulae.

Attach a label to the container prior to being filled and maintain a list of contents as waste is added to the container. Deface or remove old labels on containers used for chemical waste.

6.5 Packaging Hazardous Waste

Packaging hazardous waste is an important step in the safe management and disposal of chemical waste in labs, as it ensures that potentially harmful compounds are confined and handled in compliance with safety rules. Proper packing avoids leaks, spills, and unintentional exposure, safeguarding both laboratory staff and the environment. Before packing, waste should be classified by hazard class, such as flammables, corrosives, toxics, or reactives, and stored separately to avoid incompatible reactions.

All waste containers must be chemically compatible with their contents, appropriately sized, and clearly labeled with the full chemical names and hazard information, as per regulatory requirements. Secondary containment, such as leak-proof trays or bins, should be used for added protection during storage and transportation. Containers must be sealed securely, using tight-fitting lids, and should never be overfilled to prevent spillage. Labels should include the laboratory name, date of packaging, and the responsible person's contact information. Additionally, it is important to inspect containers regularly for signs of deterioration, leaks, or damage, and to replace them if necessary.

Adhering to these packaging guidelines not only facilitates safe and efficient waste pickup but also ensures compliance with environmental and safety regulations, thereby contributing to a sustainable and responsible laboratory waste management program. Proper packaging of waste has to be maintained by the waste generator. For laboratory chemical waste, if possible, the original chemical containers should be used for disposal. Otherwise, choose a container that has the following criteria:

- A sealable with a tight screw lid
- A waste compatible
- Not damaged or defective

6.6 Special Waste

Special waste refers to laboratory-generated materials that do not fit neatly into normal hazardous or nonhazardous waste categories but need careful treatment, management, and disposal owing to their unique qualities or potential hazards. Items in this category include pressurized gas cylinders, mercury-containing devices (e.g., thermometers and manometers), batteries, radioactive materials, and some forms of electronic trash such as circuit boards and computers. Although these items may not necessarily fit the rigorous statutory definition of hazardous waste, their disposal is subject to special standards designed to protect human health and the environment. For example, mercury-containing equipment requires collection in sealed, labeled containers to prevent vapor release, while batteries must be packaged to prevent short-circuiting and leaks. Compressed gas cylinders, whether empty or partially full, must be stored upright, properly labeled, and segregated from flammable materials before being returned to the supplier or disposed of through an authorized recycling program.

Managing special waste effectively requires laboratory personnel to understand the unique hazards associated with each type, follow appropriate segregation and labeling protocols, and coordinate disposal with approved vendors or institutional environmental health and safety (EHS) offices. By identifying and properly managing special waste, laboratories uphold their commitment to safety, environmental responsibility, and regulatory compliance.

6.7 Hazardous Waste Pickup

Hazardous waste pickup is a critical component of laboratory waste management that ensures the safe, efficient, and compliant disposal of hazardous materials generated during laboratory operations. This process involves the collection and proper handling of chemical waste that poses risks to human health or the environment including substances that are toxic, corrosive, flammable, or reactive. Laboratories typically coordinate with designated waste disposal contractors or internal EHS departments to schedule regular pickups, ensuring that hazardous waste is removed from the premises promptly. Before pickup, all waste must be appropriately labeled, segregated by hazard class, and stored in suitable containers that meet regulatory requirements.

Proper paperwork, such as garbage manifests or inventory sheets, is necessary to comply with local, national, and international hazardous waste transportation and disposal regulations. Furthermore, laboratory staff must be taught proper waste preparation procedures such as separating incompatible materials to prevent dangerous reactions. Implementing an ordered and consistent hazardous waste pickup schedule not only protects laboratory staff but also reduces the risk of environmental contamination, reduces liability, and develops a culture of safety and responsibility in the laboratory.

6.8 Recordkeeping and Compliance

Proper recordkeeping and compliance are essential components of effective hazardous chemical waste management. Regulatory agencies such as the EPA (USA), EA (UK), and Ministry of Environment (various countries) require thorough documentation to ensure that chemical waste is managed safely and legally. Key aspects (Table 6.1) include:

- Waste Manifests and Tracking. Maintaining accurate records of waste generation, storage, transportation, and disposal. This includes manifests that accompany waste shipments and documents that detail quantities, hazard classes, and final disposal methods.
- To cut down on waste and avoid accidental stockpiling or improper disposal, keep thorough inventory records for chemicals in use, storage, and disposal.
- Record staff training on the safe handling, storage, and disposal of hazardous materials. To guarantee adherence to relevant laws, training should be updated often.
- To make sure safety regulations are being followed and to identify areas for improvement and keep regular inspection records for storage facilities, trash cans, and safety equipment.
- Adherence to waste management regulations including Water Framework Directive (WFD) in the EU, COSHH in the UK, and Resource Conservation and Recovery Act (RCRA) in the United States. There might be fines, penalties, or legal action for noncompliance.

Table 6.1: Key aspects of proper recordkeeping and compliance.

Record type	Description
Waste manifests and tracking	Accurate documentation of waste generation, storage, transport, and disposal.
Inventory records	Maintain records of all chemicals in use, storage, and disposal to support waste minimization.
Training records	Document employee training on safe handling, storage, and disposal of hazardous chemicals.
Inspection reports	Regular checks of storage areas, waste containers, and safety equipment, with corrective actions.
Compliance with regulations	Adhere to national, regional, and local waste management regulations (e.g., RCRA and COSHH).

6.9 Questions and Answers

Questions		Answers
1.	What is hazardous chemical waste?	Waste that poses substantial or potential threats to public health or the environment.
2.	How should hazardous chemical waste be labeled?	With a clear label indicating "Hazardous Waste," contents, and date of accumulation.
3.	Can hazardous waste be poured down the drain?	No, it must be disposed by following proper regulatory procedures.
4.	How should waste containers be selected?	They must be compatible with the chemical waste and leak-proof.
5.	Can different types of chemical waste be mixed?	No, unless compatibility is confirmed, and mixing is approved by waste authorities.
6.	How should waste containers be kept?	Tightly closed when not in use and stored in a designated, labeled area.
7.	Who is responsible for managing chemical waste in a lab?	The lab personnel who generate the waste and designated safety officers.
8.	What PPE is required when handling hazardous waste?	Gloves, lab coat, goggles, and sometimes face shield or respirator, depending on the hazard.
9.	What should you do in case of a hazardous waste spill?	Follow the spill response procedure, use spill kits, and notify the appropriate authorities.
10.	Why is segregation of hazardous waste important?	To prevent dangerous chemical reactions and enable proper disposal.

Chapter 7
Hazard Control Measures

Building on the foundation established in previous chapters, this chapter explores the three core components of the hazard hierarchy of controls: engineering controls, administrative controls, and personal protective equipment (PPE), which together form the backbone of laboratory and workplace safety (Figure 7.1). Engineering controls, which aim to eliminate or isolate hazards at their source, are preferred over administrative measures and PPE because they provide a more effective and reliable means of protection.

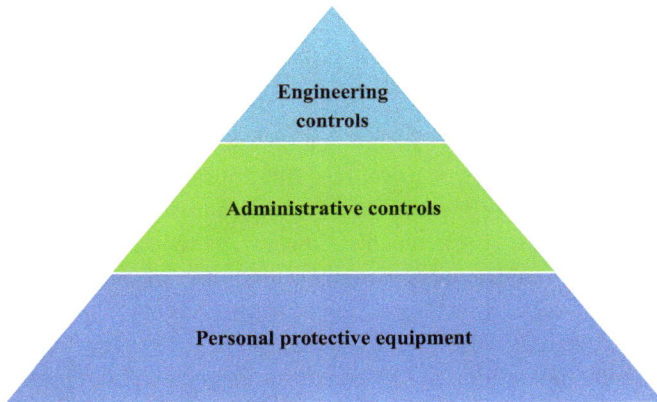

Figure 7.1: Safety three lines of defense.

However, it is important to recognize that these measures are part of a broader hierarchy of hazard control measures (Figure 7.2), in which elimination and substitution represent the most effective strategies for hazard reduction. Together, these strategies form a comprehensive approach to managing chemical risks in laboratory and industrial environments.

The type and level of control required depend on factors such as the hazard present, the level of exposure, the product's toxicity, and process specifics. Engineering controls aim to eliminate or reduce hazards at the source, such as by substituting safer chemicals, isolating the hazard, or using ventilation. Administrative controls (also known as work habits controls) include strategies like work scheduling, safe work procedures, experiment planning, and training. PPE is used when hazards cannot be fully managed through engineering or administrative means. Essential PPE includes eye protection, gloves, lab coats, and closed-toe shoes. The following sections provide detailed guidance on these controls.

https://doi.org/10.1515/9783112218105-007

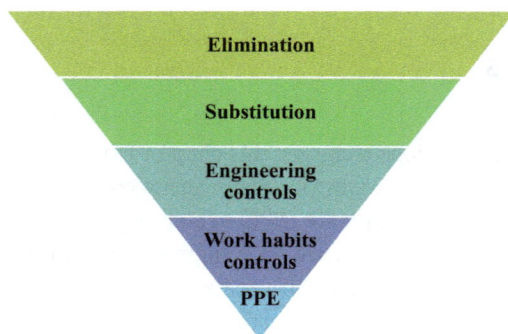

Figure 7.2: Hierarchy of hazard control measures.

7.1 Elimination or Substitution

The most effective way to control hazards is to eliminate the use of hazardous chemicals altogether, thereby removing the risk at its source. This approach not only safeguards workers' health but also reduces the potential for environmental contamination and minimizes the need for subsequent controls. However, complete elimination is not always feasible due to the requirements of certain processes, research protocols, or the lack of viable alternatives. In such cases, the next best strategy is to substitute hazardous chemicals or processes with less hazardous ones, thereby reducing the risk to acceptable levels while maintaining the necessary functionality. Examples of effective chemical substitutions include:

- Benzene, a known human carcinogen, can be replaced with toluene, which has lower toxicity and is less volatile, thereby reducing inhalation hazards.
- Carbon tetrachloride, a potent hepatotoxin and suspected carcinogen, can be replaced with dichloromethane (methylene chloride), which, while still hazardous, presents lower toxicity and is less persistent in the environment.
- Talc is primarily used in analytical chemistry laboratories, materials science labs, geology labs, and industrial hygiene labs, particularly in procedures involving powdered samples, sample preparation, lubrication, or as a drying or anti-caking agent. It is also sometimes used as a carrier or diluent in some spectroscopic and microscopic analyses. However, because talc may contain asbestos fibers or can produce fine respirable dust, which poses respiratory hazards (including lung disease), many laboratories now opt to substitute talc with chalk (calcium carbonate). Chalk is used in similar applications (e.g., as a powder medium or anti-caking agent) but is less likely to contain hazardous fibers or produce harmful dust, making it a safer alternative.
- Sandblasting, which can generate large amounts of respirable crystalline silica dust (a known cause of silicosis and lung cancer), can be substituted with steel

shot blasting, which produces significantly less dust and can often be captured more easily through local exhaust ventilation (LEV).
- Dry handling techniques (e.g., powder transfer) can be replaced with wet handling techniques (e.g., slurry preparation or wet wiping), which suppress dust emissions and reduce inhalation risks.

It is crucial to assess the suggested alternative's physical and chemical characteristics before putting it into practice to make sure that it does not create new risks or reduce the efficiency of the operation. It is important to consider elements including environmental persistence, flammability, reactivity, and regulatory compliance. In the end, substitution is a proactive measure that may greatly improve workplace and laboratory safety within the hierarchy of hazard controls. Substitution is a crucial component of an all-encompassing risk management plan when paired with engineering and administrative controls and enhanced by the proper PPE.

7.2 Engineering Control

A key process for reducing exposure to hazardous chemicals is ventilation, which is typically achieved using mechanical air handling methods. The two main types of ventilation are LEV and general (or dilution) ventilation:
- LEV captures airborne contaminants directly at the source of emission and exhausts them to a safe location or treats them with air-cleaning techniques.
- General or dilution ventilation reduces contaminant concentrations by mixing contaminated air with clean, fresh air. While it is effective for controlling heat and odors in spaces like control rooms, offices, and labs, it is not recommended for dust, mists, fumes, or highly toxic substances, especially when contamination rates are high or uneven.

Among mechanical air handling techniques, fume hoods (Figures 7.3 and 7.4) are particularly effective for protecting laboratory workers from hazardous chemicals. Chemical fume hoods have a movable front sash that acts as a protective barrier and an interior baffle that directs the airflow to maximize the capture of aerosols, fumes, and vapors. Adjust baffles as needed according to the manufacturer's manual or with guidance from the safety officer. Since most chemical fume hoods operate with a constant exhaust flow rate, lowering the sash increases the face velocity by reducing the cross-sectional opening. However, fume hoods must not be considered a substitute for proper waste disposal. Equipment inside the hood should include condensers, traps, or scrubbers to collect or contain waste and vapors, ensuring safe and efficient laboratory practices.

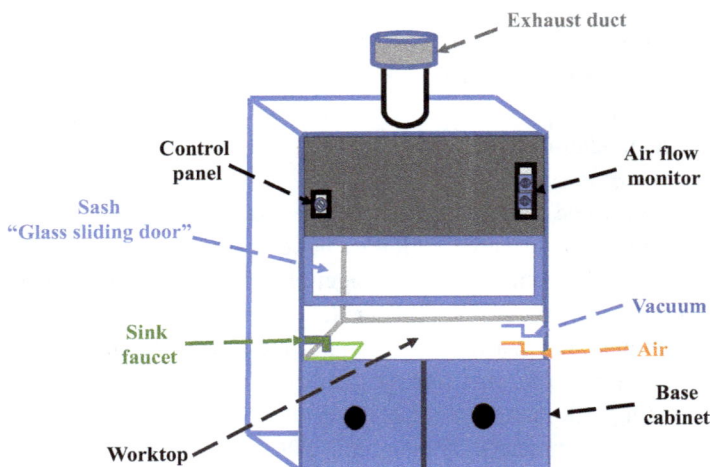

Figure 7.3: Schematic diagram for the fume hood.

Before starting work in a fume hood, always ensure that it is operational. Confirm that the local on–off switch is in the "on" position and check airflow. If no airflow gauge or velometer is available, tape a strip of 1-inch-wide tissue to the lower corner of the sash; if the tissue is gently drawn into the hood, airflow is present. A malfunctioning fume hood is more dangerous than no hood at all as it creates a false sense of safety. Never work in a fume hood that is tagged "out of service," as this could expose maintenance workers to hazardous chemicals.

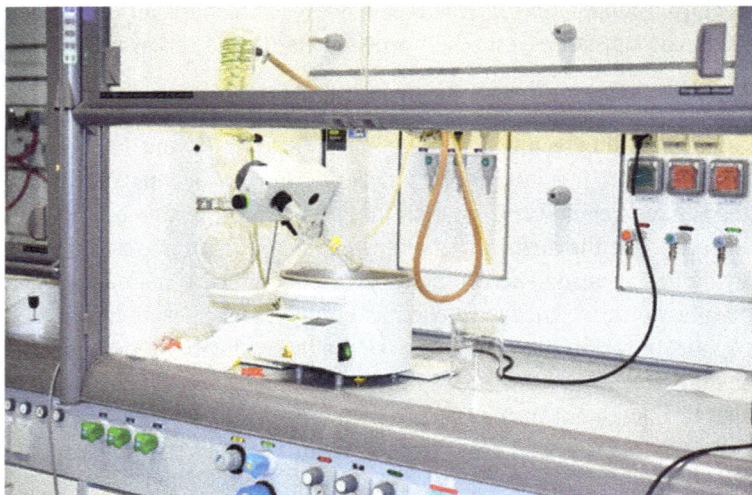

Figure 7.4: Fume hood. Image credit: https://depositphotos.com/photos/laboratory-fume-hood.html.

Recommended operating parameters:
– Typical Chemical Fume Hood. At a 30-cm sash opening, the average face velocity should be 100 fpm or 0.5 m/s; no point should be less than 80 fpm (0.4 m/s).
– Extremely Harmful Substances. There is no point below 100 fpm (0.5 m/s), with an average face velocity of 120 fpm (0.6 m/s).
– For safe functioning, the airflow should be consistent and turbulence-free.

Best practices for safe use:
– Keep the sash opening at 30 cm or less while working and fully close it when not in use. To maintain safe operation, do not exceed a sash height of 40–44 cm (16–18 inches).
– Avoid blocking air slots at the back of the hood; maintain a 3-cm clearance.
– Position the equipment at least 15–20 cm from the front edge. Use stands to lift bulky items to maintain airflow.
– Keep the hood clean and uncluttered. Store only items needed for active processes; do not use the hood for long-term storage.
– Never modify the hood's interior by adding shelves.
– Minimize foot traffic around the hood; people walking past can cause turbulence.
– Keep windows and doors near the hood closed, as drafts can disturb airflow.
– To ensure that chemical fume hoods are operating correctly and offering sufficient protection, test them at least once a year. Avoid using fans close to the hood as they can also interfere with optimum airflow. Every chemical fume hood needs to have an inspection sticker attached to it that shows the face velocity measurement and the date of the hood's inspection. Until they are fixed and re-tested, chemical fume hoods that do not pass inspection should not be utilized.
– Provide a backup containment system for containers that could shatter or leak.
– If it is necessary to put bulky equipment in the chemical fume hood, raise it about 2 inches above the ground so that air may flow underneath it.
– Ensure that all electrical devices are connected outside the chemical fume hood to avoid electrical arcing that can ignite a flammable or reactive chemical.
– Clean all chemical residues from the chemical fume hood chamber after each use.
– Keep the sash completely lowered when the chemical fume hood is not in use or when an experiment in the hood is left unattended. Lowering the sash not only provides additional personal protection but also results in significant energy conservation.
– Avoid storing chemicals and equipment in the chemical fume hood.
– Do not use perchloric acid in a conventional chemical fume hood. Perchloric acid vapors accumulate in ductwork and form perchlorate crystals that have the potential to explode, causing serious injury to personnel and damage to property.
– Never turn off the chemical fume hood.

Many types of laboratory equipment, such as gas chromatographs, atomic absorption spectrometers, and ovens, generate hazardous vapors and gases but cannot be operated inside a traditional fume hood. LEV should be used to capture and remove these emissions. Ideally, a dedicated exhaust system is recommended. If connecting to an existing hood duct, ensure that the fan capacity is increased, and airflow is balanced between all hoods. Remember that each new exhaust point requires adequate make-up air to replace the exhausted air.

General laboratory ventilation controls both air quality and quantity by continuously supplying and replacing air, thereby reducing the concentration of odorous or toxic substances. Laboratories are typically maintained at negative pressure relative to adjacent areas to prevent the spread of contaminants to other parts of the building.

7.3 Administrative Control (Work Habits Control)

Administrative controls or work habits controls are workplace policies, procedures, and practices that minimize the exposure of workers to risk conditions. Great care is needed to ensure that procedures, once adopted, are observed; particularly in the longer term, as shortcuts and nonobservance can become "custom and practice" over time and once established can be difficult to overcome.

Good and continuous housekeeping is particularly important in processes and laboratories where hazardous materials may be handled. Clear labeling, with relevant health and safety advice, careful and appropriate storage, and good work techniques all need to be addressed.

Good housekeeping can also help to minimize airborne contamination from spilled materials and waste off-cuts. The untidy workplace may also prevent access to essential system controls, such as LEV on–off switches, which could discourage their proper use, and this, in turn, affects the ventilation efficiency in the work area.

Education of employees on any health hazards in the workplace and the importance of correctly using all the control measures provided, adopting recommended operating procedures, and wearing personal protection, if required, is needed in order to minimize the risks to health.

Training of employees on the use of the appropriate control measures, operating practices, and the factors involved in the correct selection, use, and maintenance of PPE.

Good hygiene practices are also important steps workers should take to protect their health. This includes following established decontamination procedures, where applicable, regular laundering of clothing, using approved methods and facilities; good personal hygiene, such as frequent washing and showering, particularly before meal breaks.

7.4 Personal Protective Equipment

PPE (Figure 7.5) is the last control and only applicable when the preceding measures are insufficient or not reasonably practicable in achieving a satisfactory work protection environment. It is important to ensure that the protection is effective and comfortable. Regular maintenance is vital for many types of PPE if effective protection is to be obtained. PPE management programs need to be adopted whenever the option of PPE use is deemed necessary.

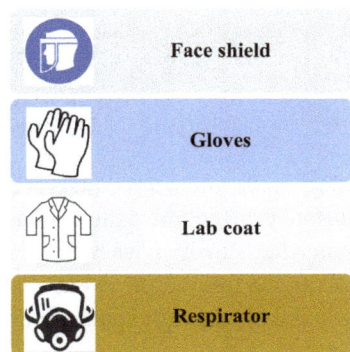

Figure 7.5: Personal protective equipment.

7.4.1 Eye and Face Protection

Eye protection should be worn in or around all laboratories when working with chemicals. The type of eye protection required depends on the risk. Safety goggles with side shields are sufficient for most situations. For more hazardous activities where there is a risk of chemical spray or explosion, safety goggles or face shields designed to protect against chemical spray should be used. This is especially true for working with corrosive chemicals.

7.4.2 Gloves and Hand Protection

The right type of gloves provides much-needed hand protection in the lab. It is recommended to wear appropriate gloves when handling hazardous chemicals, toxins, corrosive substances, or hot or cold objects. Particular attention should be paid to chemicals that have a "Skin" mark on the MSDS sheet.

When choosing a glove, consider the circumstances in which the glove will be used. The degree of protection required will depend on the hazards associated with the chemical, the type of experimental work performed, the scale, and the individual's work habits.

Disposable gloves made from suitable material are generally acceptable for routine laboratory work with small quantities of chemicals, as they offer the best combination of tactile sensitivity and barrier protection. They should be removed and replaced when contaminated. Reusable gloves should be inspected before each use, replaced each time they discolor or show signs of damage, and cleaned or disinfected after each use.

When working with chemicals, it might be more dangerous to wear the incorrect kind of gloves than to wear none at all. The glove may sustain significant harm if a chemical seeps through and is kept close to the wearer's hand for an extended period of time.

7.4.3 Aprons and Coats for the Lab

Aprons and other lab attire are worn to absorb or deflect spills and keep dangerous or corrosive materials from getting to the skin. For a normal coat, cotton is the recommended material since it is inexpensive and burns somewhat slowly. Layers of synthetic fiber are not advised since they may melt and stick to the flesh in the event of a fire. Use synthetic fabrics like Tyvek or Nomex that are resistant to chemicals or flames in high-risk scenarios.

Plastic or rubber aprons should be used when handling large amounts of concentrated acids and other corrosive substances.

7.4.4 Respiratory Protection

Respiratory protection is not normally required when working in the laboratory, due to the combination of engineering controls, such as hoods, safe working procedures, and relatively small amounts of chemicals used in the laboratory. In order to use respiratory protection, a hazard assessment must first be performed, and if the result of the assessment recommends the use of respiratory protection, then laboratory personnel in need of protection must be instructed in the proper use, care, and maintenance of respiratory equipment.

7.5 Safe Work Practices

In addition to engineering controls and personal protection equipment, proper work practices are necessary to provide a safe and healthy laboratory environment. These protocols entail carefully planning experiments, keeping clean work areas free of unnecessary chemicals, and marking each container correctly. Laboratory personnel should always read and understand the safety data sheets for the chemicals they handle and follow the correct storage procedures to prevent unintentionally combining incompatible substances.

Eating, drinking, and smoking should be strictly prohibited in lab environments. It is crucial to often wash your hands, especially before leaving the lab, to stop the spread of contaminants. Finally, putting in place a strong training program that prioritizes emergency protocols, danger awareness, and equipment usage guarantees that all staff members can operate safely and efficiently.

7.6 Emergency Showers and Eyewash Stations

Any laboratory that utilizes corrosive or other chemicals that are dangerous to the skin or eyes must have a safety shower and eyewash station (Figure 7.6) nearby and constantly available. Workers in the laboratory should be aware of the locations of the eyewash stations and safety showers as well as how to use the equipment correctly and access to these safety equipment must never be restricted. There should also be signage indicating their locations.

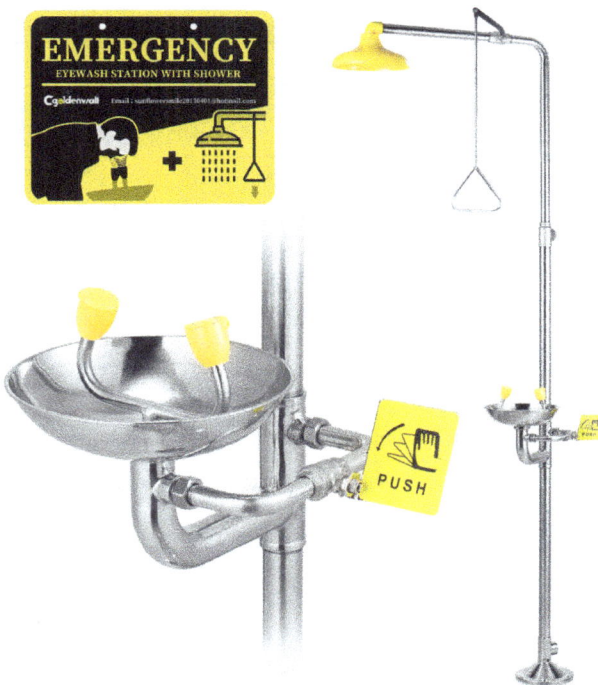

Figure 7.6: An eyewash station and a safety shower. Image credit: https://www.amazon.ca/CGOLDEN WALL-Combination-Emergency-Eyewash-Stainless/dp/B07NJTFMY4?th=1.

Dosing or drench hoses are common fittings in many labs and complement eyewashes and showers. They can be used to wash a small spot when a full shower is not necessary, to help a person who is unable to stand or is unconscious, or to irrigate under clothing before removing for a full emergency shower flush. Eyewash bottles can be used as an accessory or as a stand-alone. They provide immediate rinsing of contaminants or small particles, followed by regular 15-min rinsing at the eyewash station. Eyewashes and showers should be tested regularly by rinsing them for at least 3 min, once a week, to ensure that they are functioning properly and to prevent microbial growth or the formation of any dirt, rust, or scale.

In conclusion, effective hazard control in the laboratory requires a multifaceted approach. By applying the hierarchy of controls, from elimination and substitution through engineering controls, administrative controls, and finally PPE, laboratories can reduce risks and protect personnel. Safe work practices, proper ventilation, and readily accessible emergency equipment reinforce these measures, ensuring a safer and more sustainable laboratory environment.

7.7 Questions and Answers

Questions	Answers
1. What are hazard control measures?	Strategies used to reduce or eliminate risks associated with hazardous substances or conditions.
2. What is the first step in hazard control?	Identifying and assessing the hazard.
3. What are the five main types of hazard controls?	Elimination, substitution, engineering controls, and administrative controls and PPE.
4. What is elimination in hazard control?	Completely removing the hazardous material or activity.
5. What is substitution in hazard control?	Replacing a hazardous substance or process with a less hazardous one.
6. What are engineering controls?	Physical changes to equipment or work areas like fume hoods or ventilation systems.
7. Why is PPE considered the last line of defense?	It does not eliminate the hazard but protects the user if other controls fail.
8. What is a fume hood used for?	To limit exposure to hazardous vapors and gases by containing and exhausting them.
9. What role does safety training play in hazard control?	It ensures that workers recognize hazards and know how to respond appropriately.
10. What is the hierarchy of controls?	A ranking of hazard control methods from most effective (elimination) to least effective (PPE).

Chapter 8
Fire and Explosion Safety

8.1 Introduction

Fire can be defined as the rapid combination of a flammable substance with oxygen, with the rapid development of heat and light. When most substances burn, the actual combustion occurs when the solid or liquid evaporates or is decomposed by heat to produce gas. The visible flame is burning gas or vapor. Fire and explosions are among the most serious hazards in a chemical laboratory. Laboratories routinely handle flammable liquids, gases, reactive chemicals, and electrical equipment, all of which can contribute to fire and explosion risks if not properly managed.

8.2 Fire Safety

Understanding the fundamentals of fire is critical for successfully controlling fire threats in the laboratory. Fire is a chemical process in which a fuel source undergoes fast oxidation, resulting in heat, light, and combustion products. Fires in laboratories can swiftly grow due to the availability of combustible chemicals and reactive materials.

8.2.1 The Fire Triangle and Fire Tetrahedron

The fire triangle (Figure 8.1) illustrates the three essential elements required for a fire to occur:
- Fuel. Any combustible material such as solvents, paper, or laboratory bench surfaces.
- Oxygen. Typically, from the air but can also come from chemical oxidizers.
- Ignition Source. Heat, sparks, open flames, or static electricity.

Figure 8.1: The fire triangle. Image credit: https://firepreventionindia.com/the-fire-triangle/.

https://doi.org/10.1515/9783112218105-008

The fire tetrahedron (Figure 8.2) adds a fourth component to describe self-sustaining fires:
- Chemical Chain Reaction. The sequence of chemical reactions that propagates the fire once it starts. Removing any one of these components breaks the chain and extinguishes the fire.

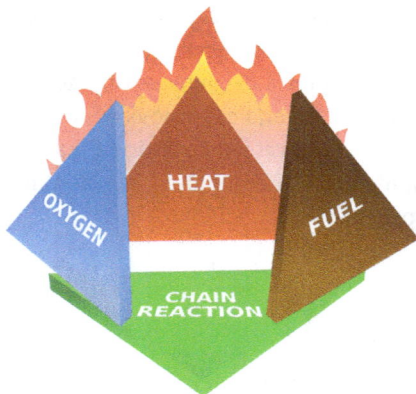

Figure 8.2: The fire tetrahedron. Image credit: https://fire-risk-assessment-network.com/blog/fire-triangle-tetrahedron/.

8.2.2 Fire Behavior in Laboratories

Laboratories present unique fire hazards due to:
- Volatile Chemicals. Many solvents (e.g., ethanol and acetone) readily produce flammable vapors.
- Confined Spaces. Fume hoods and enclosed areas can accumulate flammable atmospheres.
- Reactive Chemicals. Some chemicals (e.g., peroxides and pyrophoric materials) can ignite spontaneously or react violently.
- Multiple Ignition Sources. Bunsen burners, hot plates, electrical equipment, and static discharge.

8.2.3 Combustion Stages

Combustion is an exothermic oxidation process that produces heat and light. The chemical reaction changes depending on the kind of fuel. Under ideal conditions (complete combustion), the process yields carbon dioxide and water. However, in real fires, incomplete combustion frequently happens owing to a lack of oxygen or inadequate mixing, resulting in the generation of carbon monoxide (CO), soot (carbon particles), and other hazardous chemicals. The combustion process involves several phases, including:

- Preignition. The fuel is heated; moisture evaporates and then decomposes (pyrolysis).
- Ignition. The temperature approaches the ignition point, and combustion starts.
- Flames (Gas-Phase Combustion). Volatile gases ignite in the presence of oxygen.
- Smoldering (Surface Combustion). Occurs with little oxygen and lower temperatures but remains harmful.
- Flashover. Rapid transition to full room involvement due to buildup of heat and flammable vapors.

Fire spreads by three mechanisms known as conduction, where heat is transferred through materials such as metal surfaces. Convection is where heat is carried by gases through smoke or plumes. Radiation where the infrared heat is emitted by flames and hot surfaces. Understanding the combustion process aids in selecting appropriate extinguishment strategies and predicting fire behavior.

Fires generally progress through the following stages:
- Ignition. The point at which fuel, oxygen, and an ignition source come together.
- Growth. The fire increases in size and heat output, potentially involving additional fuels.
- Fully Developed Fire. All available fuel is involved; the fire reaches maximum intensity.
- Decay. Fuel is consumed, or oxygen is depleted, and the fire weakens.

In laboratory fires, controlling the fire at the ignition or growth stage is critical to prevent rapid escalation.

8.2.4 Key Fire Terminologies

- Combustion. The chemical process of burning in which a fuel combines with an oxidizer (usually oxygen in the air) to produce heat and light.
- Flame. The visible, gaseous component of a fire, often yellow or orange, is created by the combustion of a gas or vapor.
- Ignition. The initiation of combustion occurs when a fuel-air mixture begins to burn.
- A Flash Point. The lowest temperature at which a liquid produces enough vapor to ignite in the presence of an ignition source.
- A Fire Point. A continuous flame is the temperature at which the vapors of a liquid continue to burn.
- Autoignition Temperature. The lowest temperature at which a substance will spontaneously ignite without an external ignition source.
- Flammable Range (Explosive Range). The range of vapor concentrations in air that can ignite and burn when exposed to an ignition source.

- Lower Flammable Limit (LFL). The minimum concentration of vapor in air that can sustain combustion.
- Upper Flammable Limit (UFL). The maximum concentration of vapor in air that can sustain combustion.
- Backdraft. An explosion occurs when oxygen is suddenly introduced into a confined space filled with hot, fuel-rich gases that are below their ignition temperature.
- Flashover. A dangerous phenomenon in which all combustible materials in a room or enclosed space ignite almost simultaneously due to high temperatures.
- Pyrolysis. The chemical decomposition of a solid material (like wood or plastic) by heat, which produces flammable gases that can then ignite.
- Smoldering. A slow, flameless form of combustion that can occur in materials like fabrics or wood; it often precedes flaming combustion.
- Extinguishment. The process of stopping combustion, typically by removing one or more elements of the fire triangle (fuel, heat, and oxygen).
- Convection. Heat transfer through the movement of hot gases or liquids.
- Conduction. Heat transfer through a solid material.
- Radiation. Heat transfer via electromagnetic waves, which can ignite distant materials.
- Plume. The column of hot gases and smoke rising above a fire.
- Fire Load. The total amount of combustible material in a space, expressed in units like MJ/m^2 (megajoules per square meter).

8.2.5 Ignition Sources

An ignition source is any energy or heat source that can initiate combustion of a flammable or combustible material. Understanding ignition sources is essential for preventing fires, especially in laboratories, industrial settings, and workplaces. Different categories of ignition sources and their examples are summarized in Table 8.1.

Table 8.1: Categories of ignition sources.

Category	Examples
Open flames	Matches, candles, Bunsen burners, and welding torches
Hot surfaces	Hot pipes, machinery, and engines
Electrical equipment	Arcing, short circuits, faulty wiring, and static electricity
Mechanical sparks	Sparks from grinding, cutting, or impact
Static electricity	Discharges from equipment, clothing, or containers
Chemical reactions	Spontaneous heating and chemical decomposition
Friction	Heat from moving parts and lack of lubrication
Radiant heat	Infrared heat from sunlight or process equipment
Lightning	Natural ignition from electrical storms

8.2.6 Relative Flammability of Selected Organic Compounds

Understanding the relative flammability of organic compounds is crucial for laboratory and industrial safety. Different chemicals pose varying levels of fire risk, which is often quantified using the National Fire Protection Association (NFPA) 704 flammability ratings ranging from 0 (will not burn) to 4 (extremely flammable). Table 8.2 summarizes the relative flammability of some commonly encountered organic compounds including their flash points, qualitative flammability levels, and NFPA 704 (Figure 8.3) ratings to guide safe handling and storage.

Table 8.2: Relative flammability of selected organic compounds (with NFPA 704 ratings).

Compound	Flash point (°C)	Relative flammability	NFPA 704 flammability rating	Notes
Diethyl ether	−45	Very high	4	Extremely volatile; highly flammable
Acetone	−20	Very high	3	Common solvent; highly flammable
Hexane	−22	Very high	3	Used in extractions; very volatile
Toluene	04	High	3	Flammable; moderate vapor pressure
Ethanol	13	High	3	Highly flammable; common lab solvent
Methanol	11	High	3	Toxic and highly flammable
Benzene	−11	Very high	3	Carcinogenic; high volatility
Xylene	27	Moderate	3	Less volatile than benzene; still flammable
Isopropyl alcohol (IPA)	12	High	3	Common lab and cleaning solvent
Acetic acid	39	Moderate	2	Flammable liquid; corrosive
Phenol	79	Low	2	Limited flammability; also toxic
Naphthalene	79	Low	1	Solid at room temp; sublimes easily
Kerosene	>38	Moderate	2	Flammable; less volatile than gasoline
Gasoline	−40	Very high	3	Extremely volatile and highly flammable

8.2.7 Fire Extinguishment

Fire extinguishment is the process of stopping or suppressing a fire, thereby preventing it from spreading or continuing to burn. It is a fundamental aspect of fire safety in

0	Will not burn
1	Must be preheated before ignition can occur
2	Must be moderately heated or exposed to relatively high ambient temperatures before ignition occurs
3	Can be ignited under almost all normal temperature conditions
4	Rapidly or completely vaporizes at atmospheric pressure and normal temperatures and burns readily

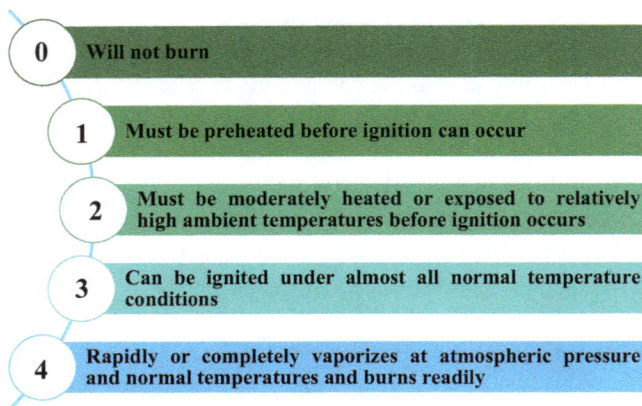

Figure 8.3: NFPA 704 flammability ratings.

industrial, laboratory, and domestic environments. Understanding how fires are extinguished requires knowledge of the basic elements that sustain combustion, often illustrated by the fire triangle (comprising fuel, heat, and oxygen) or the more comprehensive fire tetrahedron (adding the chemical chain reaction). Fire extinguishment is a critical skill in managing fire hazards. It involves understanding how fires sustain themselves and how to interrupt that process safely and effectively. By applying appropriate extinguishing methods, cooling, smothering, starvation, or chemical inhibition, fires can be brought under control and prevented from causing extensive damage or injury.

8.2.7.1 Principles of Fire Extinguishment

Fire extinguishment relies on removing or interrupting one or more sides of the fire triangle:

– Cooling. The most common method of extinguishing a fire involves lowering the temperature of the burning material below its ignition point. Water is a widely used cooling agent due to its high heat capacity and availability.
– Smothering (Oxygen Exclusion). Fires need oxygen to sustain combustion. Smothering deprives the fire of oxygen, thereby extinguishing it. Fire blankets, foam, or carbon dioxide (CO_2) are often used to achieve smothering.
– Starvation (Fuel Removal). Removing the fuel source can extinguish a fire by preventing it from sustaining itself. This can involve shutting off a fuel valve, clearing away combustible materials, or physically isolating the fire.
– Chemical Inhibition. Some fires, especially those involving flammable liquids or gases, can be extinguished by disrupting the chemical chain reaction. Dry chemical extinguishers (such as ABC powder) interrupt the reaction at the molecular level, stopping the combustion process.

8.2.7.2 Fire Classification and Extinguishing Agents

Depending on the kind of fuel used, fires are divided into several classes:

- Class A. Common flammables like fabric, paper, and wood. Foam and water work well.
- Class B. Liquids that can catch fire such as solvents, oil, or gasoline. Carbon dioxide, dry chemical, and foam extinguishers are often utilized.
- Class C. Electrical devices that have been turned on. Dry chemicals or nonconductive extinguishing agents like CO_2 are advised.
- Class D. Metals that may catch fire such as titanium, magnesium, or sodium. For these flames, special dry powders are made.
- Class F (EU) or Class K (USA). Oils and fats used in cooking. To put out these flames, wet chemical extinguishers are made especially to saponify the oil.

8.2.7.3 Safety Considerations

Selecting the correct extinguishing method is crucial. Using water on an oil fire, for example, can spread the flames, while using a water-based extinguisher on an electrical fire poses a risk of electrocution. Personnel must be trained in identifying fire classes and using extinguishers properly including the PASS (pull, aim, squeeze, and sweep) technique. Moreover, extinguishing a fire is only one part of overall fire safety. Adequate fire detection, alarm systems, evacuation plans, and maintenance of extinguishing equipment are all essential components of an effective fire safety program.

8.2.7.4 Types of Fire Extinguishers and Their Uses

Fire extinguishers (Figure 8.4) are portable devices used to suppress incipient-stage fires. They are classified based on the extinguishing agent and are labeled with the fire classes they are suitable for. Common fire extinguishers are listed in Table 8.3, and the contents of common fire extinguisher types are listed in Table 8.4.

8.3 Explosion Safety

8.3.1 Introduction to Explosion Hazards

A fast, abrupt release of energy that causes a sharp rise in pressure, temperature, or volume, usually accompanied by the creation of shock waves and extreme heat, is called an explosion. Explosions provide serious safety risks in laboratory and industrial settings because they can result in serious injuries, fatalities, and substantial property damage. Therefore, it is essential to comprehend the nature of explosions to put into practice efficient safety measures.

Figure 8.4: Fire extinguisher types. Image credit: https://depositphotos.com/photos/fire-extinguisher-types.html.

Table 8.3: Common fire extinguisher types.

Type	Mechanism of action	Suitable classes	Limitations
Water (APW)*	Cools fuel below ignition temperature	A	Not safe for B, C, D, or K fires
Foam (AFFF/FFFP)**	Smothers and cools	A and B	May conduct electricity; not for C fires
Carbon dioxide (CO_2)	Displaces oxygen and cools	B and C	Ineffective outdoors or on Class A fires
Dry chemical (ABC)	Interrupts chemical chain reaction	A, B, and C	Leaves residue; may damage sensitive equipment
Dry powder (Class D)	Forms crust to isolate burning metals	D	Ineffective on A, B, C, or K fires
Dry powder (Class D L2, Lith-Ex)	Sodium chloride or copper-based powders; specifically designed for metal fires including lithium, magnesium, and sodium–potassium alloys and also designed for use on lithium-ion batteries	D	Only for specific metals; must match the metal type
Wet chemical	Cools and reacts with oils (saponification)	K or F	Designed for commercial kitchen fires

Table 8.3 (continued)

Type	Mechanism of action	Suitable classes	Limitations
Clean agent	Chemically interrupts combustion chain reaction; leaves no residue	A, B, and C	Expensive; may have limited cooling effect

*APW stands for air-pressurized water.
**AFFF stands for aqueous film-forming foam and FFFP stands for film-forming fluoroprotein foam.

Table 8.4: Contents of common fire extinguishers.

Type	Chemical contents	Notes
Water (APW)	Water (sometimes with a corrosion inhibitor)	Pressurized air or nitrogen propels water. Simple, no added chemicals.
Foam (AFFF)	Water + AFFF concentrate (synthetic surfactants, typically fluorinated surfactants like perfluorooctane sulfonate or modern alternatives)	Diluted concentrate forms a film that floats on flammable liquids and prevents vapor release
Foam (FFFP)	Water + film-forming fluoroprotein foam concentrate (protein hydrolysates + fluorinated surfactants)	Slightly different from AFFF; forms a thicker, stable foam layer on hydrocarbons.
Carbon dioxide (CO_2)	Compressed liquid CO_2 (stored at high pressure)	Releases as cold, dense gas; no residue.
Dry chemical (ABC)	Monoammonium phosphate powder (yellowish)	Interrupts combustion chain reaction; safe on Class A, B, and C fires.
Dry powder (Class D)	Sodium chloride for magnesium, sodium, potassium, and aluminum. Copper powder for lithium fires. Graphite-based powders for some metals	Designed to smother metal fires; absorbs heat and isolates oxygen.
Dry powder (Class D L2)	Copper-based powder (often blended with other additives to improve heat absorption)	For lithium fires and other reactive metals.
Wet chemical	Potassium acetate, potassium citrate, or potassium carbonate solution	Forms a soapy film (saponification) on burning oils (Class K fires).
Clean agent	Halotron I (2,2-dichloro-1,1,1-trifluoroethane), FE-36 (hexafluoropropane), Novec 1230 (dodecafluoro-2-methylpentan-3-one), and FM-200 (heptafluoropropane)	Stored as liquid; discharged as gas. Interferes with the chemical chain reaction.

8.3.2 Types of Explosions

Explosions can be broadly classified based on their underlying cause and the nature of the energy release. Understanding these types is critical for developing appropriate safety measures, choosing equipment, and designing emergency response protocols. The most common explosion types encountered in laboratory and industrial settings include chemical explosions, physical (mechanical) explosions, dust explosions, and electrical explosions (Table 8.5).

Table 8.5: Types of explosions, their key characteristics, and common examples.

Type of explosion	Key characteristics	Common examples
Chemical	Rapid combustion of flammable materials; deflagration or detonation	Flammable gases, solvents, and reactive chemicals
Physical	Sudden mechanical failure of pressurized equipment	Gas cylinders, boilers, and reaction vessels
Dust	Fine particles suspended in air; rapid flame propagation	Grain, metal dust, and pharmaceutical powders
Electrical	Arc flashes or short circuits; ignition source	Electrical panels, batteries, and static discharge

8.3.3 Key Factors Contributing to Explosions

Several factors determine whether a fire might escalate into an explosion:
- Fuel Concentration. The mixture of flammable materials and air must fall within the fuel's LFL and UFL to sustain rapid combustion.
- Ignition Source. Sparks, open flames, hot surfaces, or even static electricity can ignite flammable mixtures.
- Confinement. When combustion occurs in an enclosed space, pressure builds rapidly, increasing the likelihood of an explosion.
- Dispersion. Especially relevant to dust explosions, where fine particles are suspended in the air, creating a highly combustible atmosphere.

8.3.4 Explosive Mixtures

An explosive mixture is a combination of a fuel and an oxidizer that, when ignited, can undergo a rapid combustion or detonation, releasing a large amount of energy in the form of heat and expanding gases. Understanding explosive mixtures is essential for preventing accidental explosions in laboratories, industrial plants, and other facilities that handle flammable or reactive chemicals.

8.3.4.1 Composition of Explosive Mixtures

Explosive mixtures require three essential components:
- Fuel. A substance that can burn or react exothermically. Examples include flammable gases (e.g., hydrogen and methane), vapors of volatile liquids (e.g., acetone and diethyl ether), and fine powders (e.g., flour and metal dust).
- Oxidizer. Usually oxygen in the air, but it can also be chemical oxidizers (e.g., nitrates and peroxides).
- Ignition Source. A spark, flame, hot surface, static discharge, or even mechanical impact.

8.3.4.2 Examples of Explosive Mixtures

Flammable Gas–Air Mixtures. These mixtures (Table 8.6) involve a flammable gas mixed with air in concentrations between the LFL and UFL. They are common in many industrial and laboratory settings.

Table 8.6: Flammable gas–air mixtures.

Fuel gas	Typical LFL (% by volume)	Typical UFL (% by volume)	Example scenario
Hydrogen	4.0	75	Laboratories using hydrogen gas (e.g., fuel cells and synthesis)
Methane	5.0	15	Natural gas leaks in pipelines or storage tanks
Acetylene	2.5	100	Welding shops or chemical labs with acetylene cylinders
Carbon monoxide	12.5	74	Incomplete combustion or leaks from furnaces
Ethylene	2.7	36	Polymerization labs or ethylene oxide sterilization processes

Flammable Vapor–Air Mixtures: Volatile liquids (Table 8.7) can evaporate and form explosive mixtures in air under certain temperature and pressure conditions.

Dust–Air Mixtures. Fine combustible powders (Table 8.8) are a specific category of explosive mixtures because they consist of fine, combustible solid particles suspended in the air, forming a cloud that can ignite and propagate flames rapidly throughout the suspended particles. When a spark or ignition source is introduced, these clouds can result in violent explosions.

Table 8.7: Flammable vapor–air mixtures.

Fuel vapor	LFL (%)	UFL (%)	Example scenario
Diethyl ether	1.9	48	Organic synthesis labs and solvent storage rooms
Acetone	2.5	12.8	Paint shops and solvent cleaning areas
Benzene	1.2	7.8	Petrochemical facilities and analytical labs
Toluene	1.2	7.1	Coatings, adhesives, and chemical processing
Ethanol	3.3	19	Alcohol distillation or storage areas

Table 8.8: Dust–air mixtures.

Combustible dust	Minimum explosible concentration (MEC)	Example scenario
Grain dust (flour, wheat, and corn)	~30–60 g/m^3	Grain elevators and food processing plants
Sugar dust	~30–50 g/m^3	Sugar refineries and confectionery plants
Metal powders (aluminum, magnesium)	~40–100 g/m^3	Metal finishing and powder metallurgy labs
Pharmaceutical powders	Varies	Drug formulation labs
Coal dust	~50–100 g/m^3	Mining operations and power plants

Reactive Chemical Mixtures. Certain chemicals (Table 8.9) can form unstable mixtures that may detonate or explode under conditions such as heat, shock, or friction. These mixtures pose significant hazards in both laboratory and industrial environments, and their handling requires strict adherence to safety protocols. Reactive chemical mixtures often involve oxidizers and fuels that can interact dangerously when mixed or stored improperly. For example, strong oxidizing agents like nitrates, chlorates, perchlorates, and peroxides can react violently with organic materials (e.g., solvents, oils, or dusts) or reducing agents (e.g., sulfides and metals). A classic example is organic peroxides, which are sensitive to heat, light, or contamination and can decompose explosively. Picric acid, an older laboratory reagent, can become sensitive to shock if it dries out or crystallizes in storage containers. Similarly, acetylides (such as copper acetylide) can form in gas lines when acetylene is stored improperly and can detonate unexpectedly. Some reactive mixtures can form accidentally during routine operations such as mixing incompatible chemicals or improper neutralization of waste. For example, adding strong acids to chlorinated solvents can generate highly reactive intermediates that may detonate or evolve toxic gases. Certain laboratory syntheses also generate intermediate compounds (e.g., nitrated organics or unstable diazonium salts) that require immediate use or safe disposal.

Table 8.9: Reactive chemical mixtures.

Chemical mixture	Example scenario
Peroxides + organic solvents	Improper storage of organic peroxides in solvent-rich environments
Nitrated compounds	Nitrocellulose in film or lab samples
Chlorates or nitrates + sulfur or charcoal	Historical gunpowder or lab mixtures

8.3.5 Explosion Risk Assessment

Explosion risk assessment is a systematic process for identifying, analyzing, and mitigating the potential hazards posed by explosive atmospheres in laboratory and industrial settings. It is a critical component of a comprehensive safety management plan, ensuring that risks are controlled and that personnel and equipment are protected from accidental explosions.

8.3.5.1 Identifying Potential Explosive Atmospheres

The first step in explosion risk assessment is to identify areas where explosive atmospheres might form. This involves analyzing the presence of:
– Flammable gases or vapors (e.g., hydrogen, acetylene, and solvents like diethyl ether, acetone, or toluene). These can leak from process equipment, storage tanks, or laboratory apparatus.
– Combustible dust clouds (e.g., grain dust, sugar dust, powdered metals, or pharmaceutical powders). These can arise during handling, processing, or cleaning activities.

Key considerations include:
– Ventilation. Poor ventilation can cause a buildup of flammable gases or vapors, increasing the danger of igniting.
– Processing Operations. Mixing, grinding, heating, and moving materials can all result in combustible atmospheres.
– Temperature and Pressure. Elevated temperatures or pressures can expand the flammability range, increasing the likelihood of an explosion.

8.3.5.2 Hazard Analysis Tools

Once potential explosive atmospheres have been identified, hazard analysis tools help systematically evaluate risks and propose control measures:

– Hazard and Operability (HAZOP)

A HAZOP is a structured, systematic technique that uses guidewords (e.g., "More," "Less," "As well as," "Reverse," "No" or "None," "Part of," and "Other than") to identify deviations from intended process conditions that could lead to hazards including explosion risks. These can be explained as follows:

"More"

This guideword examines what might happen if a process parameter (e.g., flow rate, temperature, and pressure) is higher than intended. Example: More flow might cause a tank to overfill or lead to higher vapor concentration, increasing explosion risk.

"Less"

This looks at situations where the parameter is lower than intended. Example: Less flow could cause cooling water to slow down, leading to overheating or unwanted chemical reactions.

"As well as"

This considers what happens if an additional or unintended event occurs alongside the intended process. Example: As well as unexpected impurities or byproducts in a chemical reaction that could generate hazardous gases or solids.

"Reverse"

This guideword explores the consequences if the flow or direction of a process is reversed. Example: Reverse flow could lead to backflow of flammable gases into upstream equipment, creating an ignition hazard.

"No" or "None"

The parameter is missing entirely. Example: No flow could mean a pump failure, leading to loss of cooling or unintended accumulation of reactants.

"Part of"

Only part of the system is affected or only part of the flow is delivered. Example: Part of flow could cause incomplete mixing or poor combustion control.

"Other than"

Something unexpected happens, or a different material enters the system. Example: Other than air, introduction of nitrogen might displace oxygen, affecting combustion.

– Failure Mode and Effects Analysis (FMEA)

FMEA involves analyzing each component or process step to identify potential failure modes, their causes, and the effects of those failures on the overall system. For explosion risks, FMEA might consider:

- Valve failure leading to gas release
- Static discharge igniting a dust cloud
- Temperature control failure leading to thermal runaway

Each failure mode is scored based on severity, probability, and detectability, enabling prioritization of risks and targeted control measures.

In summary, before delving into the specific types of explosions, it is crucial to recognize that explosion hazards can originate from a diverse range of materials, processes, and operational conditions. Establishing a comprehensive explosion safety program involves systematically identifying hazards, thoroughly analyzing potential ignition sources, and implementing effective control measures that address each specific risk. Table 8.10 provides an overview of key explosion types, their associated hazards, and recommended prevention strategies that laboratories and industrial facilities can adopt to reduce the likelihood of catastrophic incidents.

Table 8.10: Explosion types, hazards, and prevention strategies.

Explosion type	Hazards	Prevention strategies
Gas–air explosions	Flammable gas leaks (e.g., hydrogen and acetylene) can form explosive atmospheres in confined spaces	– Leak detection and repair – Adequate ventilation – Inerting with nitrogen or other gases
Vapor–air explosions	Volatile liquids evaporate and mix with air, forming flammable vapors (e.g., diethyl ether and acetone)	– Proper storage in explosion-proof containers – Use of spark-proof equipment – Spill control
Dust–air explosions	Combustible dusts (e.g., grain, sugar, and metals) suspended in the air can ignite explosively	– Dust collection systems – Housekeeping to prevent accumulation – Explosion venting or suppression
Hybrid mixtures	Combination of dusts and vapors/mists and increasing ignition risks (e.g., spray-painting operations)	– Control both dust and vapor hazards – Use of local exhaust ventilation – Control ignition sources
Reactive chemical mixtures	Unstable chemical combinations may detonate under heat, shock, or friction (e.g., peroxides and nitrated compounds)	– Proper chemical segregation – Temperature and pressure monitoring – Use of blast shields
BLEVE (boiling liquid expanding vapor explosion)	Pressurized liquefied gases (e.g., propane) can rupture and cause violent explosions	– Pressure relief devices – Proper tank maintenance – Safe distances from ignition sources

8.4 Questions and Answers

Questions	Answers
1. What is the fire triangle?	The three elements required for a fire: fuel, oxygen, and heat.
2. What causes most laboratory fires?	Improper handling or storage of flammable materials near ignition sources.
3. What is a flash point?	The lowest temperature at which a liquid gives off enough vapor to ignite.
4. How can you prevent flammable vapor buildup?	Use proper ventilation and keep containers sealed when not in use.
5. How are flammable chemicals stored safely?	In approved flammable storage cabinets, away from heat and incompatible substances.
6. What is a Class B fire?	A fire involving flammable liquids or gases.
7. What type of fire extinguisher is used for chemical fires?	A Class B or multiclass ABC fire extinguisher, depending on the chemical.
8. What should you do first if a fire starts in the lab?	Activate the fire alarm and evacuate the area immediately.
9. What is an explosion hazard?	A risk that a substance or condition can cause a sudden release of energy, often with a blast.
10. How can explosion risks be minimized?	Avoid incompatible chemical mixing, use blast shields, and follow safety procedures.

Chapter 9
Laboratory Glassware and Equipment Safety

Laboratory glassware and equipment constitute the foundation of experimental work in most scientific laboratories, but their usage has inherent dangers that must be appropriately controlled. Glassware, while flexible and chemically resistant, is prone to breakage and heat stress, providing risks such as cuts, burns, and chemical exposure. Similarly, laboratory equipment, which ranges from heating devices and stirrers to vacuum systems and pressure apparatus, can pose risks such as fire, electrical hazards, and mechanical accidents. Safe laboratory practices, comprehensive training, and a complete awareness of possible risks are critical for avoiding accidents and guaranteeing the safety of laboratory staff. This section provides guidelines for the safe use, handling, and maintenance of laboratory glassware and equipment, emphasizing both basic precautions and specific safety concerns.

9.1 Glassware Safety

Glassware is an essential component of most laboratory work, but it also poses specific hazards that require careful handling and awareness of best practices. Understanding the risks associated with glassware and adopting safe practices helps prevent injuries and ensures a safer laboratory environment.

9.1.1 Common Hazards of Laboratory Glassware

- Breakage. Glass can shatter or crack during use, cleaning, or storage, leading to cuts or puncture wounds (Figure 9.1).
- Thermal Shock. Rapid heating or cooling can cause glassware to crack or explode.
- Chemical Attack. Some chemicals can corrode or weaken glass (Figure 9.2) over time, increasing the risk of breakage.
- Pressure and Vacuum Hazards. Improperly using glassware with high pressure or vacuum can cause implosions or explosions.

9.1.2 General Glassware Handling Tips

Laboratory glassware must be handled carefully to avoid accidents and preserve a safe working environment. Before using any glassware, look for chips, cracks, or other defects that might compromise its integrity. If they are broken, replace or throw them away. To avoid cuts, glassware should be thoroughly cleaned and dried before

https://doi.org/10.1515/9783112218105-009

Figure 9.1: Broken glass. Image credit: https://chemjobber.blogspot.com/2019/09/were-you-ever-asked-to-pay-for-broken.html.

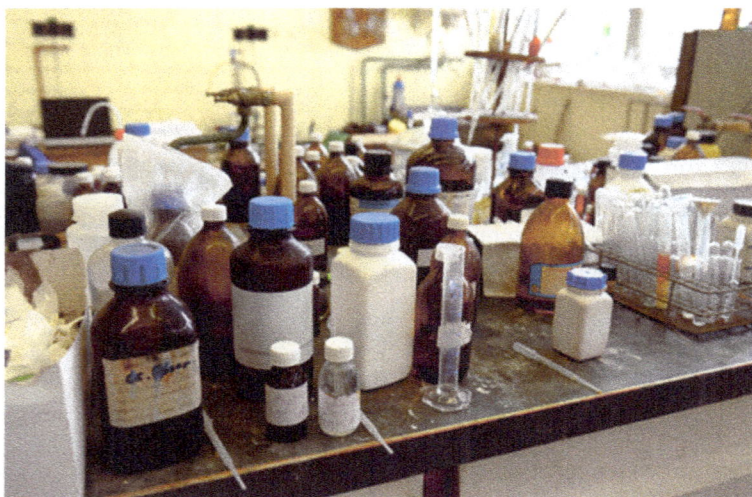

Figure 9.2: Chemically contaminated or corroded or weaken glass bottles. Image credit: https://lab-train ing.com/24417/.

use. Use heat-resistant gloves or tongs while handling hot glasses, and keep in mind that heated glass may seem cool. To lessen the risk of thermal shock, keep glassware away from cold or moist surfaces, allow it to cool gradually, and never subject it to sudden temperature changes.

Glassware should be stored securely to avoid tipping over or falling, and the right kind should be chosen for each procedure, such as borosilicate glass for heating or chemical reactions. You may greatly lower the chance of mishaps and injury in the lab by adhering to these rules.

9.1.3 Special Considerations and Best Practices

- Vacuum Systems. Only use thick-walled glassware designed for vacuum work. Always protect vacuum systems with shields or enclosures.
- Pressurized Reactions. Use proper pressure-rated equipment instead of standard laboratory glassware.
- Glass Tubing and Rods. Fire-polish or bevel the edges after cutting glass tubing to prevent cuts. Use lubrication and protective gloves when inserting glass tubing into stoppers or rubber tubing.

Understanding these key hazards and best practices helps reinforce safe laboratory habits when working with glassware. By adopting these guidelines and maintaining a cautious, deliberate approach, laboratory personnel can reduce the risk of injury and accidents while working with glass equipment. Table 9.1 summarizes common hazards, example scenarios, and recommended safe practices for quick reference.

Table 9.1: Common hazards, example scenarios, and recommended safe practices.

Hazard	Example scenario	Safe practice
Breakage	Dropping a beaker	Inspect glassware; handle with care; dispose of damaged items.
Thermal shock	Placing hot flask on a cold surface	Gradually heat/cool glassware; use appropriate support.
Chemical attack	Acid etching	Use resistant glassware; inspect regularly.
Vacuum/pressure failure	Evacuating a filter flask	Use thick-walled glassware; protective shields.
Handling cuts	Breaking a test tube	Use tongs/gloves; know first-aid procedures.

9.2 Electrical Equipment Safety

Electrical equipment in the laboratory may cause electrical shock and act as an ignition source for flammable or explosive chemicals. To minimize the possibility of either of these, several precautions can be taken:
- All laboratory receptacles and equipment should be equipped with three-prong grounded plugs.
- Equipment should be located safely to minimize the possibility of chemical spills on or under it.
- Inspect cords on a regular basis for frayed and/or damaged connections.
- Devices equipped with motors used where there are flammable vapors present should be either non-sparking induction or air-driven motors.
- On–off switches, rheostat-type speed controllers, and similar devices can produce sparks every time they are adjusted. If electrical equipment is to be used in the fume hood, all controls should be outside the hood.
- Unplug electrical equipment before making repairs or modifications.

Electrical devices such as stirrers and mixers are often operated over extended periods of time with the possibility of mechanical failure, electrical overload, or blockage of the stirrer. If they are to be left unattended, the associated equipment should be fitted with a suitable fuse or thermal protection device that will shut down the apparatus in the event of such problems.

9.3 Vacuum Pumps Safety

Working at reduced pressure carries with it the risk of implosion and the subsequent dangers of flying glass, splashing chemicals, and possibly fire. Any apparatus under reduced pressure should be shielded to minimize that risk. When using a vacuum pump (Figure 9.3):
- Place cold traps between the apparatus and the vacuum source to minimize the amount of volatile material that enters the system.
- Vent vacuum pumps to an air exhaust system, not directly into the laboratory.

9.4 Heat Source Safety

Whenever possible, use suitable electrically heated sources such as hotplates, heating mantles, or similar devices in place of gas burners as they are inherently safer. Steam baths are best for temperatures under 100 °C since they present neither shock nor spark risks and the temperature is guaranteed not to rise above 100 °C.

Figure 9.3: Vacuum pump. Image credit: https://depositphotos.com/photos/lab-vacuum-pump.html.

9.4.1 Heating Mantle Safety

Heating mantles (Figure 9.4) incorporate a heating element in layers of fiberglass cloth and do not present a risk of shock or fire if used properly. Some precautions that should be taken when using mantles include:
- Do not use if the fiberglass cloth is worn or broken, exposing the heating element.
- Be careful not to spill water or other chemicals on the mantle, as this poses a serious shock hazard. Depending on the spilled chemical, it may also present a fire or explosion hazard.
- Always use a variable transformer to control the input voltage. Never plug directly into an electrical outlet. High voltage will cause the mantle to overheat, damaging the fiberglass insulation and exposing the heater.

9.4.2 Oil Baths Safety

Electrically heated oil baths are commonly used in situations where a stable temperature is required, or a small or irregularly shaped vessel must be heated. Some precautions that should be taken when using oil baths include:

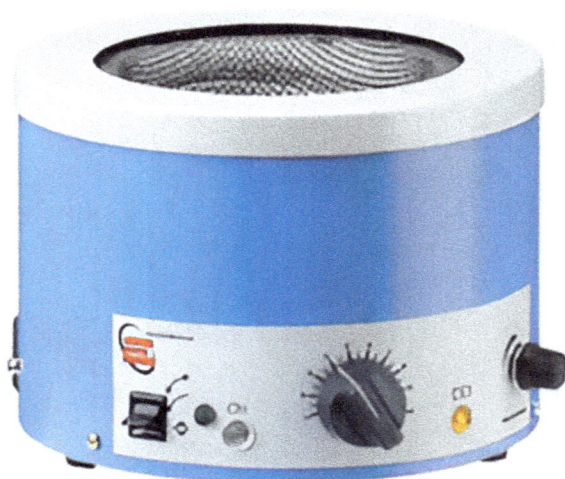

Figure 9.4: Heating mantles. Image credit: https://www.fishersci.ca/ca/en/browse/90056015/heaters-and-heating-mantles?page=1.

– Avoid spilling water or volatile substances into the bath, which may result in a splattering of hot oil or smoking or ignition of the bath.
– Saturated paraffin oil is suitable for up to 200 °C, and silicone oil should be used for temperatures up to 300 °C.
– Always monitor the temperature of the bath to ensure that it does not exceed the flash point of the oil.
– Mix well to prevent "hot spots" from forming.
– Support with a laboratory jack or similar apparatus so the bath can be lowered and raised easily without recourse to manually lifting the hot bath.

9.4.3 Oven and Furnace Safety

Ovens are most commonly used for drying laboratory glassware and chemical samples. Only laboratory-approved ovens that have the heating elements and temperature controls separated from the interior atmosphere should be used. Laboratory ovens generally vent directly into the laboratory. If toxic vapors or gases may be released while using the oven, the vapors should be vented into a fume hood. Furnaces are used for high-temperature applications. Ensure that the reaction vessels and other equipment used are designed to withstand high temperatures.

9.5 Refrigerator and Freezer Safety

Refrigerators and freezers used in the laboratory must be carefully selected for specific chemical storage needs. Commercial refrigeration units are not designed to meet the special hazards presented by flammable materials. The interior of a commercial refrigerator contains a number of electrical contacts that can generate electrical sparks. Frost-free models often have a drain, which could allow vapors to reach the compressor, and electrical heaters used to defrost the refrigerator are also a spark hazard. For these reasons, only specially designed lab refrigerators or modified commercial units should be used for cold storage of flammable chemicals.

Those classified as "flammable" do not have internal switches or unprotected wires that can act as a source of ignition. An "explosion-proof" unit has switches and wires protected both inside and outside and is suitable for use in environments where flammable vapors may reach explosion and/or ignition limits outside the refrigerator. For storage of flammable materials in most laboratories, a unit rated for "flammable storage" is sufficient. Commercial refrigerators and freezers are acceptable for the storage of nonflammable materials but must be visibly labeled as unsuitable for the storage of flammable materials.

Laboratory refrigerators should be clearly labeled as being for chemical storage only. A major concern with chemical storage refrigerators is that as tightly sealed spaces, they can allow the buildup of toxic and/or flammable vapors. Containers must be adequately sealed to minimize the likelihood of this happening. Beakers, flasks, and bottles covered with aluminum foil or plastic wrap are unacceptable for the storage of volatile chemicals in the refrigerator. Likewise, corks and glass stoppers are also inadequate. Screw top caps with a seal inside are best suited for refrigerator storage. Refrigerators should also be regularly defrosted and cleaned to minimize the accumulation of ice and hazardous vapors inside the unit. Chemicals no longer used must be disposed of as hazardous waste.

9.6 Autoclave and Centrifuge Safety

Autoclaves and centrifuges are indispensable laboratory equipment that support critical processes such as sterilization and separation. However, their powerful mechanical and thermal operations also present unique hazards that must be managed effectively. Understanding the risks associated with each device and following proper operational procedures ensures both user safety and the integrity of laboratory experiments.

9.6.1 Autoclave Safety

9.6.1.1 Hazards Associated with Autoclaves

Autoclaves (Figure 9.5) operate at high temperatures and pressures to sterilize laboratory materials, making them essential in biological and chemical laboratories. However, these same operating conditions can pose hazards including burns, scalds from steam, pressure-related injuries, and exposure to hazardous materials if not handled properly.

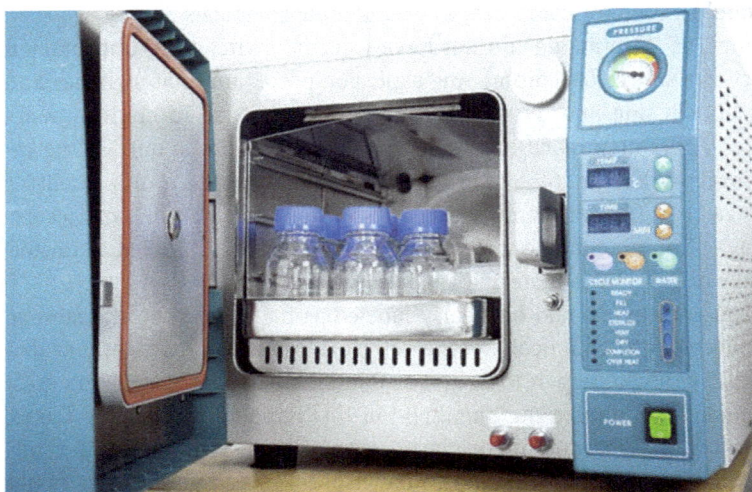

Figure 9.5: Autoclave. Image credit: https://depositphotos.com/photos/autoclave.html.

9.6.1.2 Safe Operation of Autoclaves

Before using an autoclave, inspect it for any signs of damage, leaks, or malfunction. Make sure that the door seal and locking mechanisms are in good condition and free of debris. Load items properly, never overload an autoclave, as overcrowding can lead to incomplete sterilization and increase the risk of accidents. Use autoclavable trays and secondary containment to catch spills and prevent leaks.

Always wear appropriate personal protective equipment (PPE), including heat-resistant gloves, a lab coat, and eye protection, when loading and unloading the autoclave. Stand to the side when opening the door to avoid exposure to residual steam. Allow the autoclave to depressurize and cool down fully before opening the door; opening an autoclave too soon can cause sudden steam release and burns.

9.6.1.3 Best Practices for Autoclave Use

- Use only approved materials for autoclaving. Some plastics or chemicals may degrade or release harmful gases under autoclave conditions.
- Clearly label items to be sterilized with the appropriate biohazard or chemical hazard labels.
- Follow manufacturer guidelines for routine maintenance including regular checks of safety valves, door gaskets, and temperature/pressure sensors.
- Keep an autoclave log documenting the date, cycle type, load type, and operator.
- In case of malfunction, do not attempt repairs yourself, contact trained maintenance personnel immediately.

9.6.2 Centrifuge Safety

9.6.2.1 Hazards Associated with Centrifuges

Centrifuges (Figure 9.6) operate at high rotational speeds, generating significant centrifugal force to separate sample components. The primary hazards include mechanical failure (e.g., rotor imbalance or breakage), exposure to hazardous materials (e.g., biohazards or chemicals), and ergonomic issues (e.g., repetitive motion or lifting heavy rotors).

9.6.2.2 Safe Operation of Centrifuges

Before use, inspect the centrifuge and rotor for any signs of corrosion, cracks, or mechanical damage. Always balance the centrifuge by placing tubes of equal weight opposite each other. An unbalanced rotor can cause violent shaking or catastrophic failure.

Use appropriate centrifuge tubes rated for the intended speed and compatible with the sample material. Make sure that tubes are capped tightly to prevent leaks or aerosol formation. For biohazardous materials, use sealed buckets or safety cups to contain any potential spills or aerosols.

9.6.2.3 Best Practices for Centrifuge Use

- Close the centrifuge lid before starting and do not open it until the rotor has come to a complete stop. Attempting to stop the rotor manually is extremely dangerous.
- Always adhere to the manufacturer's recommended speed and temperature limits for both the centrifuge and the rotor.
- Use appropriate PPE, such as a lab coat, gloves, and eye protection, especially when handling hazardous materials.
- Clean and disinfect the centrifuge regularly, particularly after any spills or leaks.

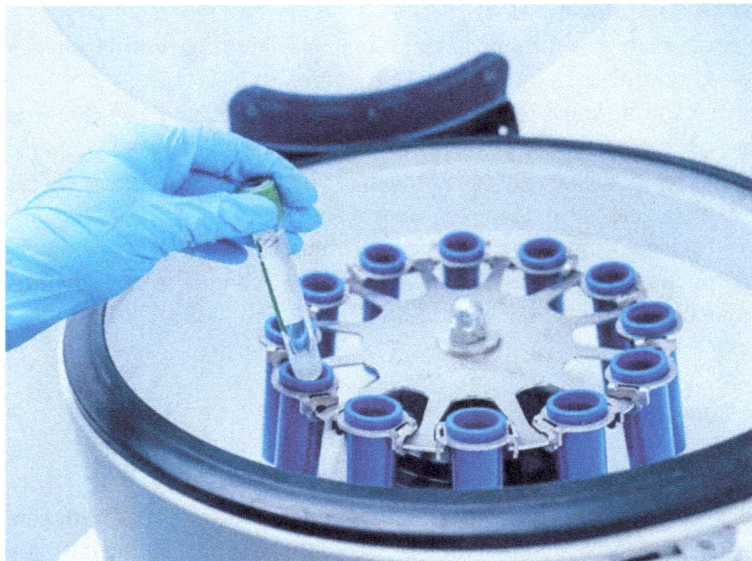

Figure 9.6: Centrifuge. Image credit: https://depositphotos.com/photos/centrifuge.html.

– Maintain a documented log of rotor usage to track the number of runs and life-span, as rotors are subject to metal fatigue and can fail if overused.
– For high-speed or ultracentrifuge use, follow any special maintenance procedures recommended by the manufacturer including periodic rotor inspections and stress testing.

9.7 Decontamination of Laboratory Equipment

Any equipment used in a laboratory that contains hazardous materials will become contaminated over time. Thus, laboratory equipment should be decontaminated prior to removal. This applies whenever equipment is transferred to another laboratory, sent out for repairing or calibration, or disposed of as waste or surplus equipment.

Decontamination includes removing all hazardous products, containers, or other potentially contaminated items such as refrigerators and cupboards. The equipment should then be visually inspected for stains, residues, or other evidence of chemical contamination, and such contamination, if found, should be removed by washing with soapy water, decontamination solution, or other necessary means.

9.8 Maintenance and Inspection

Proper maintenance and regular inspection of laboratory glassware and equipment are critical components of a comprehensive laboratory safety program. Routine checks help ensure that equipment performs reliably and that glassware remains intact and safe for use, thereby preventing accidents and ensuring accurate experimental results.

9.8.1 Routine Checks of Glassware Integrity

Before each use, laboratory personnel should carefully inspect all glassware for signs of damage including cracks, chips, or scratches. Even small defects can weaken the structure of the glass and increase the risk of breakage under normal laboratory conditions, particularly when exposed to thermal or mechanical stress. Damaged glassware should be removed from service immediately and disposed of or repaired according to laboratory protocols. Regularly inspecting glassware not only minimizes the risk of cuts and injuries but also reduces the likelihood of contamination or experimental failure due to leaks or unexpected breakage.

9.8.2 Equipment Maintenance Schedules

To guarantee optimum performance and safety, every laboratory equipment, from vacuum pumps and fume hoods to centrifuges and heating mantles, needs routine maintenance. The manufacturer's recommendations, equipment usage rates, and laboratory-specific needs should all be taken into consideration when creating maintenance schedules. Lubricating moving parts, changing worn parts, checking electrical connections, and cleaning filters and ventilation systems are a few examples of preventive maintenance activities. Completed inspections, repairs, and calibration procedures should all be documented in a maintenance journal. Following a regular maintenance plan prolongs the life of costly lab equipment, minimizes downtime, and aids in the early detection of any problems.

9.8.3 Calibration of Instruments

Valid experimental results depend on precise and trustworthy measurements. Measurements stay within allowable accuracy limits when laboratory equipment, including pipettes, pH meters, balances, and analytical instruments, is regularly calibrated. The manufacturer's instructions and all applicable legal requirements should be followed when performing calibration procedures. After every calibration event, calibration records that include the date, the person doing the calibration, and the test re-sults should be kept up to date. When an instrument fails calibration, it should be properly labeled and kept out of use until it is fixed or replaced. Labs preserve the integrity of their data and promote secure and efficient research procedures by mak-ing sure that their equipment is calibrated correctly.

In summary, systematic maintenance and inspection practices are indispensable for promoting laboratory safety, protecting personnel, and preserving the accuracy and reliability of scientific work. Establishing and enforcing these routines as part of the laboratory's standard operating procedures not only supports compliance with safety regulations but also fosters a culture of continuous improvement and risk re-duction.

9.9 Safe Storage and Disposal

To provide a safe working environment and lower the chance of contamination and accidents, laboratory glassware and equipment must be stored and disposed of prop-erly. Labs may decrease waste, prevent accidents, and adhere to safety requirements by putting into practice safe disposal procedures and efficient storage techniques.

9.9.1 Storing Glassware to Prevent Breakage

Glassware needs to be kept in spaces that are specifically made to prevent tipping or dropping. To lower the chance of breakage, glassware storage cabinets or shelves should be strong, orderly, and have cushioned or lined surfaces. Shelves should not be overcrowded because this may cause objects to move and tumble when accessible. To lessen the chance of damage from dropping, store heavy or bulky goods on lower shelves. In order to facilitate recovery and avoid stacking delicate objects, glassware should also be arranged according to size and kind. Organization and safety can be further improved by labeling shelves and utilizing the proper racks or holders for spe-cialty glassware such as burettes and pipettes.

9.9.2 Handling Broken Glassware

In the laboratory, broken glassware poses a significant risk and needs to be handled carefully. Never use your bare hands to pick up shattered glass. Rather, gather the fragments with a brush and dustpan or tongs and put them in a sturdy, impenetrable container that is marked "Broken Glass." To protect laboratory and maintenance workers from harm, this container (Figure 9.7) should be stored apart from general garbage and chemical waste. When cleaning up shattered glass, always use the proper

PPE such as gloves and eye protection. Check the area after cleaning to make sure that there are no tiny fragments behind.

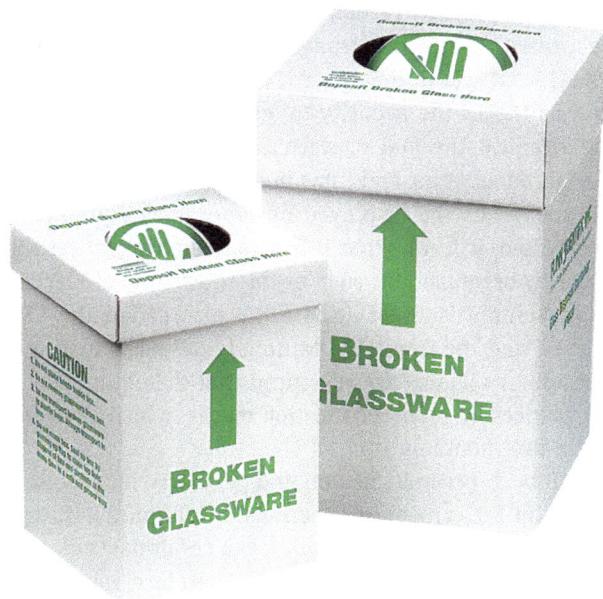

Figure 9.7: Broken glass box. Image credit: https://www.flinnsci.com/glass-disposal-container-benchtop-model/ap8829/.

9.9.3 Disposal of Damaged Equipment and Glassware

Damaged glassware and equipment that are no longer safe or functional should be disposed of promptly and properly. Place broken or irreparable glassware in a designated, puncture-resistant "Broken Glass" container for disposal. Larger equipment that is no longer functional, such as cracked beakers, damaged heating mantles, or malfunctioning centrifuges, should be tagged as "Out of Service" and removed from the lab to avoid accidental use. Follow your institution's guidelines for disposing of laboratory equipment, which may include recycling, hazardous waste procedures, or coordination with the institution's Environmental Health and Safety Office. For glassware or equipment contaminated with hazardous chemicals, decontaminate it if possible before disposal or follow local regulations for hazardous waste management.

9.10 NRTL Certification and Compliance

Following the proper protocols and receiving the requisite training is critical but so is ensuring that lab equipment satisfies nationally recognized safety standards. The Nationally Recognized Testing Laboratory (NRTL) program of the Occupational Safety and Health Administration (OSHA) is critical in this regard. OSHA has authorized independent organizations known as NRTLs the capacity to test and certify items for safety, particularly those with mechanical, thermal, or electrical dangers such as autoclaves, centrifuges, hot plates, ovens, incubators, and fume hoods.

The OSHA-approved NRTL mark must be displayed on laboratory equipment. When a piece of equipment is certified, it means that it has passed stringent testing and satisfies safety regulations set by organizations such as the International Electrotechnical Commission, Canadian Standards Association (CSA, also called the CSA Group), Underwriters Laboratories, and the American National Standards Institute. Laboratory professionals and regulatory bodies are reassured that the equipment is safe to use under normal circumstances by their certification marks, which are commonly seen on product labels or in user manuals.

Institutions have a responsibility to procure, install, and regularly inspect only NRTL-certified equipment, especially for critical applications. Additionally, during audits or safety inspections, the absence of NRTL certification can result in regulatory citations or the removal of noncompliant equipment from use. Therefore, laboratory supervisors and procurement personnel should be familiar with the NRTL program and integrate its requirements into equipment purchasing and maintenance protocols. In summary, using NRTL-certified equipment is a foundational aspect of laboratory safety compliance. It not only protects personnel and property but also ensures alignment with OSHA regulations and institutional safety policies.

9.11 Questions and Answers

Questions	Answers
1. Why is laboratory equipment safety important?	To prevent accidents, injuries, and damage to instruments or experiments.
2. What should you do before using any lab equipment?	Read the operating manual and receive proper training.
3. How often should lab equipment be inspected?	Regularly, according to manufacturer guidelines and institutional policies.
4. What is the risk of using damaged electrical cords?	Electrical shock, fires, or equipment failure.

(continued)

Questions	Answers
5. How should glassware be handled?	With care, inspect for cracks and avoid sudden temperature changes.
6. How should centrifuges be loaded?	Symmetrically, with balanced tubes to prevent vibration and damage.
7. What is an autoclave used for?	Sterilizing equipment and waste using pressurized steam.
8. What safety step is critical before opening an autoclave?	Allow pressure and temperature to return to safe levels.
9. How should lab equipment be cleaned?	According to proper protocols, using compatible cleaning agents.
10. Why should you never use equipment you are unfamiliar with?	It increases the risk of misuse and accidents.

Chapter 10
Laboratory PPE: Selection, Maintenance, and Storage

10.1 Introduction

A vital part of laboratory safety is personal protective equipment (PPE), which shields workers from physical, chemical, and biological risks. PPE is the final line of defense for lab safety, whereas engineering controls (like fume hoods) and administrative measures (like training and protocols) constitute the basis. Using PPE properly reduces the chance of mishaps and exposure to dangerous materials when working in a lab. Protecting against chemical spills and splashes, preventing the intake of harmful aerosols or vapors, preventing skin contact with poisonous or caustic chemicals, protection from physical harm (cuts, punctures, and abrasions), as well as providing fire and heat protection, are the primary goals of PPE in a chemical laboratory (Figure 10.1).

Figure 10.1: Personal protective equipment. Images credit: https://depositphotos.com/photos/personal-protective-equipment.html.

https://doi.org/10.1515/9783112218105-010

10.2 Types of PPE in the Laboratory

10.2.1 Eye and Face Protection

Types	Purpose
Safety glasses	Protect against minor splashes and flying particles
Goggles	Provide full eye coverage; necessary for high splash risks
Face shields	Protect face and neck; used alongside goggles for corrosive or reactive substances

10.2.2 Hand Protection

Glove materials	Suitable for
Nitrile	Most solvents, oils, and acids
Latex	Biological materials (but may cause allergies)
Neoprene	Acids, peroxides, and fuels
Butyl rubber	Ketones, esters, and concentrated acids
PVC	Acids, bases, and some solvents
Thermal gloves	Handling hot or cold equipment
Cut-resistant gloves	Used during glass cutting or cleanup

10.2.3 Body Protection

Item	Purpose
Lab coats	Protect skin and clothing from spills and splashes
Chemical-resistant aprons	Provide extra protection against corrosives
Flame-resistant coats	Needed when working with flammable materials

10.2.4 Respiratory Protection

Types	Use case
N95 respirators	Filtering non-oil-based particulates and aerosols
Cartridge respirators (3M 6006)	Organic vapors, acid gases, and formaldehyde
Supplied air respirators (SAR)	For highly toxic or oxygen-deficient environments

10.2.5 Foot Protection

Types	Protection from
Closed-toe shoes	Spills and dropped objects
Chemical-resistant footwear	Acids, solvents, and water
Steel-toed boots	Heavy equipment or glassware risks

10.3 Selection of Appropriate PPE and Proper Use

A comprehensive risk assessment of the activity at hand should serve as the basis for choosing PPE. The type of hazard (chemical, thermal, biological, or mechanical), possible exposure routes (inhalation, skin contact, or eye exposure), the length and frequency of exposure, and the PPE's material compatibility (e.g., chemical resistance of gloves) should all be considered in this assessment. After choosing the proper PPE, it is crucial to:
- Check PPE for wear, damage, or contamination before each usage.
- Make sure that your PPE fits comfortably and securely by wearing it appropriately.
- Steer clear of altering or abusing PPE, such as removing gloves too soon or raising eyewear.
- Before handling personal belongings, take off your gloves to avoid cross-contamination.
- Understand emergency removal procedures, such as how to safely remove contaminated gloves or extinguish burning clothing.

10.4 Maintenance and Storage

10.4.1 Cleaning and Decontamination

- Reusable PPE (e.g., lab coats and goggles) must be cleaned regularly.
- Use manufacturer instructions for washing flame-resistant coats or chemical aprons.
- Never take contaminated lab coats home for cleaning and use professional services.

10.4.2 Storage Guidelines

– Store PPE in a clean, dry, designated area, away from chemicals and direct sun-light.
– Hang lab coats on hooks or racks; do not drape them over chairs or benches.
– Keep gloves and eyewear in clean, covered containers to prevent dust and con-tamination.

10.4.3 Training and Responsibility

All laboratory personnel must be trained in:
– Proper selection and usage of PPE
– Limitations of each type of PPE
– Care, maintenance, and disposal
– Response protocols in case of PPE failure

Supervisors and instructors are responsible for ensuring PPE availability, training, and compliance. Students must report damaged or missing PPE and follow all lab safety protocols.

10.5 Limitations of PPE

While PPE significantly reduces risks, it has limitations:
– It does not eliminate the hazard; it only protects against exposure.
– It may provide only partial protection if not properly fitted or used.
– It can fail if exposed to incompatible substances, excessive heat, or wear and tear.

In conclusion, PPE serves as a critical barrier between laboratory personnel and po-tential hazards. However, its effectiveness depends not only on having the right equipment but also on making informed selections based on risk assessments, wear-ing PPE correctly, and maintaining it in good condition. Regular inspections, proper storage, and prompt replacement of damaged items are essential components of a suc-cessful PPE program. Moreover, fostering a strong safety culture, through ongoing training, supervision, and a shared commitment to best practices, empowers both stu-dents and educators to work confidently in laboratory environments. PPE should never be seen as a substitute for safe work practices or engineering controls, but rather as a vital final safeguard when other controls may not fully eliminate risk. Al-ways remember PPE is your last line of defense. Treat it with the seriousness it de-serves, select it wisely, wear it properly, and respect its role in protecting your health and safety.

10.6 Questions and Answers

Questions	Answers
1. Why is PPE important in the laboratory?	To protect workers from chemical, physical, and biological hazards.
2. What are the most common types of PPE in labs?	Gloves, lab coats, safety goggles, face shields, and respirators.
3. What type of eye protection is needed for handling corrosives?	Chemical splash goggles or a face shield.
4. Why should lab coats be worn?	To protect skin and clothing from spills, splashes, and contaminants.
5. When is a face shield required?	During procedures with a risk of splashing, explosion, or flying particles.
6. When should respirators be used?	When there is a risk of inhaling hazardous vapors, dusts, or gases.
7. Can disposable PPE be reused?	No, it should be discarded after one use or if contaminated.
8. How do you know when to replace PPE?	When it is damaged, contaminated, or after recommended use duration.
9. What is the role of training in PPE use?	To ensure proper selection, fitting, and safe usage of PPE.
10. Why must PPE fit properly?	To ensure effective protection and reduce the risk of exposure.

Chapter 11
Chemical Spill Management and Decontamination Procedures

11.1 Introduction

In laboratory settings, chemical spills (Figure 11.1) constitute an unavoidable danger. Accidents can happen even with the finest safety procedures. To reduce the negative effects of such disasters on people, property, and the environment, effective spill control and decontamination techniques are crucial. Safety is guaranteed, escalation is avoided, and laboratory activities are resumed with a prompt and appropriate reaction.

Figure 11.1: Chemical spills. Image credit: https://depositphotos.com/photos/chemical-spills.html.

11.2 Classification of Chemical Spills

Spills are generally categorized by their volume, hazard level, and potential for exposure. Types of chemical spills and their response levels are presented in Table 11.1.

https://doi.org/10.1515/9783112218105-011

Table 11.1: Types of chemical spills and their characteristics.

Spill types	Definition	Examples	Hazards	Response
Minor spills (low hazard)	Small amounts of low-toxicity chemicals, no immediate danger	Dilute acids/bases, small ethanol, or acetone spills	Minimal exposure risk	Clean by trained staff using PPE and spill kits
Major spills (high hazard)	Large volume or highly hazardous substances	Large solvent spills, concentrated acids, and toxic reagents	Fire, toxicity, and environmental harm	Evacuate and notify safety officer/ emergency response
Flammable spills	Spills of easily ignitable substances	Diethyl ether, hexane, and toluene	Fire or explosion	Eliminate ignition sources, ventilate, and use non-sparking cleanup tools
Corrosive spills	Spills of substances that burn or degrade tissues/materials	Sulfuric acid, HCl, and sodium hydroxide	Chemical burns and equipment corrosion	Neutralize (if safe), wear full PPE, and ventilate
Toxic spills	Involves harmful or lethal substances even in small amounts	Mercury, formaldehyde, and hydrogen sulfide	Poisoning via inhalation, contact, or ingestion	Evacuate, restrict area, specialized cleanup, and possible medical evaluation
Reactive spills	Spills of chemicals that react violently with air, water, or other materials	Sodium metal, peroxides, and acids with organics	Explosion and toxic gas release	Do not touch – call emergency response; isolate area
Biological/ radiochemical spills	Involves infectious agents or radioactive materials	Pathogens, P-32, and I-125	Infection and radiation exposure	Follow biosafety/ radiation protocols and report immediately

11.3 General Spill Response Principles

Determining the type and size of the spill is the first step in responding to a chemical leak effectively. If there is an imminent concern from the spill, such as fire, toxic fumes, or reactivity, leave the area. Notify others and, if required, get in touch with the appropriate people or emergency services. Use the appropriate personal protective equipment (PPE) and spill kits to clean up a small spill that is within your training level. Quickly contain the spill and, if safe, ventilate the area to stop it from spreading.

Properly dispose of contaminated materials and report the incident according to institutional protocols. Never attempt to clean up a spill unless you are trained and confident in doing so safely. These essential steps are summarized in the general response principles presented in Figure 11.2.

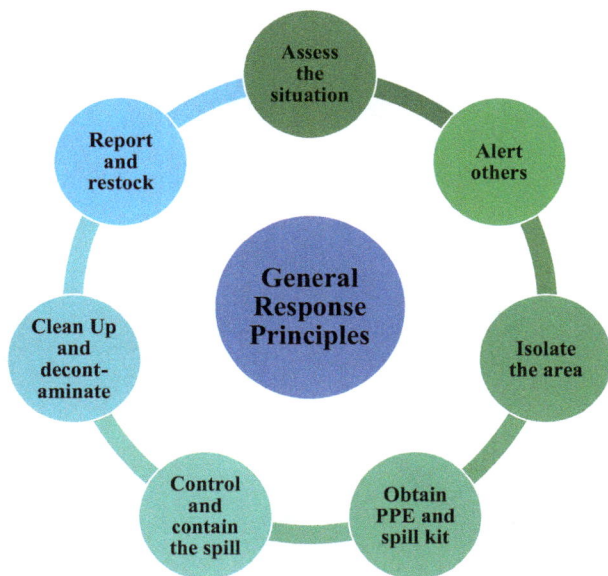

Figure 11.2: General spill response principles.

11.4 Spill Response Materials and Kits

A well-equipped spill kit (Figure 11.3) should be easily accessible and tailored to the chemicals in use. Typical contents include:
- Absorbent pads or pillows
- Neutralizing agents (e.g., sodium bicarbonate for acids and citric acid for bases)
- Chemical absorbent powders
- PPE: gloves, goggles, and disposable lab coats
- Dustpan, scoop, tongs, and plastic bags
- Waste disposal bags or containers
- Instructions for use and emergency contact list

Figure 11.3: Chemical spill kits. Image credit: https://oceansafetysupplies.com/product/chemical-spill-kit-200l/.

11.5 Response Procedures for Specific Spill Types

11.5.1 Acid and Base Spills

- Wear chemical-resistant gloves, goggles, and a lab coat.
- Neutralize the acid with sodium bicarbonate or the base with citric acid.
- Absorb the neutralized material with inert absorbent (e.g., vermiculite or paper towels).
- Scoop up and dispose of in appropriate hazardous waste containers.
- Rinse the area with water if safe to do so.

11.5.2 Solvent Spills

- Extinguish all ignition sources.
- Use non-sparking tools and adequate ventilation.

- Absorb with inert, nonflammable materials like vermiculite.
- Do not use water; it may spread the spill or react violently.
- Dispose of contaminated materials in flammable waste containers.

11.5.3 Toxic or Reactive Chemical Spills

- Evacuate the area immediately and alert emergency personnel.
- Do not attempt cleanup unless properly trained and equipped.
- Wait for the Hazardous Materials Response Team (HAZMAT) or Environmental Health and Safety (EHS) personnel.

11.5.4 Mercury Spills

- Do not use a vacuum cleaner or broom.
- Evacuate and ventilate the area.
- Use a mercury spill kit (Figure 11.4) with sulfur powder or zinc dust.
- Collect droplets with an eyedropper or tape.
- Dispose of in mercury-specific containers.

Figure 11.4: Mercury spill kits. Image credit: https://www.mscdirect.com/product/details/92977719.

11.6 Decontamination Procedures

After a spill is contained and absorbed, decontamination of affected surfaces and materials is necessary. This can be done as follows:
– Remove all absorbents and waste.
– Wash contaminated surfaces with mild detergent and water.
– Use appropriate neutralizers for lingering residues.
– Rinse the area thoroughly.
– Ventilate the space to remove vapors or airborne contaminants.
– Check for lingering contamination with pH paper, odor checks, or chemical sensors if applicable.

In conclusion, effective spill management and decontamination (Table 11.2) are vital to maintaining a safe and functional laboratory environment. Proper preparedness, prompt action, and systematic response can prevent injuries, equipment damage, and environmental contamination. All personnel must be trained in spill response protocols, understand the hazards associated with laboratory chemicals, and be capable of using spill kits and PPE appropriately. Establishing clear procedures for handling different types of spills, whether biological, chemical, or radioactive, ensures that incidents are managed swiftly and safely. Regular training sessions, spill response drills, and proper signage can significantly enhance readiness. Additionally, ensuring that emergency equipment such as eyewash stations, safety showers, absorbent materials, and neutralizing agents are accessible and well-maintained is essential. Creating a culture of safety, where vigilance and accountability are integral to daily operations, reinforces the importance of proactive spill management. By integrating spill response into broader laboratory safety protocols and encouraging continuous improvement, institutions can foster an environment that not only protects individuals but also preserves the integrity of research and the facility as a whole.

Table 11.2: Key elements of effective spill management and decontamination.

Elements	Description
Preparedness	Ensure that spill kits, PPE, and emergency equipment are stocked and accessible.
Training	Conduct regular training and drills on spill response procedures.
Hazard identification	Know the physical and chemical properties of substances used in the lab.
Immediate response	Evacuate if necessary; alert others and secure the area.
Use of PPE	Select and wear appropriate PPE before approaching the spill.
Containment and cleanup	Use absorbents, neutralizers, and appropriate tools to control and clean spill.

Table 11.2 (continued)

Elements	Description
Waste disposal	Dispose of contaminated materials in labeled hazardous waste containers.
Decontamination	Clean surfaces thoroughly to remove residues and prevent secondary exposure.
Reporting and documentation	Report the incident to supervisors and document the response steps taken.
Review and improvement	Assess the response after the incident and update protocols as needed.

11.7 Questions and Answers

Questions	Answers
1. What is the first step in responding to a chemical spill?	Alert others, evacuate if necessary, and assess the type and extent of the spill.
2. What is a spill kit?	A collection of tools and materials used to safely clean up hazardous chemical spills.
3. What items are commonly found in a spill kit?	Absorbent materials, neutralizers, gloves, goggles, disposal bags, and instructions.
4. Who should clean up a chemical spill?	Trained personnel using appropriate PPE and spill response procedures.
5. What is the purpose of a neutralizer in a spill kit?	To safely neutralize acid or base spills before cleanup.
6. How do you clean up a small, low-risk spill?	Wear PPE, contain and absorb the spill, decontaminate the area, and dispose of waste properly.
7. What should you do in case of a large or high-risk spill?	Evacuate the area and call the emergency response team.
8. What is decontamination?	The process of removing or neutralizing hazardous substances from people, surfaces, or equipment.
9. What areas should be decontaminated after a spill?	All affected surfaces, tools, and any surrounding areas that may have been exposed.
10. Why is it important to document a spill incident?	To review causes, improve safety procedures, and comply with regulations.

Chapter 12
Biological and Radiological Hazards in Chemistry Labs

12.1 Introduction

While chemistry laboratories are primarily associated with chemical risks, many also involve biological and radiological hazards, especially in interdisciplinary settings such as environmental chemistry, biochemistry, medicinal chemistry, and materials science. Failure to recognize and manage these hazards can result in serious health risks and regulatory violations.

12.2 Biological Hazards in Chemistry Labs

12.2.1 Definition and Sources

Biological hazards (biohazards) are biological materials (Figure 12.1) that can harm human health. These hazards may be present within the laboratory or may escape into the surrounding environment. In chemistry laboratories, they can arise from:
- Microorganisms used in environmental or biochemical studies such as bacteria and fungi
- Recombinant DNA technology and genetic engineering experiments
- Biodegradation studies using living organisms
- Wastewater and soil samples containing unknown microbial contaminants

Figure 12.1: Biological substances. Image credit: https://depositphotos.com/photos/biological-substan ces-bacteria.html.

https://doi.org/10.1515/9783112218105-012

12.2.2 Risk Classification (Biosafety Levels)

Biological agents are classified into biosafety levels (BSL 1–4) based on their infectivity, severity, and transmissibility (Table 12.1). Most chemistry labs will encounter BSL-1 or BSL-2 organisms.

Table 12.1: Biological agents' biosafety levels.

Level	Description	Examples
BSL-1	Minimal threat, standard microbiological practices	Nonpathogenic *E. coli*, yeast
BSL-2	Moderate risk, requires limited access and PPE	*Salmonella*, *Staphylococcus aureus*
BSL-3	High risk, controlled access, specialized ventilation	*Mycobacterium tuberculosis*
BSL-4	Life-threatening, maximum containment	Ebola virus, Marburg virus

12.2.3 Control Measures

- Engineering Controls. Biosafety cabinets, HEPA filters, and dedicated autoclaves.
- Administrative Controls. Access restrictions, standard operating procedures, and training.
- PPE. Gloves, lab coats, eye protection, and face shields.
- Disinfection and Decontamination. Use of 10% bleach, autoclaving contaminated waste.
- Proper Waste Disposal. Biohazard bags, sharps containers, and incineration of infectious waste.

12.3 Radiological Hazards in Chemistry Labs

12.3.1 Types and Sources of Radiation

Radiological hazards in chemistry labs can stem from both ionizing and nonionizing radiation. Radiation is often used in radiolabeling experiments, radioactive tracers in reaction kinetics or biological assays, and material activation and radiochemical analysis (Figure 12.2).
- Ionizing radiation:
  - *Alpha particles* from radium and uranium.
  - *Beta particles* from tritium and carbon-14.
  - *Gamma rays/X-rays* from cobalt-60 and cesium-137.
- Nonionizing radiation:
  - Ultraviolet (UV) light used in photochemistry and sterilization.

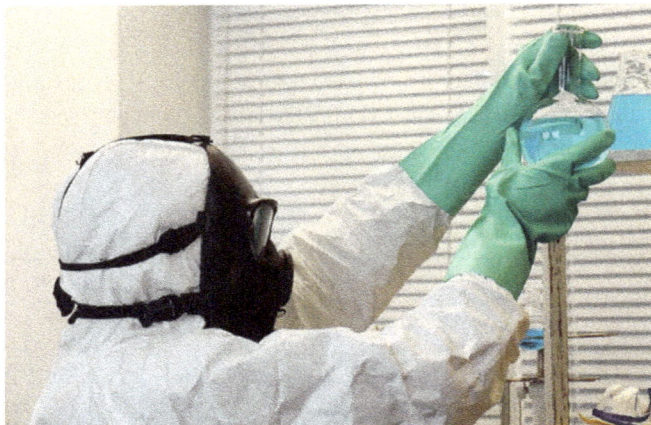

Figure 12.2: Radiological hazards. Image credit: https://depositphotos.com/photos/radiological-hazards-in-labs.html.

12.3.2 Radiation Units and Dosimetry

Dosimetry and radiation units are crucial for quantifying and evaluating ionizing radiation exposure. While the biological impact of radiation is stated in sieverts (Sv), which takes tissue sensitivity and radiation type into consideration, the absorbed dose is measured in grays (Gy). In order to ensure that doses stay below safe and regulated levels to safeguard human health, dosimetry uses specialized tools and procedures to monitor and assess radiation exposure in industrial, medical, and research contexts. Dosimeters (Figure 12.3) are required for personnel handling ionizing radiation in order to monitor exposure.

12.3.3 Control Measures

- Time, Distance, Shielding Principle:
  - Minimize exposure time.
  - Maximize distance from sources.
  - Use appropriate shielding (e.g., lead, acrylic, and concrete).
- Containment and Labeling:
  - Clearly mark radioactive materials with the international radiation symbol.
  - Use designated radioisotope work areas.
  - Store materials in shielded, labeled containers.

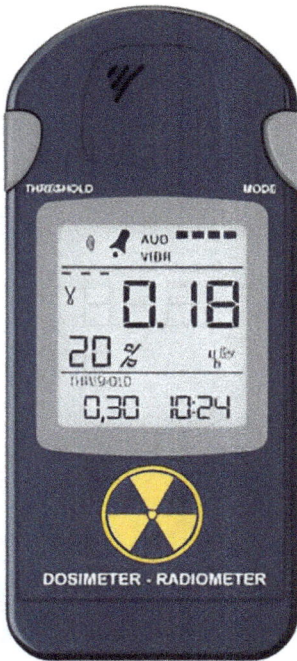

Figure 12.3: Dosimeter. Image credit: https://depositphotos.com/photos/dosimeter.html.

– Monitoring and Surveying:
  – Routine area surveys with Geiger counters (Figure 12.4) or scintillation detectors (Figure 12.5).
  – Regular wipe tests for contamination.
– Training and Authorization:
  – All users must complete radiation safety training.
  – Work must be authorized under institutional radiation safety programs.

12.4 Combined Bio-radiological Hazards

Some advanced experiments involve both biological and radiological materials such as:
– Radiolabeled DNA or RNA in genetic studies
– Radioisotope tracking of cellular uptake
– Environmental bioremediation studies involving radionuclide-contaminated organisms

Figure 12.4: Geiger counter. Image credit: https://depositphotos.com/photos/geiger-counter.html.

Figure 12.5: Scintillation detector. Image credit: https://depositphotos.com/photos/scintillation-detector.html.

12.5 Emergency Procedures

12.5.1 Biological Spills

- Don PPE, contain the spill with absorbents.
- Apply appropriate disinfectant (e.g., 10% bleach).
- Allow contact time (usually 10–30 min).
- Clean and decontaminate the area thoroughly.
- Dispose of waste in biohazard bags.

12.5.2 Radiological Spills

- Evacuate and notify radiation safety personnel.
- Restrict access and use signage.
- Monitor and decontaminate surfaces with survey meters.
- Clean using absorbents and approved procedures.
- Collect all waste in radioactive waste containers.

12.6 Waste Disposal and Compliance

To keep a laboratory safe and ecologically conscious, proper waste disposal (Table 12.2) and regulatory compliance are essential. Institutional rules and governmental requirements require that all hazardous, chemical, and biological wastes be kept, labelled, and separated. Local, national, and institutional standards, including the International Atomic Energy Agency's Safety Standards for Radiation Protection and the Centers for Disease Control and Prevention's and World Health Organization's Biosafety Guidelines, must be followed by labs. In addition to the guidelines from the Occupational Safety and Health Administration.

Table 12.2: Biological and radioactive disposal methods.

Waste type	Disposal method
Biological (BSL-1/2)	Autoclaving, incineration, and biohazard bags
Radioactive (short-lived)	Decay-in-storage, monitored disposal
Radioactive (long-lived)	Collection by licensed radioactive waste vendors
Combined waste	Must follow the most restrictive disposal pathway

In conclusion, the presence of biological and radiological hazards in chemistry labs, while less frequent than chemical hazards, poses unique and often severe risks. It is crucial to implement strict containment practices, hazard identification, personal protection, and waste management systems. Awareness and training are key to maintaining a safe laboratory environment and ensuring institutional and regulatory compliance.

12.7 Questions and Answers

Questions	Answers
1. What are biological hazards in a chemistry lab?	Risks from microorganisms like bacteria, viruses, or biological toxins.
2. What are radiological hazards?	Risks from exposure to ionizing radiation such as alpha, beta, gamma, or X-rays.
3. What PPE is needed for handling biological agents?	Gloves, lab coats, goggles, and sometimes face shields or masks.
4. What is a biosafety cabinet used for?	To provide containment and protect workers from exposure to infectious agents.
5. What equipment is used to monitor radiation levels?	Geiger counters, scintillation counters, and dosimeters.
6. How should radioactive waste be stored?	In labeled, shielded containers following institutional protocols.
7. What is a dosimeter badge?	A device worn to track cumulative radiation exposure.
8. How should biological spills be handled?	With PPE, disinfectants, and proper containment procedures.
9. How are biohazardous materials labeled?	With the universal biohazard symbol and details of the risk.
10. Why is training essential for working with bio/radiological hazards?	To ensure safe practices, proper use of equipment, and emergency preparedness.

Chapter 13
Emergency Preparedness and Response Plans

13.1 Introduction

Emergencies in chemical laboratories can arise from a variety of sources, chemical spills, fires, gas leaks, equipment malfunctions, natural disasters, or personal injuries. To minimize harm and protect people, property, and the environment, it is essential to have a well-developed and practiced emergency preparedness and response plan (EPRP). Preparedness encompasses a continuous cycle of hazard identification, risk assessment, preventive controls, and competency-based training that equips every laboratory worker, from students to senior researchers, with the knowledge and confidence to act decisively when seconds count. A robust EPRP establishes clear roles, communication pathways, and resource allocations (e.g., spill kits, fire-suppression equipment, and first-aid stations) before an emergency occurs, ensuring that response actions are swift, coordinated, and proportional to the threat. Regular drills and after-action reviews transform the plan from a binder on a shelf into a living, evolving program that integrates lessons learned, new regulatory requirements, and advances in technology. By embedding emergency preparedness into daily operations and institutional culture, laboratories not only reduce the likelihood and severity of incidents but also safeguard continuity of research, protect institutional reputation, and uphold their ethical responsibility to the broader community and environment. The importance of emergency preparedness is summarized in Figure 13.1.

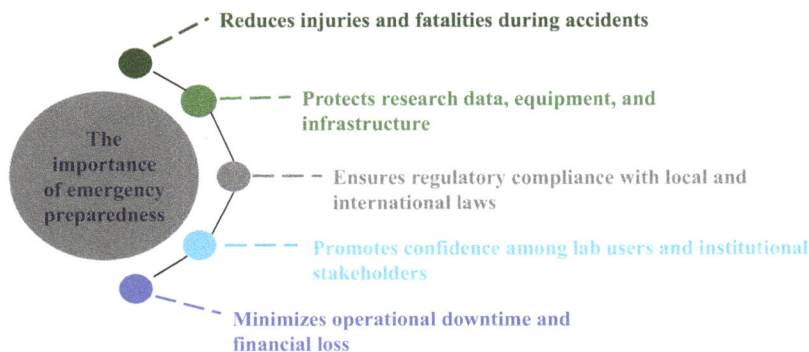

- Reduces injuries and fatalities during accidents
- Protects research data, equipment, and infrastructure
- Ensures regulatory compliance with local and international laws
- Promotes confidence among lab users and institutional stakeholders
- Minimizes operational downtime and financial loss

The importance of emergency preparedness

Figure 13.1: The importance of emergency preparedness.

https://doi.org/10.1515/9783112218105-013

13.2 Key Elements of an Emergency Preparedness and Response Plan

An effective EPRP should be comprehensive, well-structured, and tailored to the specific hazards and operational needs of the laboratory. Core elements typically include hazard identification, defined emergency roles and responsibilities, communication protocols, evacuation procedures, access to emergency equipment, and coordination with external emergency services. Together, these components ensure that laboratory personnel are prepared to respond quickly and effectively to minimize harm. The key components of an effective EPRP and their purposes are summarized in Table 13.1.

Table 13.1: Key components of an effective EPRP and their purposes.

Components	Purpose
Hazard identification	Understand all possible risks (chemical, physical, and biological)
Risk assessment	Evaluate likelihood and severity of each hazard
Emergency procedures	Define step-by-step actions for various emergency scenarios
Communication plan	Ensure quick and clear notification of all relevant personnel
Roles and responsibilities	Assign tasks to emergency coordinators, lab managers, and team members
Training and drills	Prepare all users to respond correctly under stress
Resource inventory	Maintain emergency equipment and first aid materials
Post-incident review	Learn from incidents and improve future preparedness

13.3 Types of Laboratory Emergencies and Responses

13.3.1 Chemical Spills

- Minor Spills (<100 mL of low-toxicity substances):
    - Alert people nearby.
    - Wear appropriate PPE.
    - Clean using spill kit (absorbents and neutralizers).
    - Dispose of waste properly.
- Major Spills (toxic, flammable, or large volumes):
    - Evacuate the area.
    - Notify EHS personnel.
    - Use signage and restrict access.
    - Decontaminate under supervision.

13.3.2 Fires

− Activate the fire alarm (Figure 13.2) system.
− Use a fire extinguisher only if:
  − You are trained.
  − The fire is small and manageable.
  − You have a safe escape route.
− Evacuate immediately if the fire grows or if smoke spreads.
− Shut lab doors (do not lock) to contain the fire.
− Proceed to designated assembly areas and account for personnel.

Figure 13.2: Fire alarm system. Image credit: https://depositphotos.com/photos/fire-alarm-system.html.

13.3.3 Medical Emergencies (First-Aid Kits and Immediate Medical Response)

– First-aid kits
First-aid kits (Figure 13.3) are a critical component of laboratory safety infrastructure, designed to provide immediate care for minor injuries and stabilize more serious conditions until professional medical help arrives. The availability, accessibility, and proper maintenance of these kits are essential for ensuring a timely and effective response to injuries such as cuts, chemical burns, splashes, punctures, or minor thermal burns that may occur in a laboratory setting. First-aid kits must be located in clearly marked, easily accessible areas within the laboratory, preferably near exits or designated safety stations. All personnel should be familiar with their locations and trained

on how to use the contents effectively. In larger laboratories or facilities with multiple rooms, multiple kits should be distributed to ensure no area is too far from emergency supplies. While contents may vary depending on the type of work performed in the lab, a standard kit should typically include:
- Sterile gauze pads and adhesive bandages (various sizes)
- Antiseptic wipes and hydrogen peroxide or iodine solution
- Adhesive tape and scissors
- Tweezers and disposable gloves
- Burn cream and cold packs
- Eye wash solution or single-use ampoules
- CPR face shield or mask
- Chemical splash treatment supplies (e.g., neutralizing pads)
- Emergency contact list and first-aid instruction guide
- Triangular bandages and elastic wraps
- Pain relievers and antihistamines (if allowed under institutional policy)

Figure 13.3: First-aid kit. Image credit: https://depositphotos.com/photos/first-aid-kit.html.

– Injury or exposure (cuts, burns, and inhalation of toxic fumes):
- Call emergency medical services
- Use eyewash stations or safety showers for chemical exposures
- Provide first aid if trained
- Report all incidents to the lab supervisor and institutional safety office

13.3.4 Gas Leaks and Electrical Failures

– Evacuate if gas leak (Figure 13.4), odors, or alarms are detected.
– Do not use electrical devices or create sparks.
– Shut down power (if safe) and notify facilities management.
– Await clearance before re-entering the area.

Figure 13.4: Gas leak. Image credit: https://depositphotos.com/photos/gas-leak-in-the-laboratory.html?
qview=472355262.

13.3.5 Natural Disasters (Earthquake and Flood)

In case of these kinds of disasters (Figure 13.5), do the following:
– Seek cover under sturdy furniture during earthquakes.
– Avoid low areas and chemical storage rooms during flooding.
– Secure hazardous materials and equipment during high-risk seasons.
– Follow institutional evacuation and shelter protocols.

Figure 13.5: Natural disasters: earthquake or flood. Image credit: https://depositphotos.com/photos/natural-disasters-earthquake-or-flood.html.

Table 13.2: Emergency equipment items to be checked on regular basis.

Items	Use
Fire extinguishers	Put out small fires (classified A, B, C, or D)
Eyewash stations/showers	Rinse contaminants from eyes or body
Spill kits	Contain and clean chemical spills
First-aid kits	Provide immediate care for minor injuries
Emergency lighting	Maintain visibility during power outages
Emergency contact lists	Provide rapid access to key personnel and responders
Alarm systems	Alert lab users and trigger evacuations

13.4 Emergency Equipment and Supplies

Each laboratory should maintain and regularly inspect the following items shown in Table 13.2.

13.5 Roles and Responsibilities

Clearly defined roles and responsibilities are essential for an organized and effective emergency response. Each laboratory member, from students to supervisors, should understand their specific duties during an emergency such as initiating alarms, assisting with evacuations, using fire extinguishers, or contacting emergency services. Assigning and communicating these roles in advance help reduce confusion, ensure accountability, and facilitate a coordinated response under pressure. Clear delegation of duties improves the speed and effectiveness of emergency response. Table 13.3 summarizes the roles and responsibilities of each laboratory personnel.

Table 13.3: Roles and responsibilities of laboratory supervisors, coordinators, instructors, and laboratory personnel.

Roles	Responsibilities
Laboratory supervisor	Develop and update EPRP and ensure training and compliance
Safety officer	Coordinate institutional response and documentation
Emergency coordinator	Lead evacuation and assist in communication and response actions
Instructors/researchers/students	Follow procedures, report incidents, and participate in drills

13.6 Post-emergency Actions

After an incident:
- Conduct a headcount and check in with all lab personnel.
- Provide medical support and counseling as needed.
- Document the event: cause, response, injuries, and damage.
- Perform a root cause analysis.
- Revise SOPs and the emergency plan based on lessons learned.
- Communicate changes to all lab users.

13.7 Questions and Answers

Questions	Answers
1. What is an emergency preparedness plan?	A strategy to anticipate, respond to, and recover from emergencies effectively.
2. Who is responsible for emergency preparedness?	Everyone in the lab, especially safety officers and supervisors.
3. What is the purpose of an emergency response plan?	To provide clear procedures and roles during an emergency.
4. How often should emergency drills be conducted?	Regularly, at least annually or as required by regulations.
5. What is the role of emergency contact lists?	To ensure quick communication with responders and key personnel.
6. What should you do if you discover a fire?	Activate the fire alarm, notify others, and evacuate immediately.
7. When should you attempt to use a fire extinguisher?	Only if the fire is small, you are trained, and it is safe to do so.

(continued)

Questions	Answers
8. What is a chemical spill response plan?	Procedures for safely containing and cleaning chemical spills.
9. Why is training essential for emergency response?	To ensure everyone knows how to act quickly and safely.
10. Why must emergency equipment be maintained?	To ensure it functions properly when needed.

Chapter 14
Safe Handling of Nanomaterials and Emerging Risks

14.1 Introduction

Advancements in nanotechnology have led to a surge in the use of nanomaterials in chemistry laboratories. These materials, ranging from nanoparticles to nanotubes, possess unique properties due to their nanoscale dimensions (1–100 nm), which enable innovations in medicine, energy, catalysis, and materials science. However, their novel properties also introduce uncertainties and safety concerns, especially related to toxicity, reactivity, and environmental persistence.

14.2 Nanomaterials

Materials with at least one dimension in the nanoscale range (1–100 nm) are referred to as nanomaterials (Figure 14.1). Key characteristics include:
- High surface area to volume ratio
- Quantum effects that alter optical, magnetic, and electrical behavior
- Enhanced chemical reactivity and catalytic activity

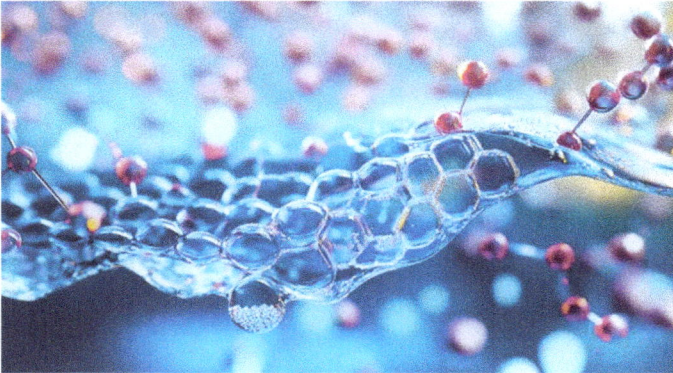

Figure 14.1: Nanomaterials. Image credit: https://depositphotos.com/photos/nanomaterials.html.

https://doi.org/10.1515/9783112218105-014

14.3 Common Types Used in Chemistry Labs

Nanomaterials have become increasingly important in modern chemistry laboratories due to their unique physical, chemical, and biological properties at the nanoscale. These materials offer enhanced surface area, reactivity, conductivity, and mechanical strength, making them valuable for a wide range of applications including catalysis, sensing, drug delivery, and materials science. The selection and use of nanomaterials in laboratory settings require a clear understanding of their types, characteristics, and associated safety considerations. Table 14.1 outlines the common types of nanomaterials (Figure 14.2) used in chemistry labs, along with their typical forms and applications.

Table 14.1: Common types of nanomaterials.

Nanomaterials	Typical application
Carbon nanotubes (CNTs)	Sensors, electronics, and structural materials
Titanium dioxide (TiO$_2$)	Photocatalysis, sunscreens, and water purification
Silver nanoparticles	Antibacterial agents, inks, and diagnostics
Graphene and derivatives	Conductive films, batteries, and drug delivery
Quantum dots	Bioimaging and photovoltaics
Iron oxide nanoparticles	Magnetic separations and biomedical applications

Figure 14.2: Common types of nanomaterials. Images credit: https://depositphotos.com/photos/quantum-dots.html.

14.4 Health and Safety Risks

The small size of nanomaterials allows them to penetrate biological membranes and be transported throughout the body. Major exposure routes and their toxicological concerns are presented in Figure 14.3.

14.5 Laboratory Risk Management

14.5.1 Hazard Identification

All nanomaterials should be treated as potentially hazardous unless proven otherwise. The risk level increases if materials are:

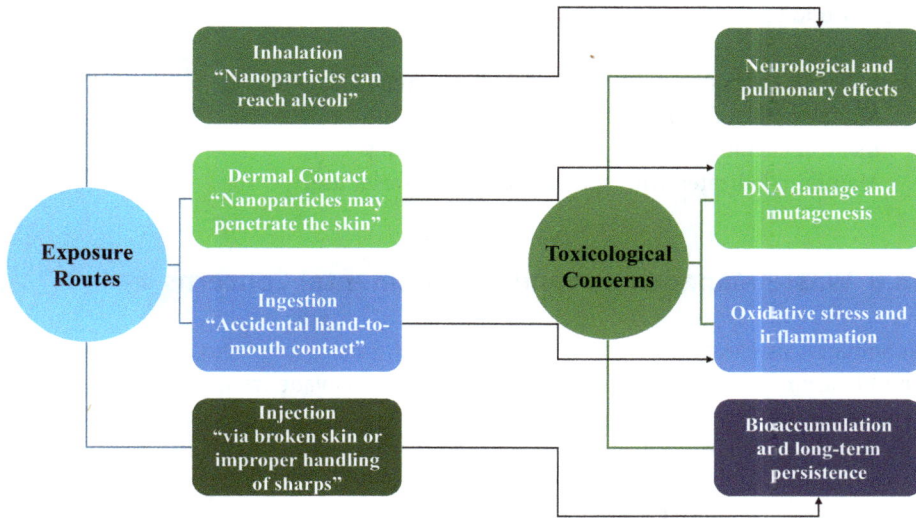

Figure 14.3: Major nanoparticle exposure routes and their toxicological concerns.

– Dry powders (easily airborne)
– Highly reactive or photoreactive
– Unknown or uncharacterized in composition

14.5.2 Engineering Controls

– Use fume hoods, biosafety cabinets, or glove boxes for handling dry nanomaterials.
– Equip ventilation systems with HEPA filters.
– Use wet methods (suspensions and slurries) instead of dry powders whenever possible.
– Implement closed systems for synthesis or scale-up experiments.

14.5.3 Personal Protective Equipment (PPE)

PPE type	Recommended features
Respirators	N95 or higher filtration rating for airborne nanoparticles
Gloves	Nitrile or chemical-resistant; double-gloving recommended
Eye protection	Safety goggles or face shields for liquids and powders
Lab coat and clothing	Nonwoven materials, disposable lab coats preferred

14.5.4 Good Laboratory Practices

- Label all containers with "Nano-" prefix and material identity.
- Minimize dust formation; avoid sweeping or compressed air.
- Use wet cleaning methods instead of dry wiping.
- Maintain dedicated workspaces for nanomaterial use.

14.6 Waste Management and Environmental Concerns

Nanomaterials can persist in air, water, and soil, raising concerns about ecotoxicity and bioaccumulation. Their small size allows them to evade traditional filtration and separation systems.

14.6.1 Waste Handling Recommendations

- Collect dry nanomaterials in sealed, labeled containers.
- Do not dispose of nanomaterials down drains or in regular trash.
- Treat liquid suspensions as hazardous chemical waste.
- Use dedicated containers for contaminated PPE and cleanup materials.

14.6.2 Environmental Compliance

While nanomaterials are often not specifically regulated, disposal must adhere to general hazardous waste regulations (EPA, REACH, and CLP). Researchers should:
- Consult institutional EHS offices for disposal guidance.
- Report significant releases or incidents involving nanoparticles.
- Track environmental emissions when manufacturing or scaling production.

In conclusion, nanomaterials hold tremendous promise across various scientific disciplines, but they also present unique health, environmental, and regulatory challenges. Effective management requires a proactive approach that includes hazard identification, implementation of appropriate controls, responsible waste handling, and continuous education. As scientific understanding advances, safety protocols must evolve to ensure that nanotechnology continues to benefit society without compromising human health or the environment. Institutions have a responsibility to:
- Keep all nanomaterials' inventories and safety information current and offer thorough instruction on the hazards and countermeasures.
- Create and implement standard operating procedures for the usage of nanomaterials.

- Keep an eye on new findings and adjust safety procedures as necessary.
- Refer to guidelines issued by agencies like NIOSH, the OECD, ISO, and under laws like REACH.

14.7 Questions and Answers

Questions	Answers
1. How can nanomaterials enter the body?	Through inhalation, skin absorption, ingestion, or injection.
2. What type of PPE is recommended for handling nanomaterials?	N95 respirators or higher, gloves, lab coats, and eye protection.
3. What engineering controls help reduce exposure to nanomaterials?	Fume hoods, glove boxes, and local exhaust ventilation.
4. How should nanomaterials be stored?	In clearly labeled, sealed containers to avoid accidental release.
5. What should be done if a nanomaterial spill occurs?	Use appropriate PPE, contain the spill, and clean up using wet methods or HEPA-filtered vacuums.
6. Why is proper training important for handling nanomaterials?	To understand risks, proper techniques and emergency procedures.
7. How is waste containing nanomaterials managed?	As hazardous waste following institutional and regulatory guidelines.
8. Can standard chemical hygiene practices apply to nanomaterials?	Yes, but with additional precautions due to their unique properties.
9. What is the role of risk assessment for nanomaterials?	To evaluate potential hazards and determine control measures.
10. Are nanomaterials visible under regular microscopes?	No, specialized equipment like electron microscopes is required.

Chapter 15
Good Laboratory Practices

15.1 Introduction to Good Laboratory Practices

Good laboratory practices (GLPs) refer to a set of principles that provide a framework within which laboratory studies are planned, performed, monitored, recorded, reported, and archived. Originally developed for regulatory compliance in research and testing laboratories, GLP has since evolved into a cornerstone of scientific integrity, accuracy, reproducibility, and safety. Laboratory work involves essential practices such as proper sample handling and responsible conduct. Given the frequent presence of hazardous chemicals, volatile solvents, and complex instrumentation, laboratories can pose significant health and safety risks. However, with appropriate organization and adherence to standardized practices, these environments can support safe and effective research. Personnel should adhere to GLPs in order to keep the laboratory safe, healthy, and productive. GLP is a quality management method that guarantees the consistency, dependability, and integrity of nonclinical safety research. Pharmaceuticals, medical gadgets, food and color additives, veterinary goods, environmental chemicals, and other topics may be investigated. Unlike standard safety rules (such as wearing gloves, goggles, or lab coats), GLP controls a larger range of laboratory activities.

It covers the qualification of personnel, validation of instruments and methods, and thorough documentation of all study phases from planning and execution to data archiving and reporting. GLP enhances the traceability and reproducibility of experimental data and supports regulatory submissions by ensuring transparency and accountability. By defining roles and responsibilities throughout the study process, GLP contributes to reducing errors, improving study reliability, and safeguarding human health, animal welfare, and the environment. The main goals of having GLP are:
- Ensure data quality and integrity
- Promote safety and environmental protection
- Facilitate reproducibility of experiments
- Enhance regulatory compliance

15.2 GLP Basic Elements

GLPs are a set of internationally recognized principles (Table 15.1) intended to ensure the quality, integrity, and reliability of nonclinical laboratory studies. GLP aims to promote consistency, transparency, and accountability in laboratory environments. These guidelines cover the organizational process and conditions under which laboratory studies are planned, performed, monitored, recorded, reported, and archived. Understanding

https://doi.org/10.1515/9783112218105-015

the basic elements of GLP is essential for maintaining scientific credibility, ensuring compliance with regulations, and safeguarding human health and the environment.

Table 15.1: Basic elements of good laboratory practices.

Elements	Description
Organization and personnel	Clear roles, responsibilities, qualifications, and training of all staff.
Quality assurance (QA)	Independent QA unit to monitor study conduct and ensure compliance with GLP.
Facilities	Adequate space and infrastructure to avoid mix-ups and ensure study integrity.
Equipment	Properly maintained and calibrated instruments suitable for their intended use.
Test and control articles	Accurate characterization, handling, and storage of substances being tested.
Standard operating procedures (SOPs)	Written procedures for routine tasks to ensure consistency and repeatability.
Study protocol	Detailed plan outlining study objectives, methods, and responsibilities.
Data collection and recording	Timely, accurate, and permanent recording of raw data and observations.
Reporting of results	Clear, complete, and accurate final reports of study findings.
Archiving and records	Proper storage of raw data, reports, and materials for future reference.
Test system management	Proper care and monitoring of biological, chemical, or physical test systems.

15.3 GLP Key Principles

The key principles of GLP provide an organized framework for performing, reporting, and archiving laboratory investigations, ensuring the creation of high-quality, traceable data. These principles encourage scientific integrity and require institutions to explicitly identify staff roles and duties in order to ensure good research administration and repeatability. GLP compliance necessitates complete adherence to all principles; partial implementation is not considered acceptable compliance. Figure 15.1 highlights the 10 key principles of GLP.

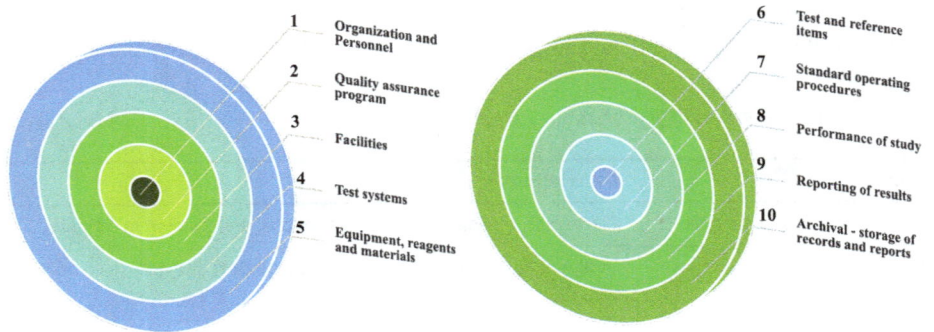

Figure 15.1: The 10 key principles of GLP.

15.4 Laboratory Conduct and Behavior

– Personal Conduct:
  – No eating, drinking, or inappropriate behavior in the lab.
  – Always wear appropriate PPE.
  – Maintain personal hygiene and handwashing routines.
– Respect for Equipment and Colleagues:
  – Use tools only for their intended purpose.
  – Do not remove items from common areas without permission.
  – Collaborate respectfully and clean up shared workspaces.

15.5 Recordkeeping and Documentation

GLP emphasizes *meticulous and traceable* documentation. Every action or observation in the lab must be recorded in a timely and legible manner:
– Use bound notebooks with numbered pages.
– Avoid white-out or erasing entries; correct mistakes with a single strike-through and initials.
– Record date, time, observations, and any deviations from standard procedure.
– Maintain electronic records securely with access logs.

15.6 Handling Chemicals and Samples

– Clearly label all containers with:
  – Chemical name
  – Concentration

- Date prepared
- Hazard warnings and preparer's initials
- Use compatible containers and proper secondary containment.
- Store chemicals by compatibility group.

15.7 Equipment Use and Maintenance

- Check equipment calibration before use.
- Report any malfunctions immediately.
- Do not operate unfamiliar equipment without training.
- Maintain logs for balances, pH meters, spectrophotometers, and other equipment.

15.8 Waste Management and Spill Response

- Segregate chemical, biological, radioactive, and sharps waste.
- Use labeled waste containers with tight-fitting lids.
- Maintain updated spill response kits and train personnel in their use.

15.9 Quality Assurance and ISO 9000 Integration

A key element of GLP is quality assurance (QA), which guarantees that all laboratory operations adhere to set standards and yield reliable, repeatable findings. Examples of QA include personnel training, standard operating procedures (SOPs), documentation, and recurring audits. Modern QA systems must incorporate ISO 9000 standards, especially ISO 9001, which describes the prerequisites for a strong quality management system. By integrating ISO principles into GLP, laboratories may comply with international best practices, which increases their legitimacy in business or regulatory contexts.

15.9.1 ISO 9000 and Laboratory Practice

The ISO 9000 family of standards (Figure 15.2) provides a framework for implementing consistent, process-driven management systems. For laboratories, ISO 9001 emphasizes on customer focus, leadership commitment, process-based approaches, and continuous improvement. ISO 9001:2015 is widely adopted in analytical, research, and industrial labs. While ISO 17025 is specific to testing and calibration labs, ISO 9001 provides a broader structure compatible with GLP principles. Adoption of the ISO 9000 system requires that a manufacturer has documentation for managing an effective quality system. Documentation should be legible, readily identifiable, easily accessible, and include revision dates. All out-of-date documentation should be removed and disposed of according to set procedures.

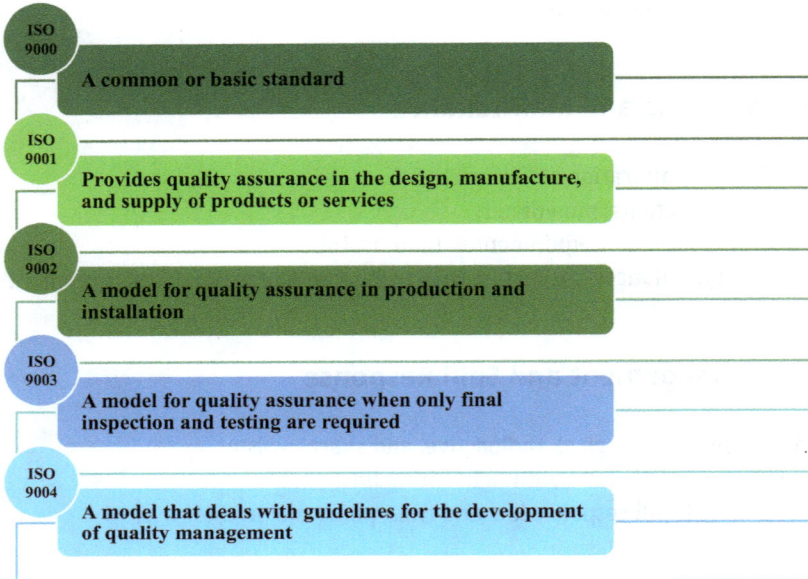

Figure 15.2: ISO 9000 standards.

15.9.2 Common Deficiencies at ISO 9000 GLP Audit

Compliance with established processes, documentation protocols, equipment standards, and personnel training is critical in laboratories and quality-controlled environments to ensure accuracy, safety, and regulatory compliance. Common nonconformities in functional areas such as documentation, equipment, materials, staff, and administration can jeopardize data integrity, pose safety hazards, and violate regulations. Table 15.2 lists common findings, organized by categories, to assist in identifying and addressing systemic weaknesses.

Table 15.2: Typical findings at ISO 9000 GLP audit to address systemic weaknesses.

Category	Common nonconformities
Documents	– Unapproved documents – Procedures not followed in practice – Misplaced documents – Incomplete approved suppliers list (ISO 9000 only) – Faulty product release protocol
Equipment	– Unqualified new/alternate equipment – Incomplete calibration – Poor performance measurement – Inadequate response to exceptions
Material	– Lack of or unclear identification – Use of unapproved sources – Poor material condition – Inadequate shelf-life control – Unclear inspection status – Nonconforming materials not controlled
People	– Lack of training – Unawareness of requirements – Noncompliance with documented requirements – Inadequate training program/records – Organizational or responsibility issues
Administration	– Incomplete/inconsistent methodologies – Nonconformance not managed (undetected, undocumented, and uncommunicated) – Ineffective corrective actions – Infrequent or ineffective management reviews – Deviation from documented procedures – Disorganized or incomplete records

15.9.3 Benefits of ISO Integration into GLP

–   Enhances document control and traceability
–   Establishes clear roles and responsibilities
–   Promotes preventive actions through risk management
–   Facilitates internal and external audits
–   Encourages standardization of procedures and training

15.10 Statistical Tools in GLP

Statistics play a vital role in validating laboratory methods, controlling quality, and ensuring data reliability. Statistical evaluation helps determine the significance, accuracy, and precision of experimental results and supports decisions during audits or investigations.

15.10.1 Key Statistical Concepts in GLP

- Accuracy and Precision:
  - Accuracy refers to the closeness of measured values to the true value.
  - Precision measures the reproducibility of results under similar conditions.
- Standard Deviation and Relative Standard Deviation:
  - Indicators of data spread and variability
- Linearity and Range:
  - Used to determine if an analytical method produces results proportional to concentration across a specific range.
- Limit of Detection and Limit of Quantification:
  - Determine the lowest concentration levels that can be reliably detected or quantified.

15.10.2 Statistical Applications in GLP

- Control Charts (e.g., Shewhart Charts):
  - Monitor process consistency and identify trends or deviations.
- Outlier Tests (e.g., Grubbs' Test):
  - Detect anomalous data points that may require further investigation.
- Regression Analysis:
  - Used to establish calibration curves and method linearity.
- Measurement Uncertainty:
  - Quantifies confidence in the final reported values.

15.11 Data Integrity and Audit Trails

Data integrity relates to the quality, completeness, and consistency of data during its entire lifespan. It is a critical GLP component, especially in the digital age, when electronic data is as prevalent as written records.

15.11.1 ALCOA and ALCOA+ Principles of Data Integrity

ALCOA (Table 15.3) is an acronym used in regulated environments, particularly in pharmaceuticals, biotechnology, and laboratory settings, to describe the principles of data integrity. These principles ensure that data generated and recorded during laboratory and manufacturing processes are trustworthy and reliable. The ALCOA+ (Table 15.4) framework outlines key attributes for maintaining trustworthy data. It includes additional principles to further reinforce data integrity.

Table 15.3: ALCOA principles.

Letter	Principle	Description
A	Attributable	It must be clear who performed an action and when it was performed.
L	Legible	Data must be readable and permanent, so it can be easily understood later.
C	Contemporaneous	Data must be recorded at the time the work is performed – not afterward.
O	Original	The first recording or a certified copy of the data must be retained.
A	Accurate	Data must be correct, with no errors or misrepresentations.

Table 15.4: ALCOA+ additional principles to further reinforce data integrity.

Additional principle	Meaning
Complete	All data (including repeat or failed results) must be included.
Consistent	Data should be recorded in a logical, chronological order.
Enduring	Data must be stored on durable media that retains the record over time.
Available	Data should be readily accessible for review and audit throughout its life.

15.11.2 Audit Trails

An audit trail is a secure, time-stamped electronic record that tracks the creation, modification, and deletion of data. It enables transparency and accountability by preserving:
- User identity
- Timestamp of activity
- Changes made
- Reason for modification (when required)

Audit trails are especially important for laboratories using Laboratory Information Management Systems, electronic lab notebooks, or other digital platforms. Regulatory agencies (e.g., FDA and EMA) require audit trails for compliance with electronic record regulations such as 21 CFR Part 11.

15.11.3 Practices to Support Data Integrity

- Train staff in ethical and proper data handling.
- Use software with audit trail capabilities.
- Prevent unauthorized access through role-based permissions.
- Conduct regular audits and data reviews.
- Avoid data transcriptions, when possible, to reduce error risk.

15.12 Ethical and Legal Considerations

- Fabrication, falsification, or plagiarism of data is unethical and legally punishable.
- Confidential information (e.g., patient samples and proprietary compounds) must be protected.
- Researchers must declare conflicts of interest and obtain ethical approvals where necessary.

15.13 Legal Compliance in Laboratory Operations

Laboratories are bound by local, national, and international laws that govern everything from waste disposal to worker safety and intellectual property. The key legal areas in GLP context are summarized in Table 15.5.

Table 15.5: The key legal areas in GLP context.

Regulatory area	Key requirements	Implications
Occupational Health and Safety Laws	– Mandate safe working conditions – Require use of PPE – Emphasize emergency preparedness	Noncompliance may lead to legal penalties, injuries, or fatalities
Environmental Regulations	– Govern storage, use, and disposal of hazardous materials (chemical, biological, and radioactive waste) – Include national and international regulations such as CEPA (Canada), EPA (USA), and the Basel Convention	Ensure environmental protection and legal compliance

Table 15.5 (continued)

Regulatory area	Key requirements	Implications
Data Protection and Privacy Laws	– Apply to labs handling personal or sensitive data (e.g., clinical samples) – Require compliance with GDPR and PIPEDA	Ensure ethical data handling and secure storage of digital records
Regulatory Compliance and Accreditation	– Relevant to labs in regulated sectors (e.g., pharmaceuticals and environmental testing) – May involve certification like ISO/IEC 17025 or FDA GLP	Subject to legal audits and inspections to ensure compliance with regulatory standards

15.14 Institutional Responsibility and Whistleblower Protection

Organizations must promote a culture of accountability and transparency. Encouraging open dialogue and ethical reflection helps prevent misconduct and fosters a safe, supportive research environment. Institutions are responsible for:
– Establishing ethical codes of conduct.
– Providing ethics training and support.
– Creating confidential channels for reporting unethical behavior.
– Protecting whistleblowers from retaliation.

15.15 Questions and Answers

Questions	Answers
1. What is the main purpose of GLP?	To ensure the quality, reliability, integrity, and reproducibility of laboratory studies.
2. Which regulatory bodies enforce GLP standards?	Organizations like the FDA (USA), OECD (nternational), and EPA (USA) enforce GLP regulations.
3. What is an SOP in GLP?	A written document that describes how to perform a specific task consistently and accurately.
4. What role does a QA unit play in GLP?	QA ensures compliance by auditing studies, facilities, equipment, and data integrity.
5. Why is documentation important in GLP?	It provides traceability, accountability, and evidence that procedures were correctly followed.
6. What is the purpose of archiving records in GLP?	To preserve study data and materials for future review, inspection, or legal use.

(continued)

Questions	Answers
7. How should deviations from SOPs be handled in a GLP-compliant lab?	They must be documented, justified, investigated, and approved by management or the study director.
8. What is ISO 17025?	ISO 17025 is an international standard for testing and calibration laboratories to demonstrate technical competence and produce valid results.
9. Can a laboratory be compliant with both GLP and ISO 17025?	Yes. Many labs implement both to meet regulatory and quality management system requirements.
10. How often should audits be conducted in ISO/GLP-compliant labs?	Internal audits should be conducted regularly (usually annually), with external audits for accreditation or regulatory compliance.

Chapter 16
Pressure and Vacuum Systems and Equipment in the Chemical Laboratories

16.1 Introduction

Laboratory experiments, whether in research or industry, rely heavily on specialized equipment to ensure accuracy, efficiency, and safety. Among the most critical yet often overlooked components are vacuum and pressure systems. These systems are essential across numerous procedures from basic separations to high-end analytical methods. This chapter provides a comprehensive overview of laboratory vacuum and pressure systems, explaining their principles, types, applications, and safety considerations. The discussion also contextualizes how such systems and equipment support both separation and analytical techniques.

16.2 Pressure and Vacuum Fundamentals

16.2.1 Definitions

- Pressure is the force exerted per unit area by a gas or a liquid. It is often measured in atmospheres, pascals, or pounds per square inch.
- Vacuum refers to a space with reduced gas pressure, lower than atmospheric pressure. Depending on the level of reduction, vacuums can be categorized as low, medium, high, or ultra-high vacuum.

16.2.2 Importance in the Laboratory

Controlling pressure and vacuum conditions is vital for:
- Enhancing reaction rates
- Improving separation efficiency
- Protecting sensitive instruments
- Ensuring sample integrity during analysis

https://doi.org/10.1515/9783112218105-016

16.3 Pressure Systems and Control

16.3.1 Compressed Gas Cylinders

These provide gases such as nitrogen, argon, or hydrogen at high pressure. Key components include:
− Pressure regulators
− Flow meters
− Flashback arrestors for flammable gases

16.3.2 Pressure Equipment

− Pressure Vessels: Used in supercritical fluid extraction or high-pressure reactions.
− Autoclaves: Sterilization under elevated pressure and temperature.
− Reactor Bombs: Small-scale reactors for synthesis under controlled pressure.

16.4 Pressure Measurement Instruments

Accurate pressure measurement is critical in laboratory operations involving vacuum systems, gas handling, and high-pressure reactions. Whether monitoring atmospheric pressure, regulating gas flow, or verifying vacuum levels, a range of instruments is used to ensure precision and safety. This section explores key pressure-measuring devices including traditional and modern instruments.

16.4.1 Mercury Barometer

A mercury barometer (Figure 16.1) monitors atmospheric pressure by raising and lowering a vertical column of mercury within a glass tube to balance the weight of the surrounding air. It consists of a glass tube with one end sealed and submerged in an open, mercury-filled basin. When air pressure rises, more mercury is forced up the tube; when pressure lowers, the mercury level decreases. The height of the mercury column, measured in millimeters of mercury, is proportional to the atmospheric pressure. Its use, advantages, disadvantages, and safety considerations are summarized in Table 16.1.

Table 16.1: Laboratory use, advantages, disadvantages, and safety considerations of mercury barometer.

Topic	Details
Laboratory use	Although largely phased out due to safety and environmental concerns, mercury barometers are still used in calibration labs and meteorology for reference measurements
Advantages	Highly accurate and stable over time No mechanical parts Long service life Suitable for calibration standards
Disadvantages	Mercury is toxic and poses a contamination risk Fragile and not portable Bulky and requires vertical installation
Safety considerations	Handle barometers carefully to prevent mercury spills Use mercury spill kits for cleanup Dispose according to hazardous waste regulations

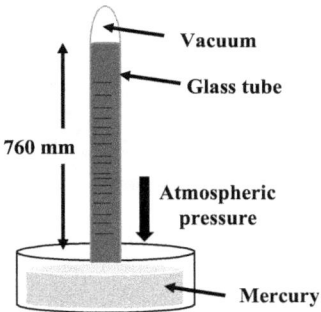

Figure 16.1: Schematic diagram of mercury barometer.

16.4.2 Aneroid Barometer

The aneroid barometer (Figure 16.2) is a compact, mercury-free instrument. It consists of a sealed, flexible metal chamber (the aneroid capsule) that expands or contracts with changes in atmospheric pressure. These movements are transmitted through a mechanical linkage to a needle on a dial. Its use, advantages, and disadvantages are summarized in Table 16.2.

Table 16.2: Laboratory use, advantages, and disadvantages of aneroid barometer.

Topic	Details
Applications in the laboratory	Atmospheric pressure tracking in labs where barometric conditions affect sensitive instruments or reactions Used in portable field kits for onsite pressure readings
Advantages	Safe, nontoxic, and compact Suitable for mobile or field work No liquids involved
Disadvantages	Less accurate than mercury under some conditions Can drift over time without recalibration Sensitive to mechanical shock or vibration

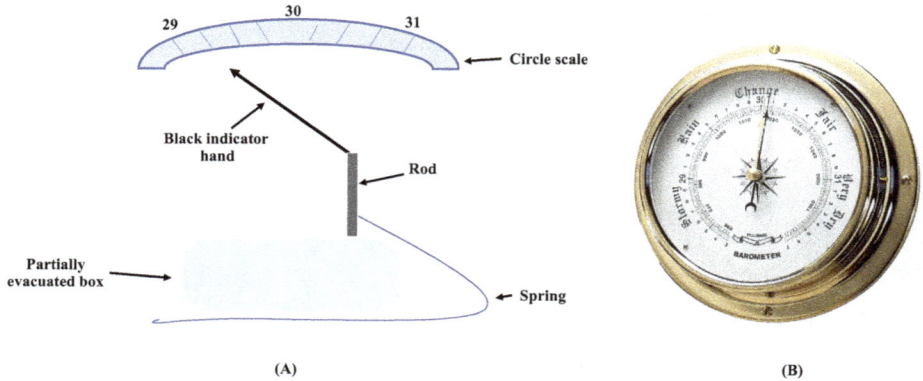

Figure 16.2: Schematic diagram of aneroid barometer (A) and aneroid barometer (B). Image credit: https://maritimeacademia.wixsite.com/my-site-1/post/aneroid-barometer.

16.4.3 Bourdon Tube Gauge

The Bourdon tube gauge (Figure 16.3) is the most common type of pressure gauge used in gas cylinders, pressurized systems, and vacuum setups. It consists of a coiled metal tube that uncoils slightly under internal pressure. This movement drives a needle to indicate pressure. Its key features and applications are summarized in Table 16.3.

Table 16.3: Key features and applications of Bourdon tube.

Topic	Details
Key features	Available in a wide range of pressure ratings (vacuum to 10,000 psi)
	Rugged, inexpensive, and simple to install
	Often includes dual-scale readings (e.g., psi and bar)
Applications	Monitoring cylinder pressures
	Regulating compressed gas systems
	Measuring pressure in filtration or reactor setups

Figure 16.3: Bourdon tube gauge. Image credit: https://www.wika.ca/111_10r_en_us.WIKA.

16.4.4 Digital Pressure Gauges

Modern laboratories increasingly use digital gauges (Figure 16.4), which convert pressure into electrical signals for display and data logging. Their key features and applications are summarized in Table 16.4.

Table 16.4: Key features and applications of digital gauge.

Topic	Details
Features	High-accuracy and multiunit display (atm, Pa, torr, and psi)
	Easy integration with computers or control systems
	Suitable for both absolute and differential pressure measurements
Applications	Vacuum line monitoring
	Instrument calibration
	Real-time pressure regulation in automated systems

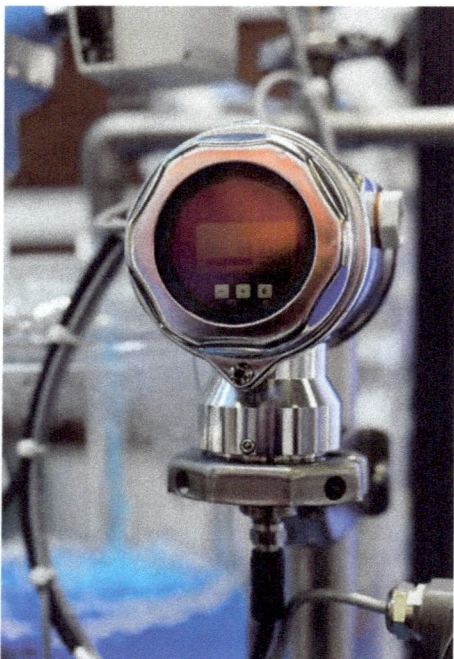

Figure 16.4: Digital gauge. Image credit: https://depositphotos.com/photos/digital-gauge.html?qview=300955126.

16.5 Vacuum Systems and Equipment

16.5.1 Vacuum Pumps

16.5.1.1 Types of Vacuum Pumps

To address the different vacuum needs in chemical laboratories, numerous types of vacuum pumps are available, each with a unique design, performance, and appropriateness for certain applications. Vacuum pump selection is determined by the required vacuum level, chemical compatibility, maintenance requirements, and budget limits. Table 16.5 outlines the essential characteristics, benefits, and limits of popular vacuum pump types such as rotary vane, diaphragm, scroll, turbomolecular, and diffusion pumps. Complementing this, Figure 16.5 provides visual representations of selected pumps to aid in understanding their structural differences.

Table 16.5: Key features, advantages, and limitations of common vacuum pumps.

Pump type	Vacuum range	Operating principle	Advantages	Limitations	Typical applications
Rotary vane pump	~1×10^{-3} to 1×10^{-1} mbar	Oil-lubricated mechanical compression via rotating vanes	High vacuum capability; suitable for solvents	Requires oil changes; not ideal for corrosive vapors	Rotary evaporators, freeze-drying, and Schlenk lines
Diaphragm pump	~1–100 mbar	Oscillating diaphragm compresses gas	Oil-free; chemical-resistant models available	Limited vacuum depth; lower flow rate	Filtration, drying, and solvent evaporation (low bp)
Scroll pump	~1×10^{-2} to 1 mbar	Interleaved spiral scrolls trap and compress gas	Oil-free; quiet; good for clean environments	Expensive; sensitive to particulates	Analytical instruments (e.g., MS and GC-MS)
Diffusion pump	~1×10^{-7} to 1×10^{-3} mbar	Vapor jet entrains gas molecules toward exhaust	Achieves ultra-high vacuum; suitable for metals	Requires backing pump; oil contamination possible	High-vacuum physics and vacuum metallurgy
Turbomolecular pump	~1×10^{-10} to 1×10^{-3} mbar	High-speed rotors impart momentum to gas molecules	Ultra-high vacuum; clean operation	Expensive; needs backing pump	Surface science and high-vacuum systems

16.5.1.2 Components of Vacuum Pumps

Vacuum pumps consist of several components that work together to create a vacuum. The main ones are:

- Pump Head. The pump head is the main body that contains the mechanism responsible for creating the vacuum. It can be made of different materials depending on the application.
- Motor. The motor provides the power to the vacuum pump, enabling the pump head to function.

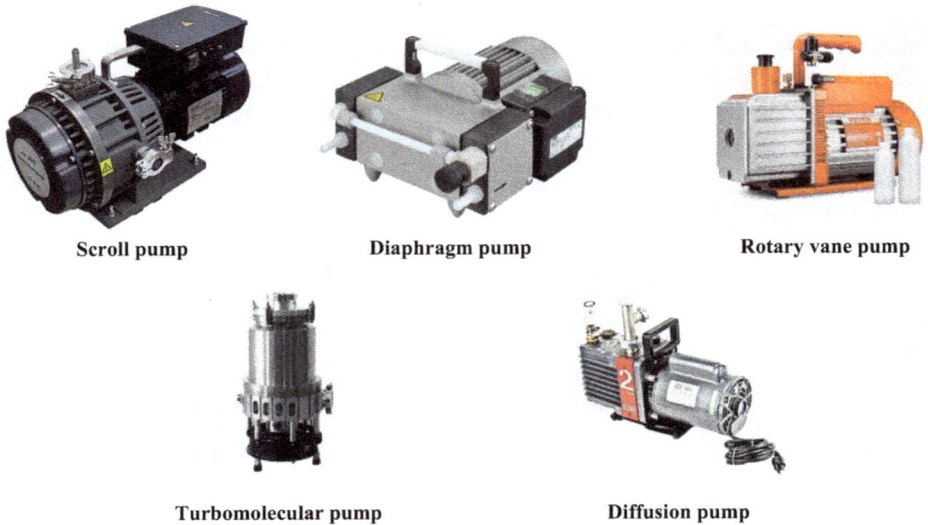

Scroll pump   Diaphragm pump   Rotary vane pump

Turbomolecular pump   Diffusion pump

Figure 16.5: Visual representations of selected vacuum pumps. Image credit: https://www.munroscientific.co.uk/vacuum_pumps.

- Inlet and Outlet Ports. The inlet and outlet ports are the openings where the pump is connected to the system and the atmosphere, respectively.
- Vacuum Gauge. The vacuum gauge is a device that measures the pressure inside the system, allowing the user to adjust the pump accordingly.

16.5.1.3 Vacuum Pump Maintenance

Maintaining a vacuum pump is critical to its good operation and lifetime. Regular maintenance can help to prevent pump failure, save downtime, and ensure the correctness of scientific studies. Here are a few maintenance tips:

- Regularly check and change the oil. Oil-powered vacuum pumps must be serviced on a regular basis. Dirty- or low-oil levels can harm the pump and cause it to fail. Consult the user handbook for the recommended oil type and frequency of changes.
- Clean pump components. Dust, dirt, and debris can collect within the pump, reducing its performance. Clean the pump head, motor, intake, and outlet ports on a regular basis with a soft brush or cloth.
- Inspect the pump for leaks. Leaks in the pump may cause a loss of vacuum pressure, impacting the experiment results. Run a leak test to detect leaks and address any concerns as soon as possible.
- Replace worn-out parts. Vacuum pump components wear down over time such as the vanes in rotary vane pumps and the diaphragm in diaphragm pumps. Replace worn components as soon as possible to prevent pump failure.

16.5.1.4 Setting Up a Vacuum Pump

Before using it, it is essential to set it up correctly. Here are the steps to follow:
-   Select the appropriate vacuum pump. Choose a vacuum pump that matches the application's requirements such as the required vacuum level, flow rate, and type of gas being removed.
-   Connect the pump to the system. Connect the pump to the system using hoses or tubing, ensuring a tight seal. Use clamps or connectors to secure the connection.
-   Connect the vacuum gauge. Connect the vacuum gauge to the pump's inlet port to monitor the pressure inside the system.
-   Turn on the pump. Turn on the pump and allow it to run for a few minutes to reach its operating temperature and create a vacuum.

16.5.2 Common Vacuum Equipment

16.5.2.1 Water Aspirators

A water aspirator (Figure 16.6) is a simple device used to generate a vacuum using the Venturi effect, a fluid dynamics principle, wherein a fluid's velocity increases as it passes through a constricted section of pipe, leading to a corresponding drop in pressure. In the case of an aspirator, water is forced through a narrow nozzle, and this rapid flow generates a region of low pressure that can draw air or gas from an attached vacuum line. The aspirator is typically connected to a tap water supply. As water flows through the device, it exits at high speed, creating a partial vacuum at a side arm, to which vacuum tubing can be attached. This vacuum is sufficient for a variety of low- to moderate-vacuum laboratory applications, particularly where high precision is not essential. Water aspirator applications, advantages, limitations, and safety considerations are summarized in Table 16.6.

Table 16.6: Applications, advantages, limitations, and safety considerations of water aspirator.

Category	Details
Applications	Vacuum filtration (e.g., Büchner funnel setups)
	Degassing liquids
	Drying of samples
	Supporting distillations of low-boiling solvents under reduced pressure
Advantages	Inexpensive and maintenance-free
	No moving parts or electrical components
	Easy to use and install on standard laboratory taps

Table 16.6 (continued)

Category	Details
Limitations	Dependent on water pressure and temperature (colder water produces better vacuum) Cannot achieve deep vacuum levels Environmental concern due to continuous water usage and waste Not suitable for air-sensitive or highly volatile substances due to poor vacuum control
Safety considerations	Use backflow preventers or vacuum traps to prevent water from entering the sample or vacuum system Never aspirate flammable or corrosive vapors Ensure that tubing and connectors are securely attached to prevent detachment under suction

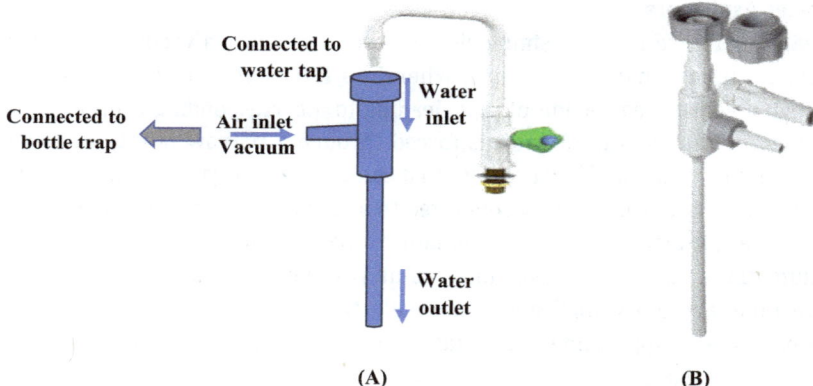

Figure 16.6: Schematic diagram of water aspirator (A) and water aspirator (B). Image credit: https://www.ubuy.com.sa/en/product/PWVEGPC-dynalon-312635-water-jet-faucet-aspirator-vacuum-pump?ref=hm-google-redirect#gallery.

16.5.2.2 Rotary Evaporator

A rotary evaporator, or rotovap (Figure 16.7), is a widely used laboratory instrument designed to gently remove volatile solvents from nonvolatile solutes under reduced pressure. It integrates vacuum, heat, and rotation to facilitate efficient and controlled solvent evaporation, making it indispensable in organic chemistry, pharmaceuticals, and food science laboratories. The principle is based on vacuum-assisted distillation, where reducing the pressure in the system lowers the boiling point of the solvent. A rotating flask increases the surface area of the solution and prevents bumping. A heated water or oil bath provides thermal energy to accelerate evaporation, while a condenser cooled with circulating fluid condenses the vapor, which then collects in a receiving flask. Rotary evaporator advantages and limitations are shown in Table 16.7.

Figure 16.7: Rotary evaporator. Image credit: https://depositphotos.com/photos/rotary-evaporator.html.

16.5.2.2.1 Main Components
A typical rotary evaporator system includes:
–   Rotating evaporation flask (round-bottomed flask)
–   Motor unit to spin the flask at variable speeds
–   Heating bath to provide controlled temperature (typically water at 30–80 °C)
–   Vertical or diagonal condenser connected to a cooling system
–   Receiving flask for collecting condensed solvent
–   Vacuum pump or aspirator to reduce system pressure
–   Vacuum control system (gauge, controller, or bleed valve)
–   A cold trap is often installed between the evaporator and vacuum pump to protect the pump from solvent vapors and extend its lifespan.

16.5.2.2.2 Applications
Rotary evaporators are employed in a wide range of laboratory processes:
–   Solvent removal after extraction or chromatography
–   Concentration of solutions
–   Crystallization of solutes
–   Recovery of expensive or hazardous solvents
–   Purification of reaction products

Table 16.7: Advantages and limitations of rotary evaporator.

Category	Details
Advantages	Faster evaporation than simple distillation
	Minimizes decomposition of heat-sensitive compounds
	Allows recovery of solvent and solute
	Safe, reproducible, and scalable for preparative work
Limitations	Limited to liquids with low-to-moderate boiling points
	Inappropriate for large-scale distillations or substances requiring high vacuum
	Equipment is relatively costly and requires routine maintenance
	Glassware is fragile and prone to breakage under vacuum if mishandled

16.5.2.3 Freeze-dryer

Freeze-drying, or lyophilization, is a vacuum-assisted drying process widely used in pharmaceutical, food, and biochemical laboratories to preserve materials without applying heat. A freeze-drying machine, or lyophilizer (Figure 16.8), combines freezing and sublimation under vacuum to remove water or solvent from sensitive samples. By lowering the pressure, the frozen solvent transitions directly from the solid to the gas phase (sublimation), preserving the structure and activity of delicate compounds. Lyophilizers require a reliable high-vacuum system, often involving rotary vane or scroll pumps, depending on the desired vacuum level. They are essential in laboratories dealing with biological samples, protein formulations, vaccines, or plant extracts. Lyophilizer features and descriptions are shown in Table 16.8.

Table 16.8: Features and descriptions of freeze-dryer.

Feature	Description
Operating principle	Removes water or solvent via sublimation under vacuum after initial freezing of the sample
Temperature control	Includes freezing chamber (−40 to −80 °C) and heated shelves for controlled sublimation
Vacuum source	Usually rotary vane or scroll pump; may include secondary cold trap to protect the pump
Chamber type	Bench-top or floor-standing; may have manifold ports for flasks or shelf trays for bulk drying
Sample capacity	Varies by model: from milliliters (lab scale) to liters (pilot and production scale)
Applications	Preservation of biological samples, pharmaceuticals, proteins, enzymes, plant extracts, and vaccines
Advantages	Preserves sample structure and activity; no thermal degradation; long-term stability
Limitations	High cost; longer process times; requires low temperatures and reliable vacuum

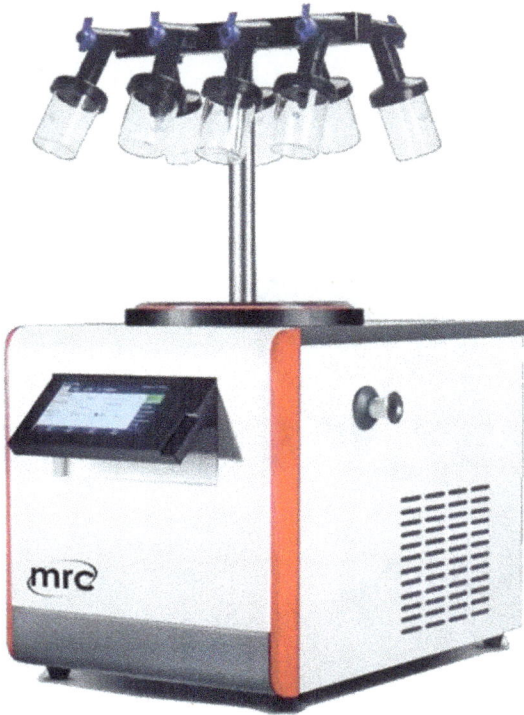

Figure 16.8: Vacuum freeze dryer. Image credit: https://www.munroscientific.co.uk/vacumm-freeze-dryer_t-type.

16.5.2.4 Vacuum Ovens

A vacuum oven (Figure 16.9) is a temperature-controlled enclosure designed to heat materials under reduced pressure, allowing for drying, curing, or degassing without exposing samples to high temperatures that could cause degradation. By reducing the pressure inside the chamber, the boiling point of liquids (especially water and solvents) is lowered, which facilitates gentle and uniform drying of thermally sensitive or reactive materials. Vacuum ovens are typically connected to an external vacuum pump (e.g., rotary vane or diaphragm), and some models include inert gas purging capabilities to maintain an oxygen-free atmosphere during heating.

16.5.2.4.1 Components
- Sealed metal or glass-lined chamber with internal heating
- Temperature controller (ambient to ~250 °C)
- Vacuum gauge and regulator
- External vacuum pump

- Inert gas inlet (optional)
- Shelving or trays for holding samples

Figure 16.9: Vacuum oven. Image credit: https://www.amazon.ca/Laboratory-Industrial-Temperature-Circulation-Precision/dp/B0CD5YDX7J.

16.5.2.4.2 Applications
- Drying of heat-sensitive chemicals, polymers, or pharmaceuticals
- Removal of moisture or solvents from hygroscopic materials
- Curing of adhesives or coatings under vacuum
- Degassing of elastomers or composites
- Thermal aging or stability testing in controlled environments

16.5.2.4.3 Advantages
- Prevents oxidation and thermal decomposition
- Gentle drying preserves sample integrity
- Enables solvent removal at lower temperatures
- Can handle flammable or sensitive materials under an inert atmosphere

16.5.2.4.4 Limitations
- High initial cost and maintenance requirements
- Limited internal volume

- Longer drying times for some substances
- Requires a reliable vacuum system and trained personnel

16.5.2.4.5 Safety Considerations
- Ensure oven is rated for the operating vacuum and temperature.
- Do not place sealed or volatile containers inside.
- Monitor temperature and pressure to avoid over-pressurization.
- Use compatible vacuum tubing resistant to heat and solvents.

16.5.2.5 Schlenk Lines
A Schlenk line, also known as a dual-manifold vacuum/inert gas line, is a glass apparatus used to handle air-sensitive compounds under an inert atmosphere (usually nitrogen or argon). It provides an efficient system for alternating between vacuum and gas flow, allowing for the manipulation, transfer, and reaction of moisture or oxygen-sensitive substances in a safe, sealed environment. The Schlenk line (Figure 16.10) typically connects to a vacuum pump and gas source, with multiple ports for attaching reaction flasks, traps, and transfer adapters.

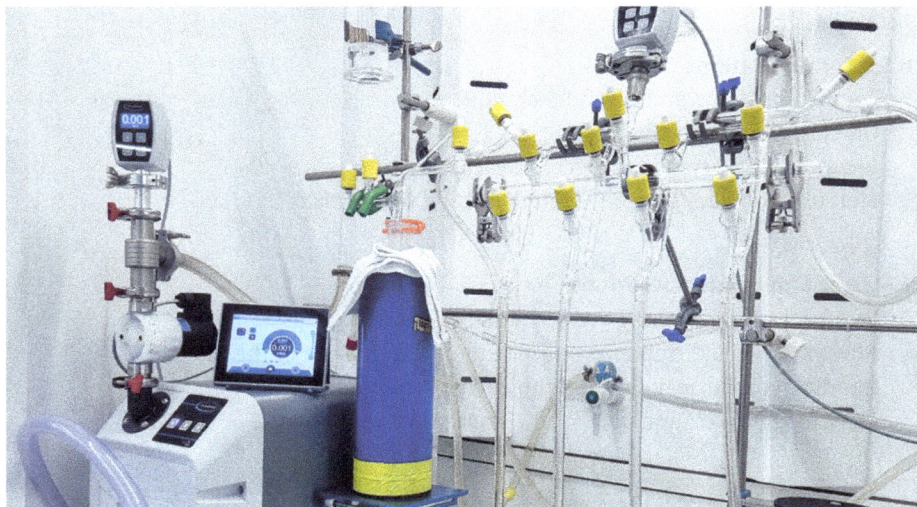

Figure 16.10: Schlenk line. Image credit: https://www.vacuubrand.com/fr/news/all-blog-posts/keeping-air-and-moisture-out.

16.5.2.5.1 Components
- Dual manifold (one line for vacuum and one for inert gas)
- Vacuum pump (often rotary vane or diaphragm)

- Cold trap (located between the pump and the manifold)
- Mercury or oil bubbler (for pressure regulation)
- Needle valves or stopcocks for controlling flow
- Connections for round-bottomed flasks and adapters

16.5.2.5.2 Applications
- Handling pyrophoric or air/moisture-sensitive compounds (e.g., Grignard reagents and organometallics)
- Degassing solvents
- Drying glassware under vacuum
- Vacuum transfer of liquids or solids
- Performing reactions under an inert atmosphere

16.5.2.5.3 Advantages
- Versatile for many types of air-sensitive operations
- Enables repeated cycling between vacuum and inert gas
- Reusable and customizable glassware
- Precise control of the atmosphere during experiments

16.5.2.5.4 Limitations
- Requires training and careful technique
- Glassware is fragile and costly
- Requires a cold trap to protect the vacuum pump
- Mercury bubblers pose environmental and health risks

16.5.2.5.5 Safety Considerations
- Ensure that all joints are greased and secured with clips.
- Always use a cold trap to avoid solvent contamination of the pump.
- Handle pyrophoric materials with extreme caution.
- Use proper ventilation and a fume hood when operating.

16.5.2.6 Vacuum Desiccators
A vacuum desiccator (Figure 16.11) is a sealed chamber designed to store or dry moisture-sensitive materials under reduced pressure, enhancing the desiccation process by accelerating the removal of water or solvent vapors. When connected to a vacuum source, the low pressure increases the vapor pressure gradient, speeding up moisture removal. It typically contains a desiccant (such as silica gel or phosphorus pentoxide) to absorb moisture and a porous platform to support the materials being dried.

Figure 16.11: Vacuum desiccator. Image credit: https://assistent.eu/en/produkt/vakuum-exsikkatoren-mit-austauschbarem-tubusdeckel-und-normschliff hahn/.

16.5.2.6.1 Components
- Thick-walled glass or polycarbonate chamber with lid
- Vacuum-tight seal (usually a grease-ring or gasket)
- Porcelain or perforated plate to hold samples
- A desiccant is placed beneath the plate
- Vacuum port for tubing connection to the pump

16.5.2.6.2 Applications
- Drying solids such as hygroscopic salts or powders
- Storage of moisture-sensitive chemicals or standards
- Cooling of materials in a dry, oxygen-free environment
- Preventing oxidation during storage

16.5.2.6.3 Advantages
- Simple and cost-effective
- Passive drying mechanism (desiccant-based)
- Useful for storage as well as drying
- No heating required, suitable for temperature-sensitive substances

16.5.2.6.4 Limitations
- Cannot dry large quantities or high-moisture materials efficiently.
- Requires regular replacement or regeneration of desiccant.
- The vacuum seal must be maintained for effectiveness.
- Brittle glass may break under vacuum if mishandled.

16.5.2.6.5 Safety Considerations
- Always inspect the glass for cracks before applying a vacuum.
- Use appropriate vacuum grease for sealing.
- Avoid drying volatile or reactive substances without containment.
- Use a secondary containment shield when working under vacuum.

16.6 Applications Across Laboratory Techniques

Vacuum and pressure systems play an important role in a variety of laboratory methods, improving separation efficiency and analytical precision. These technologies, which accelerate filtration and enable high-resolution spectroscopy, are critical for preserving ideal conditions in delicate operations. Table 16.9 outlines the applications.

Table 16.9: Applications of vacuum and pressure systems across laboratory techniques.

Category	Technique	Application/description
Separation techniques	Vacuum filtration	Enhances filtration speed for fine precipitates or viscous solutions
	Vacuum distillation	Used when substances decompose at their boiling points under atmospheric pressure (e.g., high-boiling organic compounds)
	Sublimation under vacuum	Purifies solids by direct transition from solid to gas at reduced pressure
	Supercritical fluid extraction	Requires high-pressure CO_2 to extract analytes from complex matrices
Analytical techniques	Mass spectrometry (MS)	Requires high vacuum for ion acceleration and detection. Turbomolecular pumps or ion pumps are standard
	SEM and TEM	Operate under high vacuum to allow uninterrupted electron beam travel and avoid scattering
	Gas chromatography (GC)	Pressure regulation ensures consistent carrier gas flow, impacting separation efficiency and detector sensitivity
	X-ray photoelectron spectroscopy (XPS)	Utilizes ultra-high vacuum to prevent interference with photoelectron signals

16.7 Questions and Answers

Questions	Answers
1. What is the difference between pressure and vacuum?	Pressure is force exerted by gas; vacuum is a condition where pressure is lower than atmospheric.
2. Name a common type of high-vacuum pump.	Turbomolecular pump.
3. Why are traps or filters used in vacuum lines?	To prevent contamination and protect vacuum pumps from corrosive vapors or particles.
4. What is back pressure, and how is it managed?	Resistance to flow downstream; managed using pressure relief or control valves.
5. Why is leak testing important in pressure systems?	To ensure system integrity and maintain safety and performance.
6. What is a Bourdon gauge?	A mechanical device that measures pressure using a coiled tube that flexes under pressure.
7. Why are digital pressure sensors preferred in automated systems?	They offer real-time monitoring, higher precision, and easier integration with control systems.
8. What is the main advantage of using a vacuum oven instead of a conventional drying oven?	A vacuum oven allows drying at lower temperatures by reducing pressure, lowering the solvent boiling point. This prevents thermal decomposition and oxidation of heat-sensitive materials.
9. Why is a cold trap essential when operating a Schlenk line with a vacuum pump?	A cold trap condenses and captures volatile solvents before they reach the vacuum pump, protecting it from damage and maintaining vacuum efficiency.
10. How does a vacuum desiccator enhance the drying process compared to a standard desiccator?	A vacuum desiccator lowers the pressure to increase moisture evaporation rate, enhancing drying efficiency and preserving moisture-sensitive materials without applying heat.

Chapter 17
Heating, Cooling, and Reaction Techniques in the Chemical Laboratories

17.1 Introduction

Temperature control is a critical element in laboratory operations across virtually all branches of chemistry. Whether accelerating a chemical transformation, promoting crystal formation, or preserving thermally labile compounds, the effective application of heating and cooling techniques is fundamental to both routine procedures and advanced research. This chapter explores common heating and cooling methods, the role of reaction techniques like refluxing, and the importance of thermal regulation in ensuring reproducibility, safety, and high-quality results.

17.2 Heating Techniques in the Laboratory

Heating is often required to increase reaction rates, drive off solvents, or enable phase transitions. Selecting the appropriate heating method depends on the temperature range, the nature of the substance, and the scale of the experiment.

17.2.1 Bunsen Burners

Bunsen burners (Figure 17.1) are traditional flame-based heating tools fueled by natural gas. They offer rapid, high-temperature heating, suitable for tasks such as glassware flame-drying or combustion reactions. However, open flames pose fire hazards, especially in solvent-rich environments, and are increasingly replaced by safer alternatives in modern labs.

17.2.2 Heating Mantles

Magnetic stirrer with heating mantle (Figure 17.2) combines thermal and magnetic stirring functionality, making it ideal for reactions requiring agitation designed to fit round-bottomed flasks, providing uniform heat without direct flame contact, making it preferred for organic synthesis.

https://doi.org/10.1515/9783112218105-017

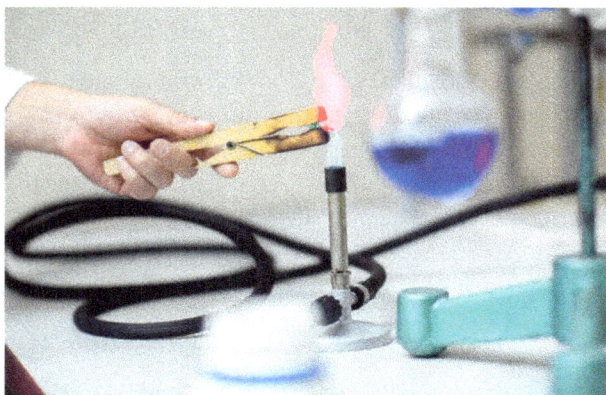

Figure 17.1: Bunsen burners. Image credit: https://depositphotos.com/photos/bunsen-burners.html.

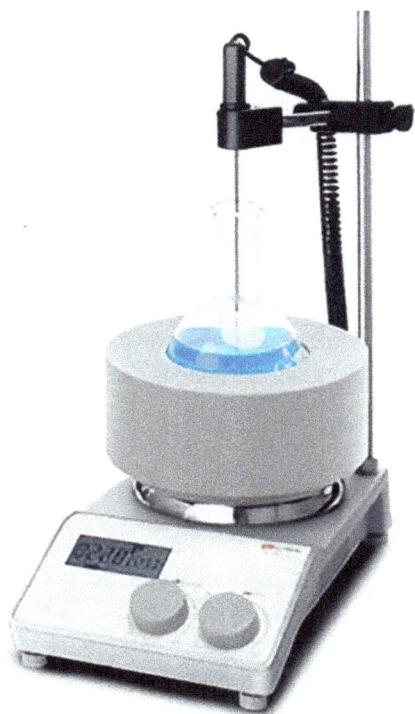

Figure 17.2: Magnetic stirrer with heating mantle. Image credit: https://www.rogosampaic.com/en/labora tory-equipment/digital-stirrer-w-heating-mantle-for-bottom-round-flasks-250-ml/.

17.2.3 Water, Oil, and Sand Baths

Water baths are commonly used for gentle heating (up to ~100 °C), ideal for enzymatic reactions or warming reagents. Oil baths extend the temperature range up to ~200–250 °C, while sand baths offer even higher temperatures and better thermal stability for high-temperature syntheses or thermolysis reactions. Sand baths also distribute heat more evenly, reducing the risk of local overheating (Figure 17.3).

Figure 17.3: Digital water–oil bath and sand baths. Images credit: https://www.rogosampaic.com/en/search/.

17.3 Cooling Techniques in the Laboratory

Cooling is equally vital in chemical laboratories to control exothermic reactions, crystallize solids, preserve temperature-sensitive reagents, or safely condense vapors.

17.3.1 Ice and Salt–Ice Baths

An ice bath (water + ice) has an approximate temperature of 0 °C. When salt (typically NaCl, $CaCl_2$, $MgCl_2$, or NH_4Cl) is added to ice, the freezing point of water decreases, resulting in colder baths (Table 17.1). These are used to delay reaction kinetics, regulate crystallization, and protect thermolabile chemicals. The salt-to-ice ratio is crucial for achieving the lowest temperature in a cooling bath. The appropriate ratio varies depending on the type of salt; however, below are some common optimal salt: ice ratios used in laboratories (Table 17.2).

Table 17.1: Ice + salt cooling bath temperature ranges.

Salt	Approx. temperature range (°C)	Notes
Sodium chloride (NaCl)	−10 °C to −21 °C	Common table salt; readily available; good for moderate cooling.
Calcium chloride (CaCl₂)	−20 °C to −40 °C	Very effective; exothermic when dissolved; widely used in refrigeration.
Magnesium chloride (MgCl₂)	−25 °C to −33 °C *(can reach −37 °C)*	Strong freezing point depression: used when lower temperature is needed.
Ammonium chloride (NH₄Cl)	−15 °C to −18 °C	Milder than CaCl₂ or MgCl₂; sometimes used in educational settings.

Table 17.2: Typical salt-to-ice ratios (by mass).

Salt	Optimal salt : ice ratio (by mass)	Resulting temperature (°C)	Notes
Sodium chloride (NaCl)	1 : 3 to 1 : 2	−10 to −21	Common and easy to handle; do not exceed 1:1 or melting slows down.
Calcium chloride (CaCl₂)	1 : 3 to 1 : 1	−20 to −40	Exothermic dissolution; use cautiously and stir constantly.
Magnesium chloride (MgCl₂)	1 : 3 to 3 : 10	−25 to −37	Highly effective; too much can saturate and reduce cooling efficiency.
Ammonium chloride (NH₄Cl)	1 : 3	−15 to −18	Mild effect: higher ratios are generally ineffective.

17.3.2 Dry Ice Baths

Dry ice (solid CO_2) sublimates at −78.5 °C under atmospheric pressure. It should be transported from styrofoam containers to the laboratory using designated dry ice buckets (Figure 17.4). When mixed with various solvents, it creates cooling baths whose temperatures depend on the solvent's freezing point and compatibility with CO_2. These cooling mixtures are commonly used for conducting low-temperature reactions, quenching strongly exothermic processes, and condensing volatile vapors during sublimation or vacuum distillation. Table 17.3 provides approximate temperatures for dry ice alone and in combination with different solvents.

Figure 17.4: Dry ice bucket. Images credit: https://www.sigmaaldrich.com/SA/en/product/sigma/cls432129.

Table 17.3: Dry ice cooling baths–solvent combinations.

Dry ice (CO_2) with:	Approximate temperature (°C)	Notes
Dry ice alone	−78.5 °C	Sublimates directly to gas; limited contact cooling; inefficient alone.
Acetone	−78 °C	Most common; efficient thermal transfer; flammable.
Isopropanol	−78 °C	Similar to acetone; slightly less volatile; flammable.
Methanol	−97 °C	One of the coldest baths; toxic and flammable; use with care.
Ethanol	−72 °C	Safer alternative to acetone; commonly used.
Diethyl ether	−116 to −100 °C (under vacuum)	Extremely cold; highly flammable; hazardous.
Chloroform	−61 °C	Moderately cold; toxic; limited use due to safety concerns.
Acetonitrile	−41 °C	Useful for moderate cooling; limited by its own freezing point.
Ethyl acetate	−77 to −84 °C	Good solvent for low-temperature baths; flammable.
Hexane	−78 to −90 °C	Excellent for deep cooling; very flammable.
Carbon tetrachloride	−23 °C	Low cooling; toxic; rarely used today.
Pyridine	−42 °C	Toxic, used when basic solvent is required.

17.3.3 Liquid Nitrogen Baths

Liquid nitrogen (–196 °C) is used for cryogenic applications. Reactions sensitive to thermal decomposition, air, or moisture are often conducted at such low temperatures. Careful handling using special dewars (Figure 17.5) is required to avoid thermal burns and pressure buildup due to rapid evaporation. Table 17.4 summarizes the approximate temperatures of liquid nitrogen alone and with different solvents.

Table 17.4: Liquid nitrogen cooling baths–solvent combinations.

Liquid nitrogen with:	Approximate temperature (°C)	Notes
Liquid nitrogen alone	–196 °C	Boiling point at 1 atm; very cold; poor thermal contact without a liquid medium.
Ethanol	–115 °C	Common LN$_2$ bath; remains liquid; good for ultra-cold reactions.
Acetone	–95 °C to –100 °C	Useful for deep cooling; flammable; efficient thermal transfer.
Isopropanol	–90 °C	Slightly warmer than acetone; safer and more stable.
Pentane	–130 °C	Extremely cold bath; flammable; volatile; useful in cryogenics.
Toluene	–95 °C	For reactions needing aromatic solvents at low temperatures.
Methanol	–97 °C	Similar to acetone; toxic; excellent for –97 °C control.
Liquid nitrogen + slush bath (in vacuum)	–210 °C (solid N$_2$)	Requires reduced pressure; rarely used; for ultra-low temperature experiments.

17.4 Reaction Techniques: Managing Thermal Energy

Beyond basic heating and cooling, certain laboratory setups are specifically designed to allow reactions to occur safely and effectively under controlled thermal conditions.

17.4.1 Refluxing

Refluxing is a technique that involves heating a reaction mixture while continuously condensing the solvent vapor and returning it to the reaction vessel This process maintains a constant temperature, typically at the boiling point of the solvent, allowing the reaction to proceed over extended periods without loss of solvent. A typical setup for a reflux system is illustrated in Figure 17.6.

Figure 17.5: Liquid nitrogen dewars. Images credit: https://www.rogosampaic.com/en/search/.

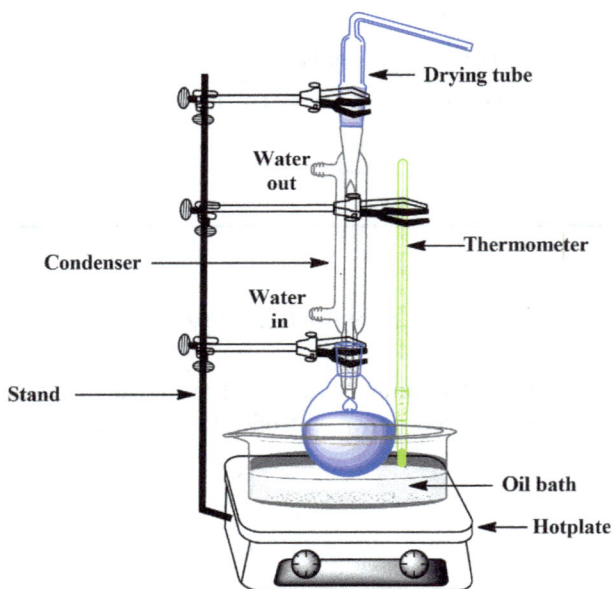

Figure 17.6: Schematic diagram for reflux system.

17.4.2 Inert Atmosphere Reactions

Some reactions are sensitive to moisture or oxygen. These are performed under an inert atmosphere, typically nitrogen or argon, using Schlenk lines or glove boxes (Figure 17.7) or septa-gas purging. Maintaining an anhydrous, oxygen-free environment is essential to prevent degradation or hazardous side reactions. Common examples of inert atmosphere reactions performed under nitrogen (N_2) or argon (Ar) to prevent unwanted reactions with air (O_2) or moisture (H_2O) are summarized in Table 17.5.

Figure 17.7: Portable glove box. Credit image: https://www.ebay.com/itm/250975392004.

Table 17.5: Examples of inert atmosphere reactions.

Reaction type	Reason for inert atmosphere
Grignard reactions	Grignard reagents react violently with water and oxygen.
Organolithium chemistry	Organolithium compounds are pyrophoric and hydrolytically unstable.
Transition metal catalysis	Catalysts (Pd^0 and Ni^0) are air-sensitive and degrade in oxygen.
Metal hydride reductions	$LiAlH_4$ reacts violently with moisture, releasing hydrogen gas.
Air-sensitive synthesis	Transition metal halides can hydrolyze or oxidize.
Coordination chemistry	Oxidation in air ruins complex integrity.
Organometallic complex formation	Intermediate oxidation states are unstable in air.
Polymerization reactions	Catalysts are highly sensitive to O_2 and H_2O.
Anionic polymerization	Terminated instantly by trace water or CO_2.
Alkali metal dissolution	Metallic sodium and solvated electrons react with air/moisture.

17.4.3 Sealed Tube or Pressure Tube Reactions

Sealed pressure tubes or bottles (Figure 17.8) are designed for reactions involving volatile or gaseous reagents. These systems (Table 17.6) permit heating above the solvent's normal boiling point and are crucial for procedures such as polymerization, hydrothermal synthesis, and catalyst evaluation. They ensure safe containment of gases (e.g., H_2, CO, O_2, and N_2), minimize solvent evaporation during extended or high-temperature reactions, allow the generation of autogenous pressure in aqueous media, and facilitate microwave-assisted reactions at elevated temperatures.

Table 17.6: Examples of sealed tube and pressure reactions.

Reaction/system	Why sealed/high pressure is needed
Hydrogenation reactions (e.g., catalytic hydrogenation of alkenes or alkynes)	Requires H_2 gas under pressure (1–100 atm) to drive reduction reactions.
Carbonylation reactions (e.g., Pd-catalyzed carbonylation)	Requires CO gas under pressure (2–30 atm) for efficient C–C bond formation.
Hydrothermal reactions (e.g., synthesis of zeolites and metal oxides)	High temperature (>100 °C) + water → pressure builds in sealed tubes (autogenous pressure).
Microwave-assisted reactions in sealed vessels	Sealed tube prevents solvent evaporation at elevated temperatures under microwave heating.
Supercritical CO_2 reactions	Supercritical CO_2 forms above 31 °C and 73 atm; used in green chemistry and catalysis.
Decarboxylation reactions (e.g., of β-keto acids)	Volatile CO_2 formed; sealed tubes prevent loss and control pressure buildup.
High-pressure polymerizations	Requires ~1,000–3,000 atm for free-radical initiation of ethylene polymerization.
Formation of sensitive intermediates (e.g., diazonium salts or azides at high concentration)	Pressure buildup helps trap gaseous or unstable intermediates for further use.
C–H activation reactions	Many C–H activations (e.g., with Ru or Rh catalysts) require pressure or sealed systems for efficiency and safety.

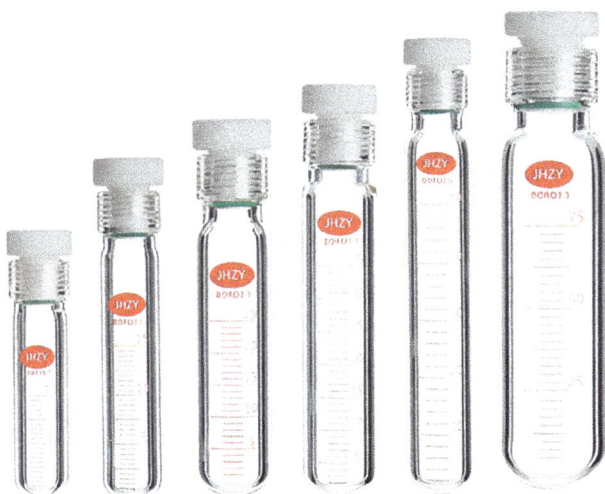

Figure 17.8: Screw pressure bottles. Credit image: https://www.aliexpress.com/item/1005004984740092.html.

17.5 Questions and Answers

Questions	Answers
1. What is the purpose of a reflux setup in chemical reactions?	To heat a reaction while preventing solvent loss by condensing and returning vapors.
2. What is the typical temperature of a dry ice–acetone bath?	Approximately –78 °C.
3. Why are salt–ice baths used instead of plain ice?	They lower the freezing point of water, allowing subzero temperatures.
4. What is the function of a heating mantle?	To provide uniform, controlled heating for round-bottomed flasks.
5. Which solvent with dry ice can reach approximately –97 °C?	Methanol.
6. Why is refluxing preferable for prolonged heating of reactions?	It maintains constant temperature and prevents solvent evaporation.
7. How does a sealed tube enable higher reaction temperatures?	By preventing solvent evaporation and allowing pressure buildup.
8. What is the purpose of an oil bath in a lab?	To provide even, adjustable heating to maintain a precise temperature.

(continued)

Questions	Answers
9. What role does autogenous pressure play in hydrothermal synthesis?	It allows reactions to proceed at high temperatures and pressures without external gas.
10. What precaution should be taken when using flammable solvents in cold baths?	Perform the procedure in a fume hood and avoid open flames or sparks.

Chapter 18
Analytical Techniques in Chemistry Laboratories

18.1 Introduction

Analytical chemistry is the backbone of scientific investigation, providing essential tools for the identification, quantification, and characterization of chemical substances. It supports diverse fields including pharmaceuticals, environmental monitoring, food safety, forensic analysis, and materials science. Analytical techniques are broadly classified into classical (wet) methods and instrumental methods. This chapter discusses the principles, instrumentation, and applications of the most commonly used analytical techniques in chemistry laboratories.

18.2 Classical Analytical Methods

18.2.1 Gravimetric Analysis

Gravimetric analysis is a classical quantitative analytical technique used to determine the amount of an analyte based on its mass. It involves the conversion of the analyte into a stable, pure, and easily weighable solid form, which is then isolated and measured. The technique is valued for its high precision and accuracy, especially when carried out under controlled conditions. Despite the development of modern instrumental methods, gravimetric analysis remains a fundamental tool in analytical chemistry, particularly for standardization, validation, and educational purposes.

18.2.1.1 Principles and Types of Gravimetric Analysis
Gravimetric methods are based on the conservation of mass. They require converting the analyte into a known chemical form and weighing it to determine its concentration. The major types include:
- Precipitation Gravimetry. The analyte is converted into a sparingly soluble precipitate (e.g., chloride as $AgCl$).
- Volatilization Gravimetry. The analyte is separated by heating and collected as a vapor (e.g., water as vapor or CO_2 from carbonates).
- Electrogravimetry. The analyte is deposited on an electrode via electrolysis (e.g., copper deposited on a cathode).
- Thermogravimetric Analysis (TGA). Monitors changes in mass with temperature to study decomposition or oxidation.

https://doi.org/10.1515/9783112218105-018

18.2.1.2 General Procedure
A typical gravimetric analysis consists of the following steps shown in Figure 18.1.

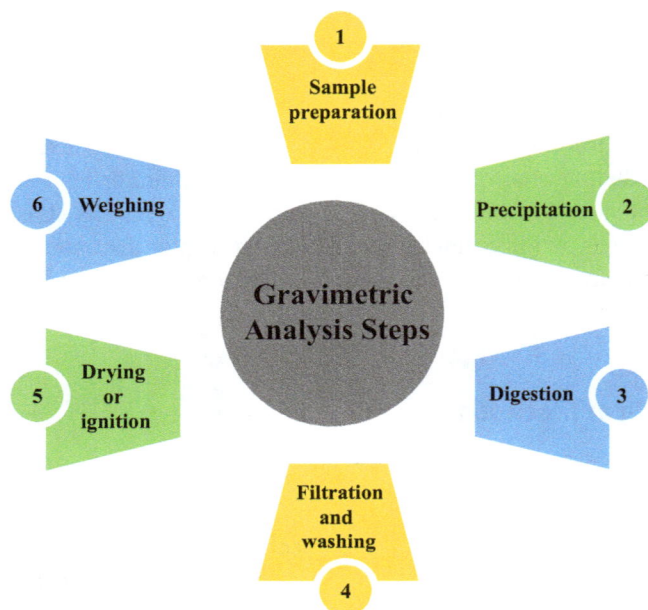

Figure 18.1: A typical gravimetric analysis steps.

18.2.1.3 Sample Preparation Techniques
Accurate gravimetric analysis requires that the sample analyzed be representative of the original material. This is especially important when analyzing heterogeneous solid samples such as ores or soils. Two widely used techniques for solid sample homogenization are coning and quartering and rolling and quartering:

- Coning and Quartering. This technique (Figure 18.2) is commonly used to reduce the size of a bulk sample while preserving its representativeness. This process can be repeated as often as required until a suitable sample size for analysis is obtained. It involves the following steps:
    - Spread the bulk solid sample into a conical pile on a clean surface.
    - Flatten the cone into a circular, uniform layer.
    - Divide the flattened sample into four equal quarters using straight lines.
    - Discard two opposite quarters.
    - Combine the remaining two quarters, mix, and repeat the process until the desired sample size is obtained.

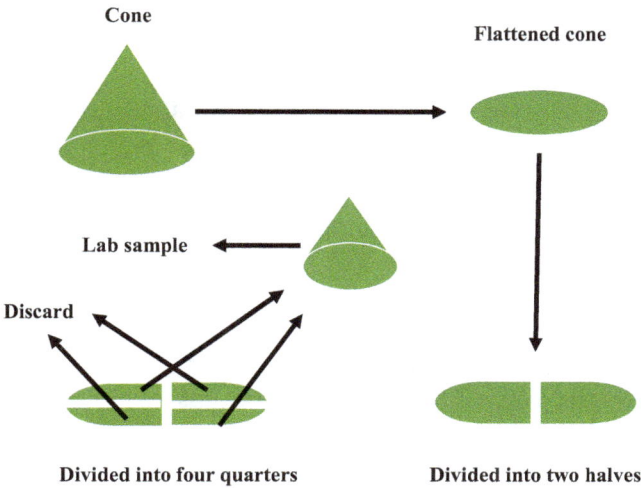

Figure 18.2: Coning and quartering.

– Rolling and Quartering: A sample preparation method is particularly effective for fine or granular powders. It begins with forming a representative cone using the coning method, which is then flattened on a flexible sheet. The sample is mixed by repeatedly folding the sheet, pulling one corner to its opposite, then another, ensuring thorough mixing. The number of rolling depends on sample size, particle size, and physical properties. Once mixed, the four corners of the sheet are lifted to collect the sample at the center. The material is then flattened into a circular layer and quartered, and the desired sample size is obtained from the two selected quarters (Figure 18.3). This method helps prevent segregation due to particle size differences.

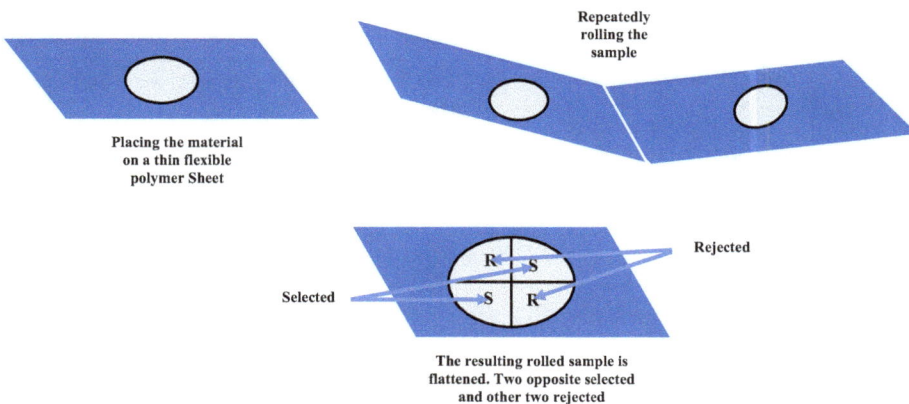

Figure 18.3: Rolling and quartering.

18.2.1.4 Requirements for Effective Gravimetric Analysis

For accurate results, the following criteria should be met:

- The precipitate must be pure, insoluble, and easily filterable.
- The product must have a known composition and stable form (e.g., $BaSO_4$ and AgCl).
- The reaction should proceed to completion and avoid coprecipitation.
- Losses during filtration, washing, or transfer must be minimized.

18.2.1.5 Applications of Gravimetric Analysis

Gravimetric analysis is used in various contexts such as:

- Determining sulfate as barium sulfate ($BaSO_4$)
- Estimating chloride as silver chloride (AgCl)
- Determining water of hydration or moisture content by mass loss
- Standardizing solutions in volumetric titrations

18.2.1.6 Example Problems and Solutions

Key equations and formulas used in solving problems are summarized in Table 18.1, and selected problems with their solutions are presented in Table 18.2.

Table 18.1: Key equations and formulas used in solving gravimetric problems.

No.	Formula/calculation	Description/use
1	Moles (n) = mass/molar mass n = mass (g)/ molar mass (g/mol)	Converts the mass of a substance (e.g., precipitate or residue) into moles.
2	Mass of analyte = mass of precipitate × (molar mass of analyte/molar mass of precipitate)	Used to calculate the mass of a specific ion or element in a precipitate (e.g., Cl^- in AgCl and SO_4^{2-} in $BaSO_4$) using stoichiometric ratios.
3	% Analyte = (mass of analyte/original sample mass) × 100	Determines the percentage composition of the analyte in the original sample.
4	x (in hydrate) = moles of water lost/moles of anhydrous salt	Used to determine the number of water molecules in a hydrated salt (e.g., in $CuSO_4 \cdot xH_2O$).
5	Mass of original compound = moles of precipitate × molar mass of original compound	Helps back-calculate the mass of the original compound that produced the precipitate (e.g., finding NaCl from AgCl precipitate).

When solving gravimetric problems:
Step 1: Convert mass of precipitate to moles.
Step 2: Use mole ratio to find moles of analyte.
Step 3: Convert moles of analyte to mass.
Step 4: Use the mass of analyte to find % composition or mass of starting material.

Table 18.2: Gravimetric problems and solutions.

Problem	Description	Key data and calculation steps	Answer
1	Sulfate determination as $BaSO_4$	Sample mass = 0.500 g Mass of $BaSO_4$ = 0.684 g Molar mass: $BaSO_4$ = 233.39 g/mol and SO_4^{2-} = 96.06 g/mol Mass of SO_4^{2-} = (0.684 × 96.06)/233.39 = 0.2814 g % SO_4^{2-} = (0.2814/0.500) × 100	56.28% SO_4^{2-}
2	Water of hydration in $CuSO_4 \cdot xH_2O$	Initial mass = 2.000 g Anhydrous $CuSO_4$ = 1.276 g Mass of water lost = 0.724 g Moles: $CuSO_4$ = 1.276/159.61 = 0.00799 mol H_2O = 0.724/18.02 = 0.0402 mol x = 0.0402/0.00799	$x \approx 5$
3	Phosphorus as $Mg_2P_2O_7$	Sample mass = 0.300 g Mass of $Mg_2P_2O_7$ = 0.200 g Molar mass: $Mg_2P_2O_7$ = 222.57 g/mol, P = 61.94 g/mol Mass of P = (0.200 × 61.94)/222.57 = 0.0557 g % P = (0.0557/0.300) × 100	18.57% P
4	Chloride determination via AgCl	Mass of AgCl = 0.430 g Molar mass: AgCl = 143.32 g/mol Moles of AgCl = 0.430/143.32 = 0.00300 mol Moles of NaCl = 0.00300 mol Mass of NaCl = 0.00300 × 58.44 = 0.1753 g	175.3 mg NaCl
5	Mass of Cl^- from AgCl precipitate	Mass of AgCl = 0.287 g Molar mass: AgCl = 143.32 g/mol, Cl^- = 35.45 g/mol Mass of Cl^- = 0.287 × (35.45/143.32) = 0.071 g	0.071 g Cl^-

18.2.2 Volumetric (Titrimetric) Analysis

18.2.2.1 Introduction

Volumetric analysis (Figure 18.4) is a chemical technique that involves reacting a substance with a solution of known concentration to determine its quantity (concentration). It is also known as titrimetric analysis since it frequently uses titration. It is rapid, precise, and cost-effective.

Advance technology automatic titrator (Figure 18.5) instruments for dosing chemical volumetric or quantitative analysis in industrial, medical, pharmaceutical, nutraceuticals, and food beverage are also used.

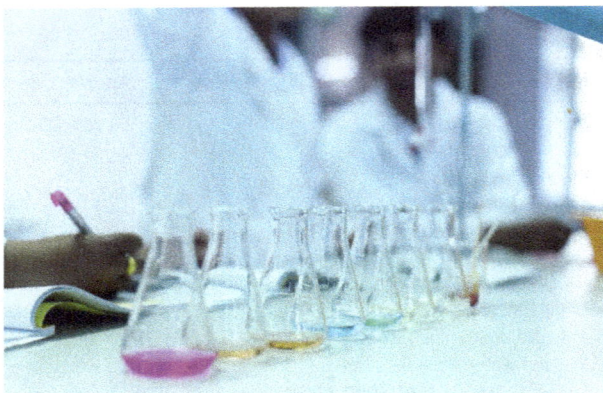

Figure 18.4: Volumetric analysis. Image credit: https://depositphotos.com/photos/titration.html.

Figure 18.5: Automatic titrator. Image credit: https://depositphotos.com/photos/automatic-titrator.html.

18.2.2.2 Definitions of Terms

– Units of Volume

Liters (L): A volume equal to 1,000 m^3 or mL

Milliliters (mL or cm^3): One-thousandth (0.001) liters

Cubic centimeters (cm^3): Can be used interchangeably with milliliters without effect.

– Titration

Titration (Figure 18.6) is the quantitative measurement of a material using a reagent. Typically, this is achieved by adding a known concentration of reagent to a solution of the substance until the reaction between the two is pronounced complete, at which point the volume of the reagent is measured.

–   Back Titration
    A process by which an excess of the reagent is added to the sample solution, and this excess is then determined with the second reagent of known concentration.
–   Standard Solution
    A reagent of known composition is used in the titration. The accuracy with which the concentration of a standard solution is known sets a limit on the accuracy of the test. Standard solutions are prepared by carefully measuring a quantity of a pure compound and calculating the concentration from the mass and volume measurements or carefully dissolving a weighed quantity of the pure reagent itself in the solvent and diluting it to an exact volume or by using a pre-standardized commercially available standard solution.
–   Equivalence Point
    The point at which the standard solution is chemically equivalent to the substance being titrated. The equivalence point is a theoretical concept. We estimate its position by observing physical changes associated with it in the solution.
–   End Point
    The point at which the physical changes arising from alterations in the concentration of one of the reactants at the equivalence point become apparent.

18.2.2.3 Typical Physical Changes During Volumetric Analysis
–   Appearance or change of color due to the reagent, the substance being determined, or an indicator substance
–   Turbidity formation results from the formation or disappearance of an insoluble phase
–   Conductivity changes in a solution
–   Potential changes across a pair of electrodes
–   Refractive index changes
–   Temperature changes

18.2.2.4 Tools of Volumetric Analysis
Pipettes, burettes, and volumetric flasks are standard volumetric equipment. A volumetric apparatus calibrated to contain a certain volume is designated as TC (to contain) and an apparatus calibrated to deliver a certain amount, TD (to deliver). Only clean glass surfaces will support a uniform film of liquid; the presence of dirt or oil tends to cause fractures in this film. The appearance of water cracks is a sure indicator of an unclean surface.

Volumetric glassware is carefully cleaned by the manufacturer before it is labeled so that they have meaning; the equipment must be equally clean when in use. As a

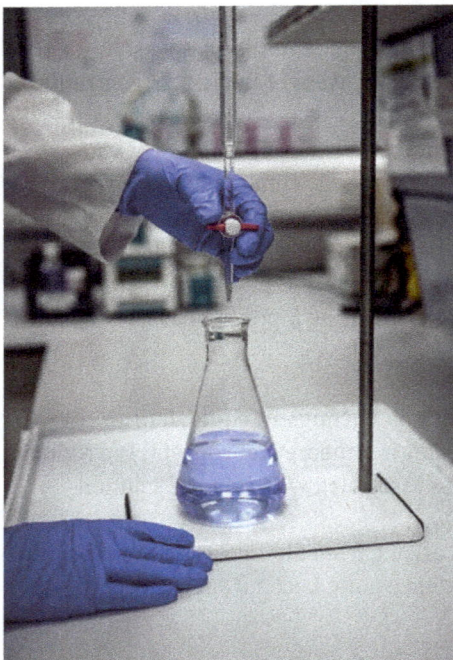

Figure 18.6: Laboratory titration technique or process. Image credit: https://depositphotos.com/photos/titration.html.

general rule, the heating of calibrated glass equipment should be avoided. Cooling too fast can permanently distort the glass and cause a volume change.

– Volumetric Flasks

The volumetric bottles are calibrated to contain a certain volume when filled to the line incised on the neck. Before use, volumetric bottles should be washed with detergent and, if necessary, a cleaning solution. Then they should be carefully and repeatedly rinsed in distilled water; they rarely need to be dried. However, if drying is needed, it is best to achieve this by clamping the flasks in the inverted position and using a mild vacuum to circulate air through them.

Direct preparation of a standard solution requires introducing a known mass of solute into a volumetric flask. To reduce the possibility of loss during transportation, insert a funnel into the neck of the vial. The funnel is then washed free of solids. After introducing the solute, fill the flask about half full and rotate the contents to obtain the solution. Add more solvent and mix well again. Bring the solution level to approximately the mark and allow time for it to drain. Then, use the dropper to make final additions to the solvent as necessary. Cap the vial firmly and turn it over frequently to ensure uniform mixing.

– Pipettes

Pipettes are designed for the transfer of known volumes of liquid from one container to another. Pipettes that deliver a fixed volume are called volumetric or transfer pipettes. Other pipettes, known as measuring pipettes, are calibrated in convenient units so that any volume up to maximum capacity can be delivered. The following instructions pertain specifically to the manipulation of transfer pipettes, but with minor modifications, they may be used for other types as well. Liquids are usually drawn into pipettes through the application of a slight vacuum. Use mechanical means such as a rubber suction bulb. The pipette can be used as follows:

- Clean the pipette by rinsing it with distilled water.
- Drain completely and then rinse three times with the solution to be used in the analysis.
- Keep the tip of the pipette below the surface of the liquid.
- Draw the liquid up in the pipette using the pipette bulb.

For proper measurement proceed as follows:

- Disconnect the rubber suction bulb when the liquid is above the calibration mark and place the index finger of the hand to prevent liquid from leaving the pipette.
- Release the pressure on the index finger to allow the meniscus to approach the calibration mark.
- At the mark, apply pressure to stop the flow of liquid and drain the drop on the tip by coming into contact with the wall of the liquid-retaining container.
- Transfer the pipette to the container to be used and release the pressure on the index finger. Let the liquid drain completely. Remove the last drop by touching the wall of the container.

– Burettes

Burettes, like measuring pipettes, deliver any amount up to their maximum capacity. Burettes of the traditional type must be filled in manually. The other side-armed burettes are gravity-filled. For more accurate work, Schellbach burettes are used. These have a white background with a blue bar and are readable at the lowest magnification point. When using unstable reagents, a burette with a reservoir bottle and pump can be used. Before being placed in service, a burette must be scrupulously clean. In addition, it must be established that the stopcock is liquid-tight. Grease films that appear unaffected by cleaning solution can be rinsed with organic solvents such as acetone. Although thorough washing with detergent should do the cleaning.

18.2.2.5 Basic Principles of Titration

In a typical titration:

- A solution of known concentration (called the titrant) is slowly added from a burette to another solution (called the analyte) until the chemical reaction between them is complete.
- The point at which the reaction is complete is known as the equivalence point.
- A chemical indicator is often used to signal that the equivalence point has been reached. The moment the indicator changes color is called the endpoint. For example, when titrating hydrochloric acid (HCl) with sodium hydroxide (NaOH), the reaction is: $HCl + NaOH \rightarrow NaCl + H_2O$

Phenolphthalein (a common indicator) changes from colorless to pink at the end point.

18.2.2.6 Types of Titrations

- Acid–Base Titration: Involves neutralization between an acid and a base.
    Example: HCl versus NaOH
    Indicators: phenolphthalein and methyl orange
- Redox Titration
  Based on oxidation–reduction (electron transfer) reactions.
    Example: Potassium permanganate ($KMnO_4$) titration to determine Fe^{2+}.
    Indicators may be self-indicating (like $KMnO_4$, which is purple and fades when reduced).
- Complexometric Titration
  Used to determine metal ions.
    Example: EDTA titration for water hardness (Ca^{2+} and Mg^{2+}).
    Indicator: Eriochrome Black T (blue to pink at endpoint).
- Precipitation Titration
  Involves formation of a solid (precipitate).
    Example: Determining chloride ions (Cl^-) using silver nitrate ($AgNO_3$)
    Indicator: Potassium chromate (forms red precipitate at end point in Mohr method)

18.2.2.7 Preparing and Using Standard Solutions

A standard solution is one whose concentration is accurately known.

- Primary standard: A pure, stable substance that can be weighed directly (e.g., KHP and Na_2CO_3).
- Secondary standard: A solution whose concentration is determined by titration with a primary standard (e.g., HCl and NaOH).

18.2.2.8 Calculations in Volumetric Analysis

– Some basic equations
  Moles = molarity × volume (L)
  $M_1V_1 = M_2V_2$ (used when the reaction is 1:1)
  % Purity = (mass of pure substance/total mass of sample) × 100
  Mass = moles × molar mass
– Titration steps
    – Write the balanced chemical equation.
    – Calculate moles of titrant used.
    – Use the mole ratio to find moles of analyte.
    – Convert moles to concentration or mass.

18.2.2.9 Common Sources of Error and Tips

Source of error	How to avoid
Incorrect burette reading	Read at eye level; remove air bubbles
Contamination of glassware	Rinse with distilled water or solution
Overtitration (overshooting end point)	Add titrant slowly near the endpoint
Wrong indicator	Choose based on the type of titration

18.2.2.10 Volumetric Analysis Practice Questions and Solutions

Q. no.	Question	Solution steps	Answers
1	25.00 mL of NaOH was titrated with 0.100 M HCl. Volume of HCl used = 20.00 mL. Find the molarity of NaOH.	Moles of HCl = 0.100 × 0.02000 = 0.00200 mol NaOH: HCl = 1:1 ⇒ moles of NaOH = 0.00200 Molarity = 0.00200/0.02500	0.0800 M NaOH
2	0.245 g of Na_2CO_3 (M = 106 g/mol) was dissolved in 250 mL. Find the molarity.	Moles = 0.245/106 = 0.00231 Molarity = 0.00231/.250	0.00924 M
3	0.100 M HCl was used to neutralize 50.00 mL of $Ca(OH)_2$. Volume of HCl = 40.00 mL. Find molarity of $Ca(OH)_2$.	Moles of HCl = 0.100 × 0.04000 = 0.00400 $Ca(OH)_2$:HCl = 1:2 ⇒ Moles of $Ca(OH)_2$ = 0.00400/2 = 0.00200 Molarity = 0.00200/0.05000	0.0400 M $Ca(OH)_2$
4	0.100 M $KMnO_4$ was used to titrate 25.00 mL of Fe^{2+}. It required 30.00 mL of $KMnO_4$. Find molarity of Fe^{2+} (reaction ratio 1:5).	Moles of $KMnO_4$ = 0.100 × 0.03000 = 0.00300 Fe^{2+}:$KMnO_4$ = 5:1 ⇒ Moles of Fe^{2+} = 0.00300 × 5 = 0.0150 Molarity = 0.0150/0.02500	0.600 M Fe^{2+}

(continued)

Q. no.	Question	Solution steps	Answers
5	25.00 mL of NaOH was titrated with 0.0500 M H_2SO_4. Volume used = 15.00 mL. Find NaOH molarity.	Moles of H_2SO_4 = 0.0500 × 0.01500 = 0.00075NaOH:H_2SO_4 = 2:1 ⇒ Moles of NaOH = 0.00075 × 2 = 0.00150 M = 0.00150/0.02500	0.0600 M NaOH
6	You used 0.02500 mol of EDTA to titrate a metal ion solution. Volume = 50.00 mL. Find molarity.	M = 0.02500/0.05000	0.500 M
7	0.321 g of pure KHP (M = 204.22 g/mol) was titrated with NaOH. Volume used = 25.00 mL. Find NaOH molarity.	Moles of KHP = 0.321/204.22 = 0.00157 Molarity = 0.00157/0.02500	0.0628 M NaOH
8	0.287 g AgCl formed from Cl^-. M (AgCl) = 143.32 and $M(Cl^-)$ = 35.45. Find mass of Cl^-.	Mass Cl^- = 0.287 × (35.45/143.32) = 0.071 g	0.071 g Cl^-
9	Vinegar sample contains 4.00% acetic acid (M = 60.05). A sample of 10.00 mL is titrated with 0.100 M NaOH. Volume used = 25.00 mL. Verify % acetic acid.	Moles of NaOH = 0.100 × 0.025 = 0.00250 Acetic acid = 0.00250 mol × 60.05 = 0.1501 g % = (0.1501/10.00) × 100	1.50% (sample is diluted)
10	A 250-mL solution contains 0.730 g of Ca^{2+}. Find molarity (M = 40.08).	Moles = 0.730/40.08 = 0.0182 M = 0.0182/0.250	0.0728 M Ca^{2+}

18.3 Spectroscopic, Microscopic, and Chromatographic Techniques

In modern chemistry laboratories, a variety of analytical techniques are used to identify, characterize, and quantify substances with high precision and accuracy. These methods are broadly categorized into spectroscopic, microscopic, and chromatographic techniques, each offering unique insights into the physical and chemical properties of materials.

Spectroscopic techniques (Figure 18.7A–C) involve the interaction of electromagnetic radiation with matter to provide information about electronic transitions, molecular vibrations, nuclear environments, and atomic composition. Techniques such as ultraviolet–visible (UV-vis) spectroscopy, infrared spectroscopy (IR), nuclear magnetic resonance (NMR), atomic absorption spectroscopy (AAS), mass spectrometry (MS), and X-ray diffraction (XRD) are widely used in both qualitative and quantitative analysis. These methods are fundamental for identifying functional groups, determining molecular structures, assessing sample purity, and measuring concentrations.

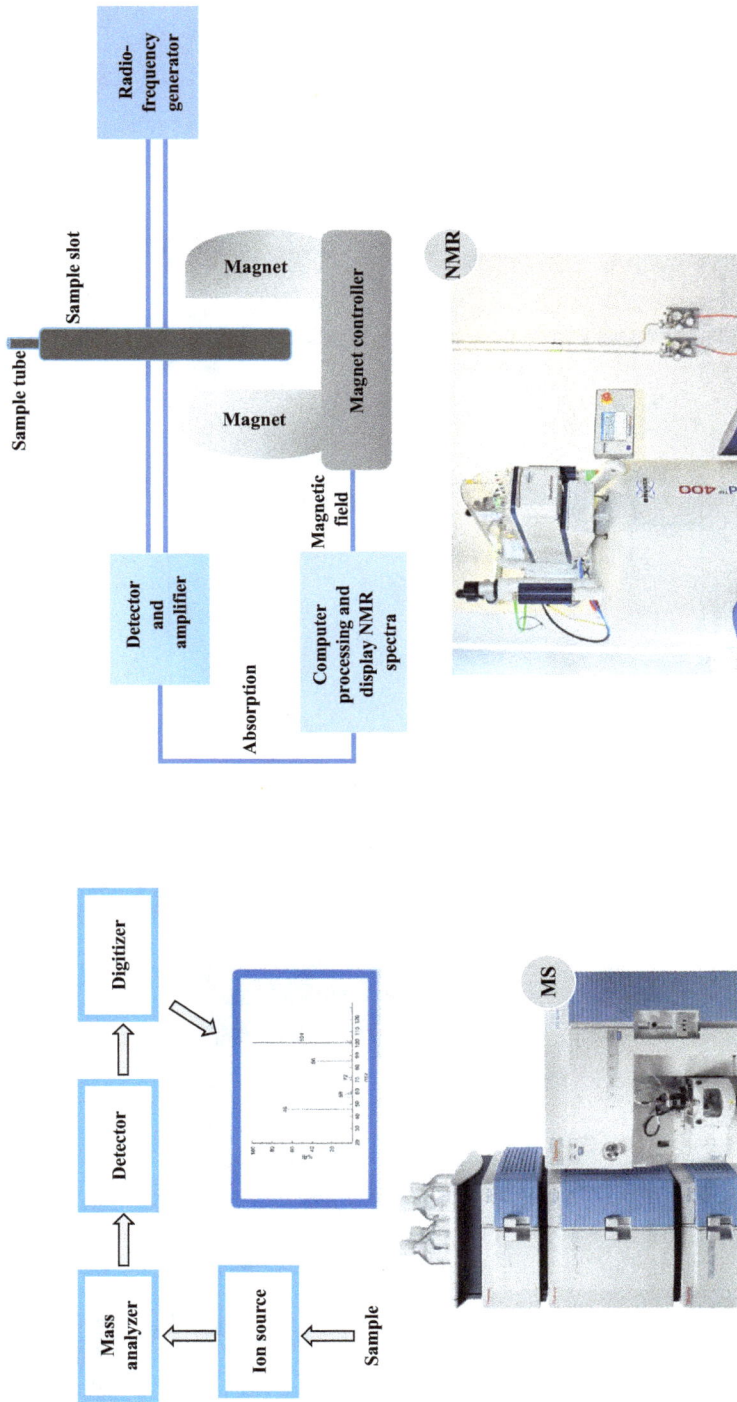

Figure 18.7: (A) Nuclear magnetic resonance (NMR) and mass spectrometry (MS). (B) Ultraviolet–visible (UV-vis) spectroscopy and infrared spectroscopy (IR). (C) Atomic absorption spectroscopy (AAS) and X-ray diffraction (XRD).

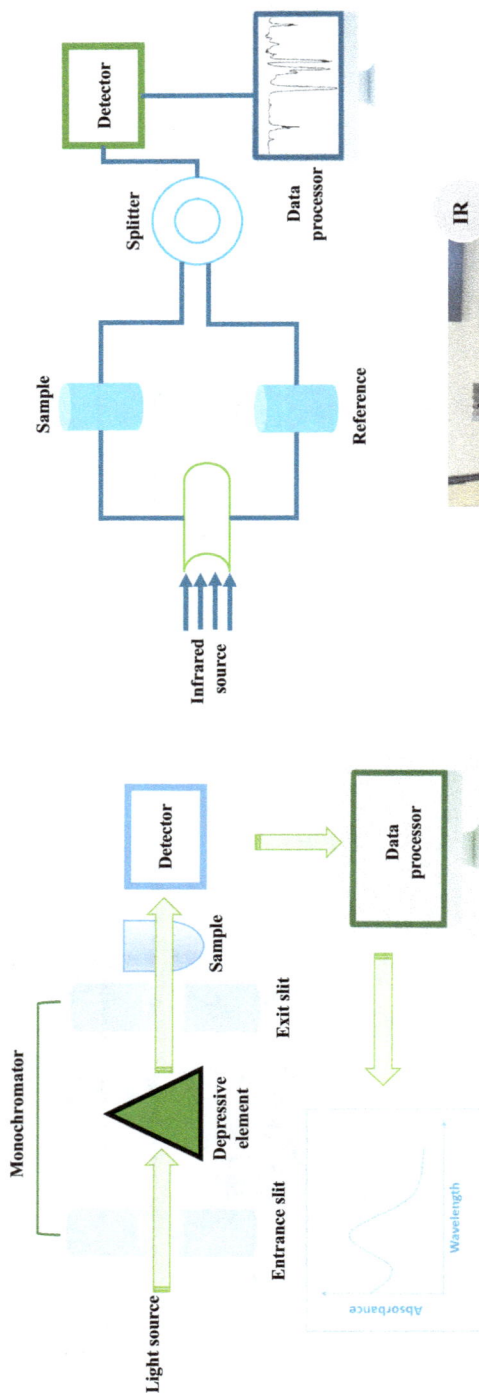

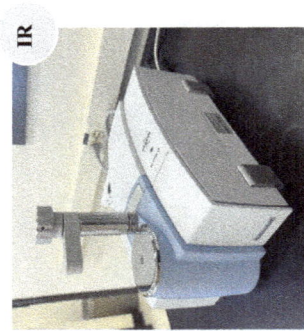

Image credit: https://www.webassign.net/ncsumeorgchem1/lab_2/manual.html.

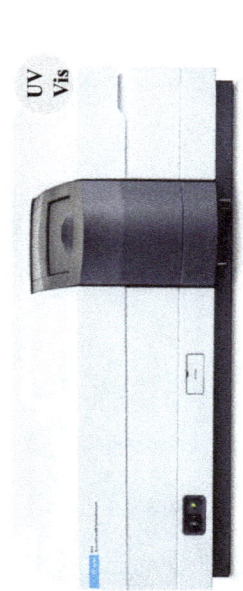

Image credit: https://www.agilent.com/en/product/molecular-spectroscopy/uv-vis-uv-vis-nir-spectroscopy/uv-vis-uv-vis-nir-systems/cary-4000-uv-vis#zoomELIBRARY_1028398.

Figure 18.7 (continued)

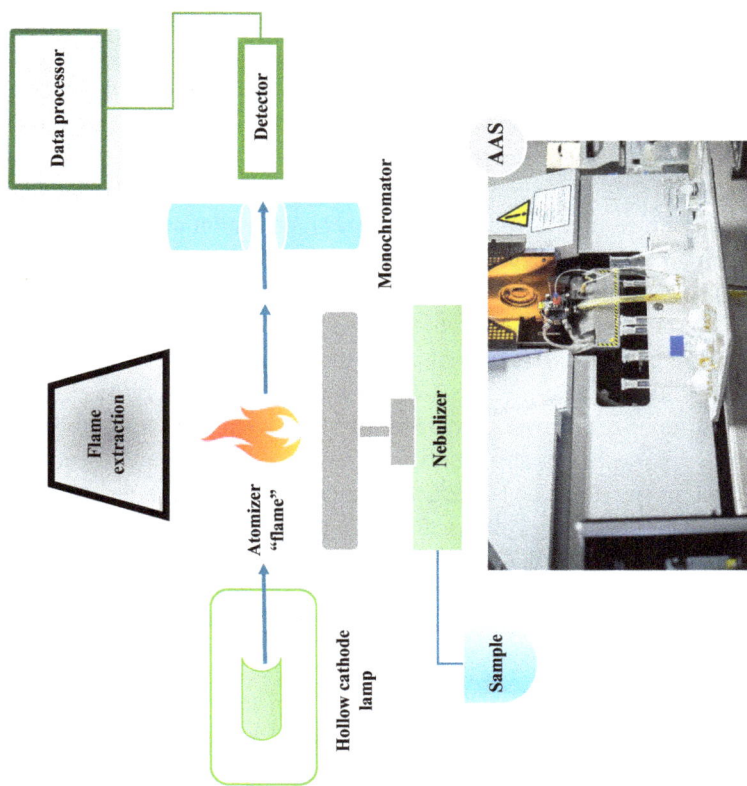

Image credit: https://www.azom.com/article.aspx?ArticleID=18556.

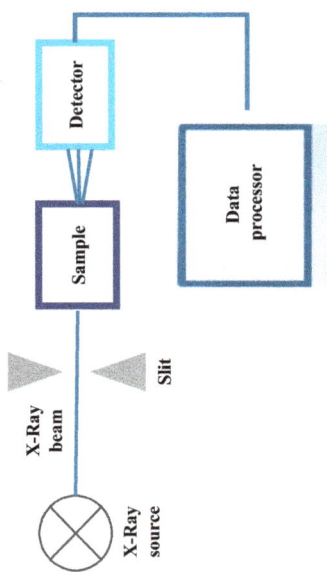

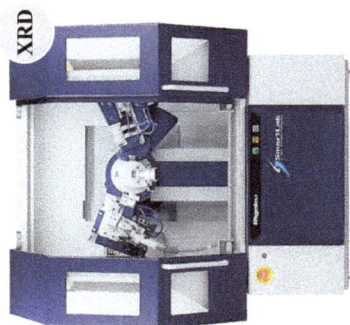

Image credit: https://rigaku.com/products/x-ray-diffraction-and-scattering/xrd.

Figure 18.7 (continued)

Microscopy or microscopic (Figure 18.8A–C) techniques are essential for investigating the physical characteristics of materials. Scanning electron microscopy (SEM) and transmission electron microscopy (TEM) provide high-resolution images of surfaces and internal structures, respectively, and are vital tools in material science and nanotechnology, while atomic force microscopy (AFM) provides three-dimensional surface mapping at the nanoscale. These techniques complement spectroscopic methods by revealing morphological information that spectroscopy alone cannot provide.

Chromatographic techniques (Figure 18.9A–C) are analytical and separation procedures that rely on variations in compound mobility within a stationary phase as a function of the mobile phase. Thin-layer chromatography (TLC), gas chromatography (GC), and liquid chromatography (LC)/high-performance liquid chromatography (HPLC) are commonly used to identify, separate, and occasionally quantify components in complicated mixtures. These approaches are very beneficial for preparing samples for spectroscopic examination, and they are critical in industries such as pharmaceutical development, forensic research, and environmental monitoring.

Together, these analytical tools (Tables 18.3–18.5) provide a comprehensive approach to chemical analysis, ensuring accuracy, specificity, and reliability in scientific research and industrial applications.

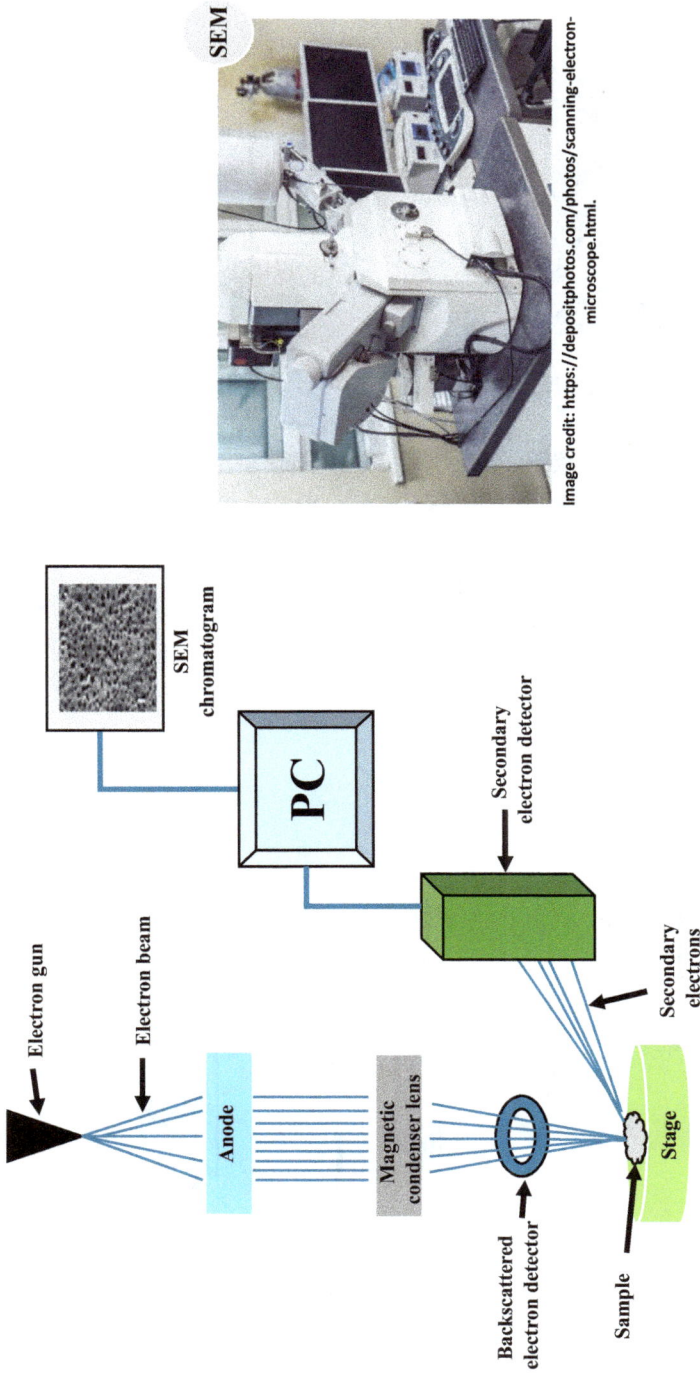

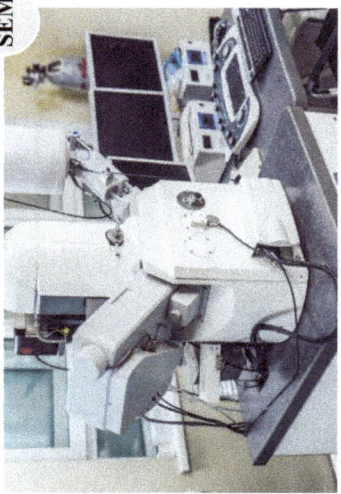

Figure 18.8: (A) Scanning electron microscopy (SEM). (B) Transmission electron microscopy (TEM). (C) Atomic force microscopy (AFM).

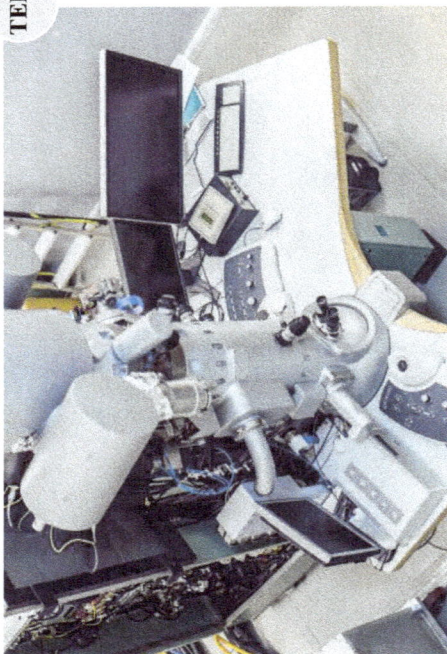

TEM

Image credit: https://depositphotos.com/photos/transmission-electron-microscope.html.

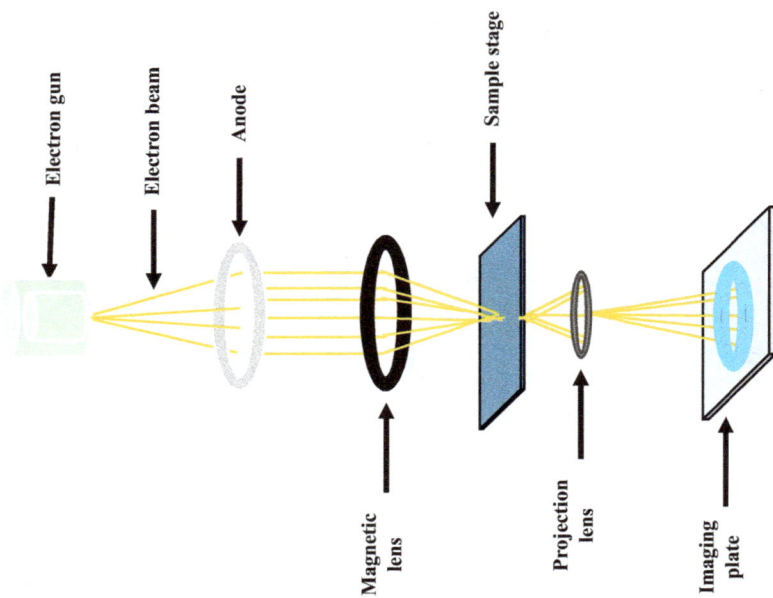

Electron gun

Electron beam

Anode

Magnetic lens

Sample stage

Projection lens

Imaging plate

Figure 18.8 (continued)

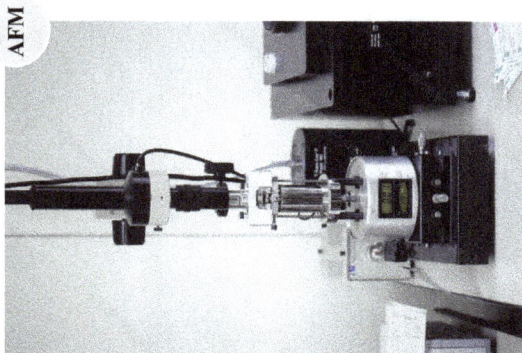

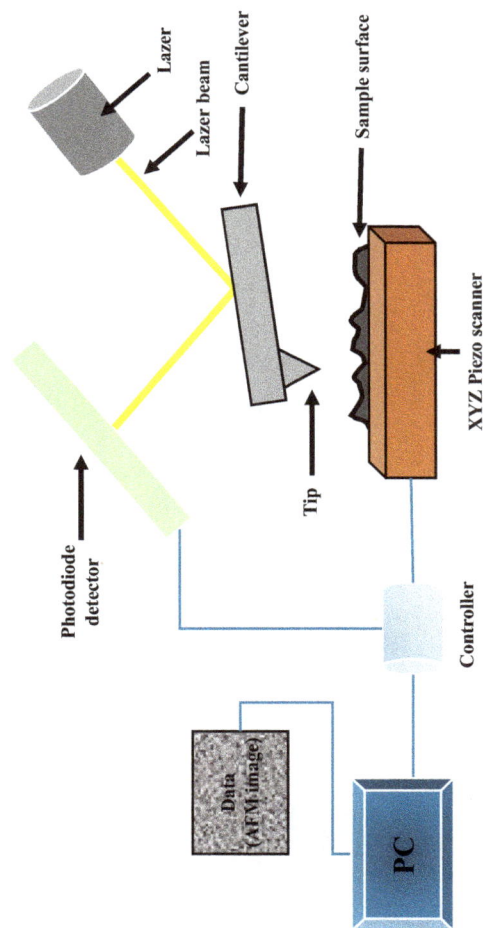

Image credit:
https://www.sciencephoto.com/media/230986/view/atomic-force-microscope.

Figure 18.8 (continued)

Image credit: https://www.azolifesciences.com/article/Thin-Layer-Chromatography-%28TLC%29-An-Overview.aspx

(A)

Figure 18.9: (A) Thin-layer chromatography (TLC). (B) Gas chromatography (GC). (C) Liquid chromatography (LC)/high-performance liquid chromatography (HPLC).

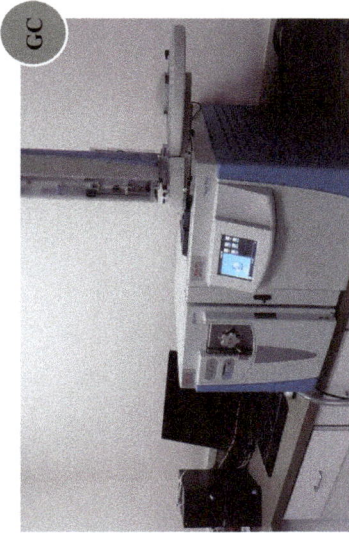

Image credit: https://www.cytogene.in/gas-chromatography-gc.

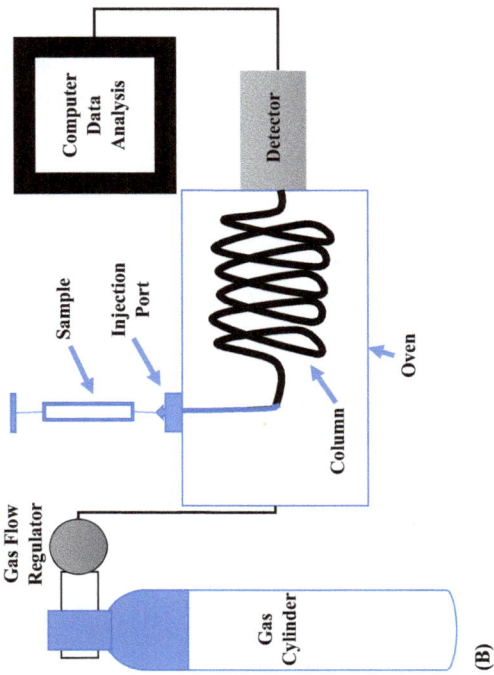

(B)

Figure 18.9 (continued)

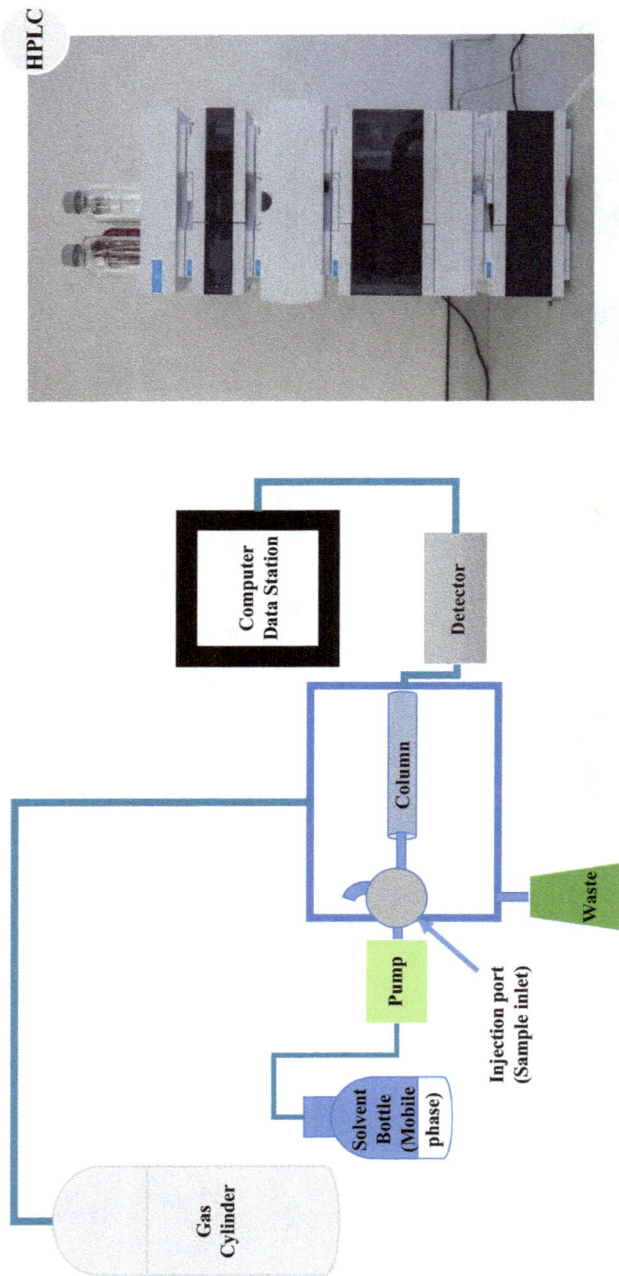

Figure 18.9 (continued)

Table 18.3: Spectroscopic techniques.

Technique	Type of radiation used	Measured quantity	Analyzed component	Typical application	Key features/notes
UV–visible (UV-vis)	Ultraviolet and visible light (200–800 nm)	Absorbance (A)	Electronic transitions (e.g., $\pi \rightarrow \pi^*$ and $n \rightarrow \pi^*$)	Concentration of colored or UV-absorbing compounds; monitoring reaction kinetics	Based on Beer–Lambert law ($A = \varepsilon cl$); simple and fast; widely used in routine analysis
Infrared (IR)	Infrared radiation (400–4,000 cm^{-1})	IR absorption	Molecular vibrations (bonds and functional groups)	Identifying functional groups in organic molecules; characterizing polymers	Produces a fingerprint spectrum; specific absorption bands for each functional group
Nuclear magnetic resonance (NMR)	Radiofrequency waves in a magnetic field	Resonance frequency (chemical shift)	Environment of nuclei (mainly ^{1}H and ^{13}C)	Structural determination of organic compounds; molecular dynamics	Highly detailed; nondestructive; useful for purity and structure verification
Atomic absorption spectroscopy (AAS)	Element-specific light (e.g., hollow cathode lamp)	Light absorption by atoms	Metal ions (e.g., Fe, Zn, and Pb) in solution	Detection and quantification of trace metals in water, food, and biological samples	Very sensitive; requires flame or graphite furnace; limited to metals only
Mass spectrometry (MS)	Ionizing radiation/electric/magnetic fields	Mass-to-charge ratio (m/z)	Molecular ions and fragments	Identifying unknown compounds; determining molecular weight and structure	Very sensitive and accurate; often coupled with chromatography; destructive technique
X-ray diffraction (XRD)	X-rays (commonly Cu Kα)	Diffraction pattern	Crystalline structures and lattice parameters	Identifying solid phases and crystal structure in materials	Ideal for solid samples; provides atomic-level structure; not suitable for amorphous samples

Table 18.4: Microscopic techniques.

Technique	Type of input/ interaction	Analyzed component	Typical application	Key features/notes
Scanning electron microscopy (SEM)	Electron beam	Surface morphology and topography	Imaging surfaces of materials, particles, and polymers	High magnification and depth of field; provides 3D-like surface images
Transmission electron microscopy (TEM)	Electron beam through thin sample	Internal structure at nanometer scale	Atomic-scale imaging of crystals, nanoparticles, and biological specimens	Extremely high resolution; requires ultra-thin samples and vacuum
Atomic force microscopy (AFM)	Probe tip scans surface	Surface topology and mechanical properties	Nanoscale surface imaging and material stiffness testing	Works in air or liquid; provides 3D topography; highly sensitive to surface interactions

Table 18.5: Chromatographic techniques.

Technique	Phase type	Principle of separation	Analyzed component	Typical application	Key features/notes
Thin-layer chromatography (TLC)	Solid stationary and liquid mobile	Polarity-based adsorption	Organic compounds (qualitative)	Monitoring reaction progress; checking purity	Quick, simple, and inexpensive; R_f values help compare components
Gas chromatography (GC)	Gas mobile, solid/ liquid stationary	Volatility and interaction with column	Volatile organic compounds	Environmental, forensic, and petrochemical analysis	High resolution; requires volatile samples; often coupled with MS
Liquid chromatography (LC)/high-performance liquid chromatography (HPLC)	Liquid mobile, and solid stationary	Polarity, size, or ion exchange	Nonvolatile, thermally unstable compounds	Pharmaceutical analysis, food safety, and biochemistry	High precision and sensitivity; automated; useful for thermally labile substances

18.4 Questions and Answers

Questions	Answers
1. What law is the basis for concentration measurements in UV-vis spectroscopy?	Beer–Lambert law ($A = \varepsilon c l$)
2. Which technique identifies functional groups based on molecular vibrations?	Infrared (IR) spectroscopy
3. What nucleus is commonly analyzed in ^{1}H NMR spectroscopy?	Hydrogen nucleus (^{1}H)
4. Which spectroscopy technique is best for detecting trace metals in water?	Atomic absorption spectroscopy (AAS)
5. What is the principle of separation in gas chromatography (GC)?	Volatility and interaction with stationary phase
6. Which microscopy technique gives surface morphology using electrons?	Scanning electron microscopy (SEM)
7. What is the purpose of TLC in chemical analysis?	To monitor reaction progress or check compound purity
8. What is titration used for in volumetric analysis?	To determine the unknown concentration of a solution by neutralization or redox reaction
9. Which spectroscopy is nondestructive and gives 3D surface topography at nanoscale?	Atomic force microscopy (AFM)
10. 25.00 mL of NaOH was titrated with 0.0500 M H_2SO_4. Volume used = 15.00 mL. Find NaOH molarity.	Moles of H_2SO_4 = 0.0500 × 0.01500 = 0.00075 NaOH:H_2SO_4 = 2:1 ⇒ Moles of NaOH = 0.00075 × 2 = 0.00150 M = 0.00150/0.02500 = 0.0600 M NaOH

Chapter 19
Separation Techniques in Chemical Laboratories

19.1 Introduction

Separation procedures are essential for experimental activities in chemical laboratories. Whether in academic research, industrial analysis, or quality control, the ability to extract pure chemicals from mixtures is crucial for understanding chemical behavior and producing correct results. Separation techniques are deeply ingrained in the chemical workflow, starting with the extraction of natural molecules and concluding with the purification of created compounds.

Chemical mixtures can include components with varying physical or chemical properties, such as boiling point, solubility, particle size, polarity, or magnetic susceptibility. These characteristics can be employed in specialized separation methods to selectively separate the desired compounds.

Filtration, for example, can remove a solid from a liquid; distillation can separate a volatile liquid from a nonvolatile residue; and ion-exchange chromatography (I-EC) can separate specific ions. Some procedures involve physical processes, such as centrifugation or evaporation, while others employ chemical reactions, such as precipitation or extraction.

The discovery and improvement of separation methods have significantly increased the efficiency and scope of laboratory operations. Traditional processes, such as decantation and crystallization, remain important due to their simplicity and efficacy. At the same time, more sophisticated procedures, such as column chromatography and electrophoresis, provide better resolution for separating complicated mixtures.

Membrane-based methods, pressure-assisted systems, and magnetic separation are becoming increasingly popular, particularly in green chemistry, biotechnology, and environmental applications that demand accuracy and sustainability.

By mastering the above-mentioned procedures, chemists can not only isolate pure compounds but also ensure the accuracy and repeatability of their experimental findings.

19.2 Filtration

19.2.1 Introduction

Filtration is a physical process by which a solid material is removed from a substrate (liquid or gaseous) in which it is suspended. It is done by passing the mixture to be processed through one of the many available filter media. These are of two types: surface filters and depth filters. The surface filter traps particles larger than the filter

https://doi.org/10.1515/9783112218105-019

pores or mesh dimensions on the surface of the filter; all other smaller particles pass through. The filter is typically made of paper, fabric, or other types of membranes.

A depth filter consists of either multiple layers or a single layer of a medium having depth, which retains particles within its structure. There are two types of depth filters. The first type is the deep bed filter, which involves filtration vertically through a packed bed of granular or fibrous material. The second type is the thick media filter. This alternative form of depth filter is predominantly composed of a replacement filter cartridge. There are four primary ways in which filtration is commonly used. These include:

- Solid–Liquid Filtration. The separation of solid particulate matter from a carrier liquid.
- Solid–Gas Filtration. The separation of solid particulate matter from a carrier gas.
- Liquid–Liquid Separation. A special class of filtration resulting in the separation of two immiscible liquids, one of them is water, by means of a hydrophobic medium.
- Gas–Liquid Filtration. The separation of gaseous matter from a liquid in which it is usually, but not always, dissolved.

Filtration is commonly employed in labs to remove solid impurities from liquids and collect precipitated or recrystallized particles from solutions. This separation may be performed via basic gravity filtration, which employs gravitational force to draw the liquid through the filter medium. However, vacuum filtration, which employs a lower pressure behind the filter, can significantly speed up the process. This creates a pressure differential, which, when combined with gravity, speeds up the filtering process and increases efficiency, especially when dealing with microscopic particles or huge amounts of liquid.

19.2.2 Filtration Methods

There are two primary methods of filtration used in the laboratory: gravity filtration and vacuum filtration (also known as suction filtration). In gravity filtration (Figure 19.1), the liquid (filtrate) passes through the filter medium primarily due to the force of gravity, aided by capillary action between the liquid and the stem of the funnel. This method is typically used when a slower, gentler filtration is sufficient, especially for hot solutions to prevent premature crystallization. This method is slow but preferable to vacuum filtering due to better retention of fine particles and less cracking or tearing of filter paper. It is also considered the fastest and most preferred method for filtering gelatinous precipitates because these precipitates tend to clog and pack the pores of the filter medium much more readily during vacuum filtration.

Vacuum filtration (Figure 19.2), on the other hand, works by establishing a pressure differential across the filter media, which speeds up the filtration process. This is performed by attaching the filtration device to a vacuum source, which clears the area un-

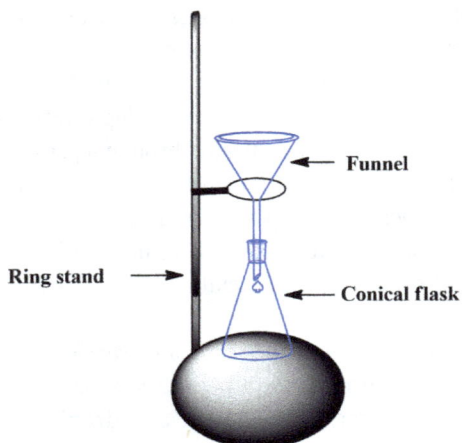

Figure 19.1: Schematic diagram for gravity filtration setup.

derneath the filter. The ensuing pressure differential causes air pressure to force the liquid through the filter faster than gravity filtration. This approach is very useful for collecting precipitates or when rapid filtering is required. A water aspirator is normally used to produce the vacuum, although a vacuum pump can be used if a larger vacuum is required, as long as it is protected by suitable traps to avoid contamination or damage.

It is essential to install a water trap between the filtration flask and the vacuum source to protect the system from backflow. Once the vacuum is applied, the combination of reduced pressure beneath the filter and atmospheric pressure above significantly enhances the rate and efficiency of filtration.

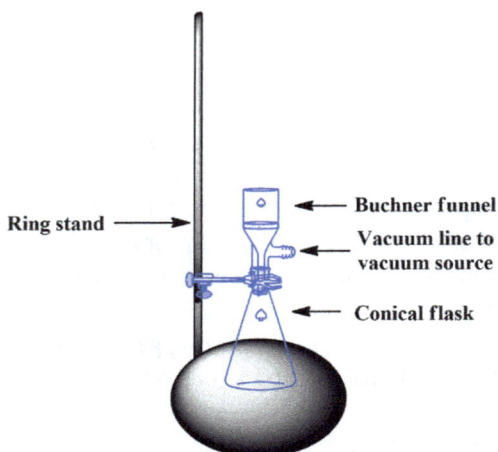

Figure 19.2: Schematic diagram for vacuum filtration setup.

The choice between gravity and vacuum filtration depends on several factors, including:
- The volume and viscosity of the liquid
- The nature and quantity of the solid
- The need for speed or delicacy in the filtration process
- The temperature and volatility of the solution
- The required purity of the filtrate or residue

19.2.3 Filter Media

19.2.3.1 Paper

There are several grades of filter papers (Table 19.1) developed to fulfill a variety of laboratory requirements. These include qualitative grades for general use, quantitative (low-ash or ashless) grades for gravimetric analysis, toughened grades for improved wet strength and chemical resistance, and glass-fiber filters for fine particle retention and high-temperature applications. When choosing filter paper, consider porosity, retention capacity, chemical compatibility, and the type of waste or residue. Filter papers are available in a range of pore sizes and diameters, allowing users to match the filtration speed and clarity required for a specific application. The goal is to complete the filtration as efficiently and rapidly as possible, while minimizing clogging and maintaining the integrity of the filter medium throughout the process.

Table 19.1: Comparison of filter paper grades and applications.

Filter paper type	Key characteristics	Typical applications
Qualitative grade	Medium retention and flow rate; not ash-free; cellulose-based	Routine qualitative analysis; general-purpose lab work
Quantitative (ashless)	Very low ash content (<0.01%); designed for precise residue analysis	Gravimetric analysis; ignition residue studies
Hardened grade	Chemically strengthened for high wet strength; acid- and tear-resistant	Filtration of acidic/basic solutions; vacuum filtration
Hardened ashless	Combines ashless properties with high strength; low contamination risk	High-precision gravimetric work involving corrosives
Glass-fiber filters	Noncellulose; borosilicate glass fibers; excellent flow and fine particle retention	Airborne particle sampling; hot/viscous liquids
Fast flow grade	Large pore size; fast filtration but lower resolution	Coarse precipitates; high-viscosity solutions
Slow flow grade	Small pore size; retains fine particles; slower flow	Fine precipitate collection; high-clarity filtration

19.2.3.2 Membrane Filters

Membrane filters are thin, ultraporous synthetic materials with certain pore sizes (Table 19.2). They are effective in removing pollutants, such as particulates, pathogens, and specific gas or liquid impurities, from fluids that flow through them because they trap particles larger than their designated pore size. Membrane filters are composed of nylon, polycarbonate, hydrophilic and hydrophobic polytetrafluoroethylene (PTFE), cellulose acetate, and mixed cellulose esters.

Depending on the need, membrane filters can provide a filtrate that is extremely clean or even sterile when chosen properly. To meet particular needs for performance and chemical compatibility, they come in a variety of pore sizes and materials. Millipore membrane filters, for instance, are made to keep all particles bigger than the filter's rated pore size on the membrane's surface. This makes it simple to see, count, or analyze the pollutants that are retained.

This surface retention mechanism sharply contrasts with that of depth filters, which trap impurities both on the surface and within the internal matrix of the filter. As a result, membrane filters offer greater precision and reproducibility in applications where particle size control and retention are critical.

Table 19.2: Comparison of membrane filters and depth filters.

Feature	Membrane filters	Depth filters
Filtration mechanism	Surface filtration: particles retained on the surface	Depth filtration: particles retained within the filter matrix
Pore size uniformity	Precisely defined, uniform pore sizes	Varying pore sizes throughout the filter material
Retention efficiency	High precision; sharp particle size cutoff	Lower precision; gradual particle retention
Clogging tendency	Higher, particles accumulate on surface	Lower, distribution of particles within depth reduces blockage
Flow rate	Lower (due to tight pore structure)	Higher (more open structure)
Sterility/ precision use	Suitable for sterile filtration and analytical applications	Suitable for bulk filtration or clarification of turbid solutions
Reusability	Typically single use	May be reusable (depending on material and application)
Common materials	Cellulose esters, PTFE, Nylon, and polycarbonate	Glass fiber, polypropylene, cellulose fiber
Typical applications	Sterile filtration, microbiological analysis, particle counting	Prefiltration, large volume filtration, and viscous or particulate-rich fluids

19.2.3.3 Fritted-Glass Filters

Alternatively, funnels with fritted-glass plates (also called sintered-glass discs) can be used for solids filtering. Because these glass filters are composed of tiny, fused glass particles and come in a variety of porosity grades (Table 19.3), the filtration needs may be taken into consideration when choosing the right pore size. Fritted-glass funnels are superior to conventional filter paper in a number of ways, including increased chemical resistance, reusability, and uniform pore structure, all of which reduce frequent problems like ripping, clogging, or filtrate loss.

Table 19.3: Porosity grades of fritted-glass filters.

Designation	Pore size range (µm)	Typical applications
Extra coarse	170–220	Rapid filtration of very coarse particles
Coarse	40–60	Filtration of coarse suspensions and precipitates
Medium	10–15	Standard laboratory filtration of precipitates
Fine	4.0–5.5	Fine precipitate collection and clarification
Very fine	2.0–2.5	Fine particulate removal; analytical applications
Ultrafine	0.9–1.4	Sterile filtration; removal of very fine solids

19.2.4 Common Filtering Accessories in Laboratories

19.2.4.1 Vacuum Filtration Accessories

Accessory	Function
Büchner funnel	Holds filter paper and retains solids during vacuum filtration
Filter flask (suction flask)	Thick-walled flask with sidearm to connect to a vacuum source
Rubber adapter (filter cone)	Seals the Büchner funnel to the filter flask to prevent leaks
Vacuum tubing	Connects flask to vacuum source; must be thick-walled to withstand pressure
Water aspirator pump	Generates vacuum using flowing water (Venturi effect)
Vacuum pump	Electrically driven device for stronger, controlled vacuum filtration
Trap flask (vacuum trap)	Protects the vacuum source from backflow of liquid
Clamp and ring stand	Holds the apparatus securely during filtration

19.2.4.2 Gravity Filtration Accessories

Accessory	Function
Glass or plastic funnel	Holds filter paper for gravity-based filtration
Filter paper	Medium through which liquid passes, retaining solids
Funnel support (ring or funnel holder)	Holds funnel in place above receiving container
Beakers/Erlenmeyer flasks	Collect the filtrate during gravity filtration
Watch glass	Covers funnel to prevent contamination or evaporation

19.2.4.3 Specialized Filtering Accessories

Accessory	Function
Fritted-glass funnels	Replace filter paper; used for finer or chemically resistant filtrations
Glass-fiber pads	Used with certain filter holders for particulate retention
Membrane filter holders	Support membrane filters in sterile or analytical filtration systems
Syringe filters	Small disposable filters for quick filtration of small-volume solutions
Filter crucibles (e.g., Gooch crucible)	Porous-bottom crucibles for gravimetric analysis
Desiccators (postfiltration)	Preserve dried precipitates in low-humidity environment

19.2.5 Operations Associated with the Filtration Process

Regardless of whether gravitational or vacuum filtration is used, three processes must be performed: decantation, washing, and precipitation transfer.

19.2.5.1 Decantation

When a solid is easily deposited on the bottom of the liquid and has little or no tendency to remain suspended, it can be easily separated from the liquid by carefully pouring the liquid so that no solid is transferred to it. This process is called decantation. To decant a liquid from a solid:

- Hold the beaker that has the mixture in it in one hand and have a glass stirring rod in the other.
- Incline the beaker until the liquid has almost reached the lip.
- Touch the center of the glass rod to the lip of the beaker and the end of the rod to the side of the container into which you wish to pour the liquid.
- Continue the inclination of the beaker until the liquid touches the glass rod and flows along it into the second container.

19.2.5.2 Washing

Washing aims to remove the excess liquid phase, and any soluble impurities present in the precipitate. Use a solvent that is miscible with the liquid phase but does not dissolve a significant amount of precipitate. The solids can be washed in the beaker after decanting the supernatant liquid phase. Add a small amount of washing liquid and mix it well with the precipitate. Allow the solid to settle. Decant the washing liquid through the filter. Allow the precipitate to settle, with the beaker slightly tilted, so that the solid accumulates in the corner of the beaker under the drain. Repeat this procedure several times. Multiple washes with small volumes of liquid are more effective in removing soluble contaminants than a single wash using the total volume.

19.2.5.3 Transfer of the Precipitate

Most of the precipitate is removed from the beaker into the filter using a stream of washing liquid from the washing bottle. The mixing rod is used to direct the flow of liquid into the filtration medium. The last traces of precipitate are removed from the walls of the beaker with a mixing rod.

19.3 Recrystallization

19.3.1 Introduction

Recrystallization is a typical process for purifying organic molecules that are solid at room temperature. The impure compound is dissolved in a hot solvent and then cooled to form crystals. The solvent used may be a single one or a combination, depending on a number of certain criteria. If crystals develop fast, contaminants may become trapped within them, rendering the purification procedure useless. However, when crystallization proceeds slowly, only the pure substance may form solid crystals, leaving impurities in solution. This slower, more selective process is commonly known as crystallization. This may be easily accomplished in the laboratory using the setting shown in Figure 19.3.

19.3.2 Solvent Requirements

In general, solvents should:
- Not react with the material to be crystallized
- Form desirable, well-formed crystals
- Be easily removed from the purified crystals
- Have high solvency for the desired substance at high temperatures and low solvency for that substance at low temperatures

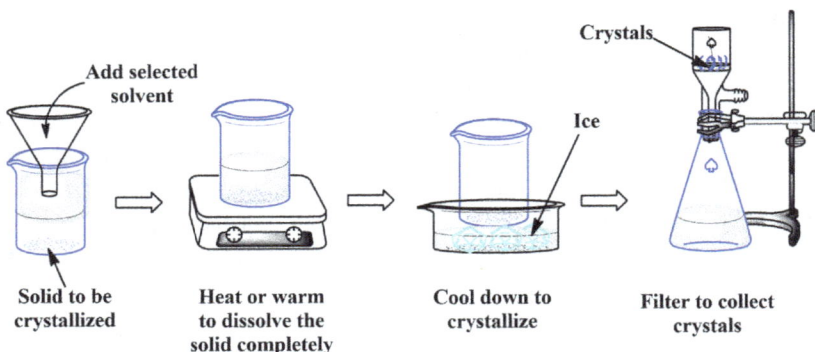

Figure 19.3: Schematic diagram for crystallization steps.

19.3.2.1 Solvency

Beyond normal temperature, the substance that needs to be purified should be very slightly soluble in the solvent; nevertheless, beyond the boiling point, it should be considerably soluble. The chemical structures of the solute and the solvent, as well as the temperature, influence how soluble a solute is in a solvent. As the temperature rises, the solute's solubility in a solvent often increases, and in certain cases, the increase is rather significant. This is the foundation for the purifying process known as recrystallization. The solubility of a solvent is determined by both temperature and the chemical composition of the solvent and the solute.

If the chemical has been published in the literature, the citation will give details about its solubility in common solvents. Polar organic compounds frequently dissolve in polar solvents such as water, low molecular weight alcohols, or a mixture of the two. Nonpolar chemicals dissolve in nonpolar organic solvents like benzene and hexane.

Solubility is based on the fundamental premise that similar chemicals dissolve in related substances. This is not always true since the entire molecule must be evaluated before concluding. Stearic acid, a long-chain carboxylic acid, behaves as a nonpolar substance rather than a polar one because the nonpolar half of the molecule outweighs the polar carboxyl group.

In general, the following points should be considered:

– A useful solvent is one that will dissolve a great deal of the solute at high temperatures and very little at low temperatures.
– If a solvent dissolves too much solute at low temperatures, it is not suitable.
– If too much solvent is required to dissolve the solute even at its boiling point, it may be possible to recrystallize several grams, but extremely large volumes of solvent would be required to recrystallize several hundred grams.
– Quick tests of solubility are sometimes unreliable because some solutes dissolve very slowly in boiling solvents. A quick observation may be misleading and cause you to reject the solvent as being unsatisfactory.

–   The suitability of a solvent depends upon the establishment of equilibrium. The maximum solute will dissolve when equilibrium has been attained between the dissolved and solid solutes.

19.3.2.2 Volatility
The volatility of a solvent determines the ease or difficulty of removing any residual solvent from the crystals that have formed. Volatile solvents may be removed easily by drying the crystals under a vacuum or in an oven (although oven drying is not the preferred way). Solvents with a high boiling point should be avoided, if possible. They are difficult to remove, and the crystals usually must be heated mildly under a high vacuum to remove such solvents.

19.3.3 Recrystallization of a Solid

19.3.3.1 General Procedure of Recrystallizing
–   Select the most desirable solvent.
–   Add the determined volume of solvent to the flask (no more than two-thirds the volume of the flask) and heat. Add a few boiling stones if required.
–   Add the minimum amount of hot solvent to the solute slowly to dissolve it. Boil, if necessary, to dissolve all the solute.
–   Allow it to stand and cool. Cool or chill rapidly in a cooling bath for small crystals, but cool slowly to get large crystals.
–   Isolate the crystals by gravity or suction filtration. Concentrate the mother liquor to get more crystals.
–   Dry the crystals in a warm oven, or better, in the desiccator because overheating in the oven may cause the crystals to degrade.

19.3.3.2 Decolorization
Colored contaminants may sometimes be removed by adding finely powdered decolorizing charcoal, which adsorbs the contaminants. Soluble contaminants, not adsorbed, remain in solution in the mother liquor filtrate. Decolorization can be achieved as follows:
–   Place the substance to be purified in a suitably sized flask.
–   Add a determined volume of solvent.
–   Add decolorizing carbon, 1% by weight of solute, if needed.
–   Boil until all crystals have dissolved.
–   Filter as quickly as possible through a fluted filter in the funnel. Stemless funnels are best to use because there is no stem in which crystallization can take place and clog the system. If necessary, warm the filter funnel to prevent the crystallization of hot filtrate in the funnel.

- Collect the filtrate in a flask; allow it to stand and cool.
- Dry the crystals.

19.4 Extraction

19.4.1 Introduction

Extraction refers to the process of removing a solute from a mixture using a solvent. A simple setup is shown in Figure 19.4. The solute must dissolve more easier in the solvent than in the mixture for the extraction to be effective. Water, water-miscible solvents (like ethanol or acetone), and water-immiscible solvents (like diethyl ether or dichloromethane) can all be used as extraction solvents. The solvent is chosen based on the solute's solubility, as well as the experiment's specific criteria, such as polarity, density, and ease of separation.

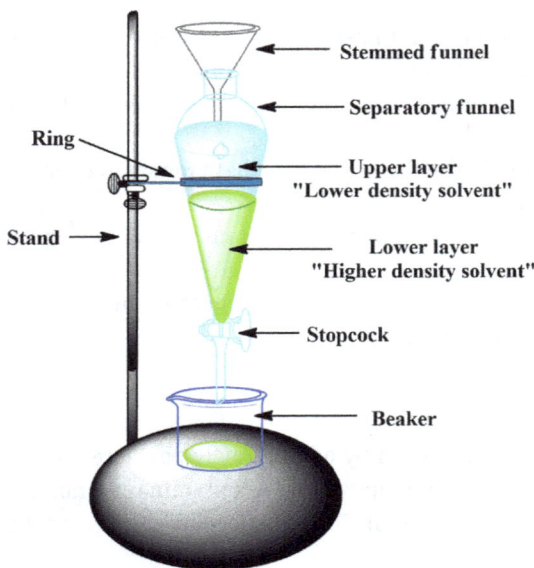

Figure 19.4: Schematic diagram for extraction setup.

19.4.2 Extraction of Solute Using Immiscible Solvents

A solute may be soluble in many immiscible solvents. When a solution of that solute in one of two immiscible solvents is shaken vigorously with the other immiscible solvent, the solute will be distributed between the two solvents in such a manner that the ratio of the concentrations in moles per liter of the solute is constant. This ratio is called the distribution coefficient, and it is independent of the volumes of the two solvents and the total concentration of the solute.

This type of extraction transfers a solute from one solvent to another. It can be used to separate reaction products from reactants and to separate desired substances from others in the solution. The separatory funnel is used for this purpose. Immiscible solvents, which are incapable of mixing to attain homogeneity and will separate from each other into separate phases, must be used. Miscible solvents, which are capable of being mixed in any ratio without separation into two phases, cannot be used. Multiple extractions with smaller portions of the extraction solvent are more effective than one extraction with a large volume. The choice of the extraction solvent determines whether the solute remains in the separatory funnel or is in the solvent which is drawn off. The solvent that has the greater density will be the bottom layer and the less dense extraction solvent remains in the separatory funnel, and the denser extraction solvent is drawn off if the extraction solvent of higher density is used. The extraction can proceed as follows:

- Use a clean separatory funnel.
- Pour the solution to be extracted into the funnel, which should be large enough to hold at least twice the total volume of the solution and the extraction solvent.
- Pour in the extraction solvent, and close with the stopper.
- Shake the funnel gently.
- Invert the funnel and open the stopcock slowly to relieve the pressure built up.
- Close the stopcock while the funnel is inverted and shake again.
- Repeat steps as needed.
- Place the funnel in the ring stand support and allow the two layers of liquid to separate. Remove stopper closure.
- Open the stopcock slowly and drain off the bottom layer.
- Repeat the operation, as needed, with fresh extraction solvent as many times as desired.
- Combine the lower layers that have been drawn off.

19.4.3 Extraction Procedures in Laboratories

19.4.3.1 Using Water

Water is a polar solvent, and polar substances are soluble in it; examples of such substances are inorganic salts, salts of organic acids, strong acids and bases, low molecular weight compounds, carboxylic acids, alcohols, polyhydroxy compounds, and amines. Water will extract these compounds from any immiscible organic solvents containing them.

19.4.3.2 Using Dilute Aqueous Acid Solution

Dilute hydrochloric acid (typically 5–10% HCl) is used to extract basic organic compounds, such as amines, cyclic nitrogen-containing heterocycles, and alkaloids, from an immiscible organic solvent. The acid reacts with the basic compound to form a water-soluble hydrochloride salt, which partitions into the aqueous phase. After the extraction, the organic layer is washed with water to remove any residual acid and water-soluble impurities. This method effectively separates basic components from neutral or acidic ones in a mixture. For example, if aniline ($C_6H_5NH_2$) is dissolved in ether and treated with HCl:

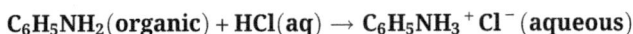

$$C_6H_5NH_2(\text{organic}) + HCl(\text{aq}) \rightarrow C_6H_5NH_3{}^+Cl^-(\text{aqueous})$$

The resulting anilinium chloride is water-soluble and moves into the aqueous layer.

19.4.3.3 Using Dilute Aqueous Basic Solution

Dilute aqueous sodium hydroxide (e.g., 5% NaOH) or sodium bicarbonate (e.g., 5% $NaHCO_3$) solutions can be used to extract acidic organic compounds from an immiscible organic solvent. The base reacts with the acidic solute (such as a carboxylic acid or phenol), converting it into its water-soluble sodium salt, which transfers to the aqueous phase. After the extraction, the organic layer is washed with water to remove residual base and other water-soluble impurities. This technique is widely used to separate acidic components from neutral or basic compounds in an organic mixture. For example, if benzoic acid is dissolved in diethyl ether and treated with aqueous $NaHCO_3$:

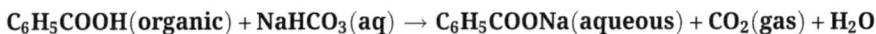

$$C_6H_5COOH(\text{organic}) + NaHCO_3(\text{aq}) \rightarrow C_6H_5COONa(\text{aqueous}) + CO_2(\text{gas}) + H_2O$$

19.4.3.4 Considerations in the Choice of Extraction Solvent

Several important factors must be considered when selecting a suitable solvent for the extraction of aqueous solutions. Table 19.4 presents characteristics of commonly used extraction solvents. Key considerations include:
-   "Like dissolves like." Polar solvents tend to dissolve polar substances, while nonpolar solvents dissolve nonpolar substances.
-   Organic solvents are generally effective at dissolving organic solutes, especially neutral compounds.
-   Water efficiently dissolves inorganic compounds and salts of organic acids and bases due to its high polarity.
-   Organic acids that are soluble in organic solvents can be transferred to the aqueous phase by converting them into their water-soluble salts using basic solutions such as sodium hydroxide (NaOH), sodium carbonate (Na_2CO_3), or sodium bicarbonate ($NaHCO_3$).

Table 19.4: Common extraction solvents organized by polarity, density, and boiling point.

Solvent	Polarity	Density (g/mL)	Boiling point (°C)	Remarks
Water	Highly polar	1.00	100	Often the aqueous phase
Methanol	Highly polar	0.79	64.7	Miscible with water
2-Butanol	Polar	0.81	99.5	Miscible with water; dries easily
Ethyl acetate	Moderately polar	0.90	77.1	Absorbs water; commonly used for polar to moderately polar organics
Diethyl ether	Slightly polar	0.71	34.6	Highly volatile; forms peroxides; lighter than water
Diisopropyl ether	Slightly polar	0.73	68.5	Forms peroxides; lighter than water
Petroleum ether	Nonpolar	~0.65–0.70	35–60 (range)	Poor for polar solutes; mixture of alkanes
Benzene	Moderately nonpolar	0.88	80.1	Both toxic and carcinogenic; forms emulsions; restricted use
Chloroform	Nonpolar	1.49	61.2	Moderately to highly toxic; denser than water; dries easily; can form emulsions
Methylene chloride	Nonpolar	1.33	39.6	Denser than water; dries easily; may form emulsions
Tetrachloromethane	Nonpolar	1.59	76.7	Toxic; very dense; restricted use

19.4.3.5 Peroxides in Ethers and Their Removal

Ethers, such as diethyl ether, are prone to forming organic peroxides when exposed to air and light over time. These peroxides are shock-sensitive and can lead to explosions, especially during concentration, distillation, or evaporation. Therefore, it is critical to test for peroxides before using or purifying ethers. One simple test involves using aqueous potassium iodide (KI), which changes color in the presence of peroxides due to the oxidation of iodide to iodine. Peroxide test procedure (KI test method) is summarized as follows:

- Prepare a 10% aqueous solution of potassium iodide (KI).
- Add 1 mL of the KI solution to a small sample of the ether (in a test tube or small vial).
- Allow the mixture to stand for 1 min.

If a yellow to brown color appears during the test, it indicates the presence of peroxides, resulting from the oxidation of iodide to iodine (I_2). In such cases, the ether must not be distilled or evaporated. Instead, it should be disposed of following institutional hazardous waste disposal guidelines.

To safely handle ethers that may contain hazardous peroxides, various chemical and physical treatments can be applied to decompose or remove these unstable compounds. Table 19.5 summarizes the methods for peroxide removal from ethers.

Table 19.5: Methods for the removal of peroxides from ethers.

Method	Reagents used	Procedure summary	Notes/precautions
Activated alumina column	Activated alumina, 5% aqueous $FeSO_4$ (for elution)	Pass ether through a column of wet activated alumina; elute with 5% $FeSO_4$ if needed.	Do not allow alumina to dry out. Use in a fume hood.
Storage over alumina	Activated alumina	Store ether in a sealed bottle over a layer of activated alumina.	Passive method; slow but safe for long-term storage.
$FeSO_4$ solution wash	100 g $FeSO_4$ + 42 mL conc. HCl + 85 mL H_2O	Shake ether with this freshly prepared solution.	Some ethers form aldehydes that must be removed.
Oxidation and base wash (if aldehydes)	1% $KMnO_4$, 5% NaOH, water	Wash ether after $FeSO_4$ treatment: (1) with 1% $KMnO_4$, (2) then with 5% NaOH, (3) then water.	Removes aldehydes and any resulting acids.
Triethylenetetramine wash	Cold triethylenetetramine (25% w/w of ether)	Shake ether with cold TETA, let settle, and then separate layers.	Useful for hard-to-remove peroxides; handle TETA with care.
CuCl reflux (for water-soluble ethers)	0.5% CuCl (by mass)	Reflux ether with 0.5% CuCl, and then distill the ether afterward.	Effective for water-miscible ethers like dioxane.

19.4.3.6 Breaking Emulsions (Dealing with Emulsions in Liquid–Liquid Extractions)

Emulsions can form during liquid–liquid extraction when the difference in densities between the aqueous and organic layers is too small, preventing clear phase separation. These emulsions can be problematic, but several strategies can be employed to break them effectively. One common approach involves modifying the density of one or both layers:

- To increase the density of the organic layer, a heavier organic solvent such as carbon tetrachloride (CCl_4) can be added.
- To decrease the density of the aqueous layer, the addition of low-density solvents like pentane may help.
- Alternatively, "salting out" can be used: adding saturated solutions of sodium chloride (NaCl) or sodium sulfate (Na_2SO_4) increases the density and ionic strength of the aqueous layer. This reduces the solubility of organic compounds in water, thus promoting phase separation.

If modifying densities does not suffice, one or more of the following methods may be helpful:
- Add a few drops of silicone-based defoamer.
- Add a few drops of dilute acid (only if chemically permissible).
- Centrifuge the emulsion in a suitable tube to force separation.
- Filter the mixture by gravity or using a Büchner funnel connected to an aspirator or vacuum pump.
- Add a few drops of a detergent solution to disrupt interfacial tension.
- Allow the emulsion to stand undisturbed for an extended period.
- Place the emulsion in a freezer, which may enhance separation upon cooling.

19.5 Distillation

19.5.1 Introduction

Distillation is the process of heating a liquid to produce vapors that may be cooled, collected, and separated from the original liquid. It may also be described as the process of separating a mixture's components according to their boiling points. In both situations, the liquid is vaporized, then recondensed and collected in a container. A residue is the liquid that remains in the flask but does not evaporate. The resultant condensed vapor is known as the distillate. This procedure of purifying and separating liquids is based on volatility differences.

19.5.2 Simple Distillation

An experimental setup for simple distillation is shown in Figure 19.5. The glass equipment may be standard and require corks or may have ground-glass fitted joints.
    Please check the following for a correct setup:
- The distilling flask should be large enough to accommodate twice the volume of the liquid to be distilled.
- The thermometer bulb should be slightly below the side-arm opening of the flask to have the correct temperature reading.
- The glass-to-glass or glass-to-cork connections should be firm and tight.
- The flask, condenser, and receiver should be clamped independently in their proper relative positions on a steady base.
- The upper outlet for the cooling water exiting from the condenser should point upward to keep the condenser full of water.

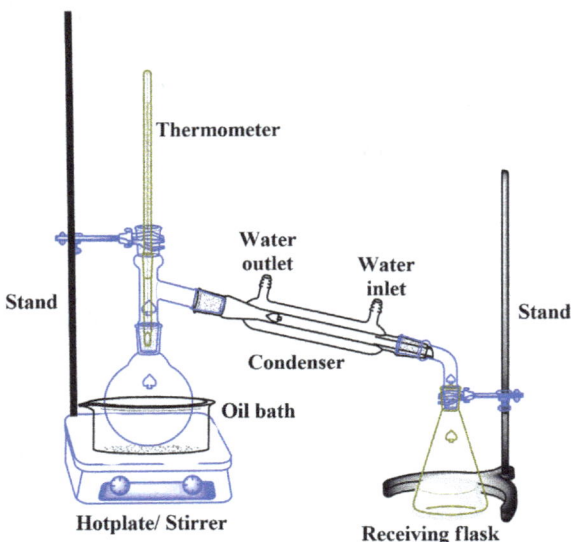

Figure 19.5: Schematic diagram for simple distillation setup.

The following procedure should be adhered to:
- Pour the liquid into the distilling flask with a funnel that extends below the side arm.
- Add a few boiling stones to prevent bumping.
- Insert the thermometer.
- Open the water valve for condenser cooling.
- Heat the distilling flask until boiling begins; adjust the heat input so that the rate of distillate is a steady two to three drops per second.
- Collect the distillate in the receiver.
- Continue distillation until only a small residue remains. Do not distill to dryness.

This type of distillation is used to test liquid purity or remove solvents. A pure liquid distills at a constant boiling point, with identical composition in the distillate and residue. Nonvolatile solutes remain in the residue, while the volatile component vaporizes and condenses. When a nonvolatile solute is present, the vapor (head) temperature matches that of the pure liquid, but the pot temperature rises due to reduced vapor pressure. As distillation continues, solute concentration increases, further raising the pot temperature.

When evaporating to recover solutes, avoid complete dryness to prevent decomposition. For separating two liquids of different volatilities, collect the distillate in fractions:
- The first fraction contains mostly the more volatile component.
- The last contains mostly the less volatile one.
- Intermediate fractions contain mixtures.

19.5.3 Azeotropic Distillation

Azeotropic mixtures boil at a constant temperature and retain the same composition in vapor and liquid, making them inseparable by simple distillation. For instance, ethanol cannot be purified beyond 95.5% by fractional distillation due to its azeotrope with water (bp 78.07 °C), which is slightly lower than pure ethanol (78.37 °C). To obtain absolute ethanol, the azeotrope is broken by distilling it with benzene, which forms a volatile azeotrope with water and removes it. A simple azeotropic distillation setup is presented in Figure 19.6.

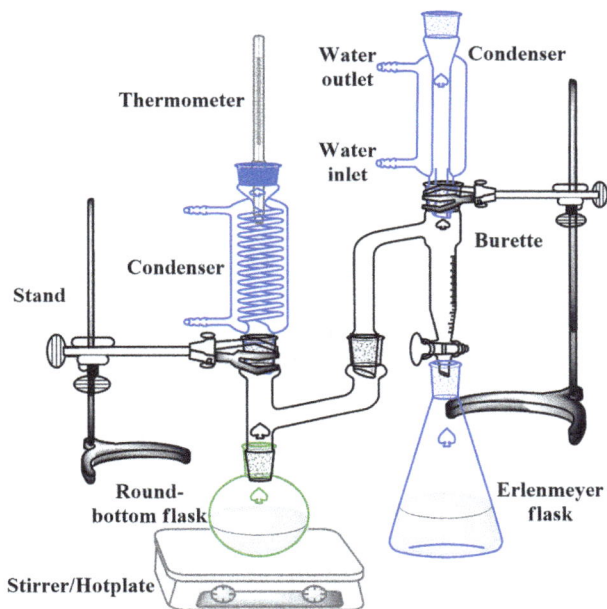

Figure 19.6: Schematic diagram for azeotropic distillation setup.

19.5.4 Fractional Distillation

Fractional distillation is a physical method of separating components in a mixture, relying on the differences in their boiling points. By monitoring the boiling points of the fractions collected, along with other physical properties, this technique can also aid in identifying the constituents of an unknown mixture.

In a two-component mixture, the substance with the lower boiling point will vaporize first and ascend the fractionating column. Over time, it will be replaced by the higher-boiling component. When the column functions efficiently, the lower-boiling substance is distilled at a constant temperature, followed by the higher-boiling component, which also emerges at a steady temperature.

A basic setup is illustrated in Figure 19.7. The fractionating column is designed to perform multiple theoretical distillations in a single continuous process, eliminating the need to collect and redistill individual fractions. The procedure for performing fractional distillation includes the following steps:

- Choose an appropriate type of fractionating column.
- Assemble the distillation apparatus.
- Open the inlet valve to allow cooling water to flow.
- Apply heat to the mixture.
- Ensure a large volume of condensate continuously returns through the column.
- Carry out the distillation slowly to maximize separation efficiency.

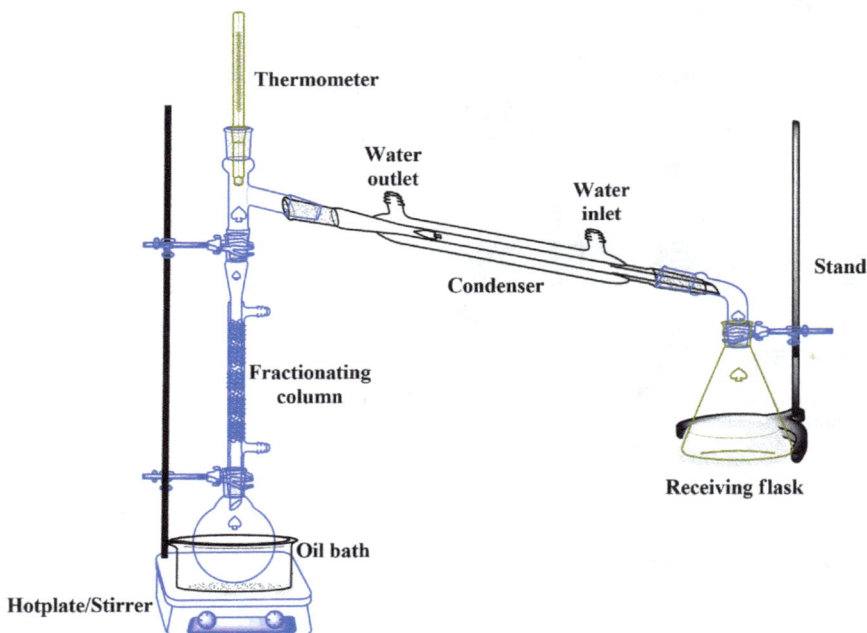

Figure 19.7: Schematic diagram for fractional distillation setup.

19.5.5 Vacuum Distillation

Many compounds are heat-sensitive and degrade before reaching boiling points, making atmospheric pressure ineffective for distillation. Vacuum distillation offers a solution by allowing distillation at a lower pressure, resulting in much lower boiling temperatures for compounds. Because a liquid's boiling point is proportional to the system's pressure, reducing pressure reduces it, and increasing pressure raises it.

The typical apparatus setup is illustrated in Figure 19.8. The general requirements for a vacuum distillation setup include:

- Efficient vacuum sources, such as water aspirators, can drop system pressure to that of water vapor. Mechanical oil-sealed vacuum pumps are also efficient; however, to prevent contamination, a cold trap containing dry ice or liquid nitrogen should be fitted.
- Safety traps protect the vacuum source and manometer against fluid contamination or backflow. The traps must be appropriately put in the system.
- A pressure gauge is essential to monitor system pressure. To prevent damage (e.g., manometer breakage), air should be introduced gradually when releasing the vacuum.
- A manostat (pressure regulator) helps maintain a constant pressure by automatically adjusting needle valves to either admit air or seal the system in response to vacuum fluctuations.
- A capillary air inlet or a spin bar is often used to promote even boiling and prevent bumping.
- Vacuum distillation flasks are specially designed to reduce contamination from frothing and splashing during boiling.
- Heating sources such as oil baths, water baths, or electric mantles are used to gently heat the liquid.
- Specialized distillation heads are required to collect fractions without disrupting the ongoing distillation process.

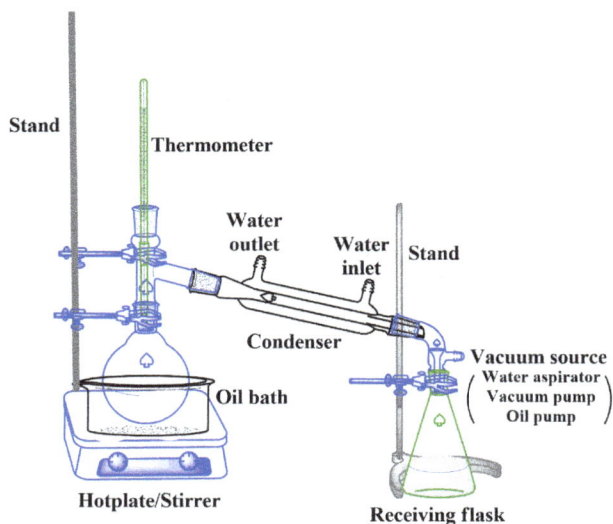

Figure 19.8: Schematic diagram for vacuum distillation setup.

19.5.6 Steam Distillation

Steam distillation is a technique used for the separation and purification of organic compounds through volatilization. It is particularly effective for substances that are either insoluble or only slightly soluble in water. When steam is introduced into a mixture of water and the target compound, the compound volatilizes and codistills with the water vapor. Upon condensation, the organic compound typically separates from the aqueous phase due to its low water solubility.

A key advantage of steam distillation is that many compounds distill at temperatures lower than their normal boiling points, often below 100 °C. This makes the method especially useful for high-boiling compounds that would otherwise decompose at their boiling points under atmospheric pressure.

The standard apparatus for steam distillation is depicted in Figure 19.9. The procedure may be carried out as follows:
-   Place the compound or mixture into a distillation flask along with a small amount of water. Connect the apparatus, ensuring that cooling water flows through the condenser. A Claisen flask may be used in place of a standard round-bottomed flask. The Claisen head helps prevent contamination of the distillate from splashing or spattering.
-   If direct steam is not available from a laboratory line, use an external steam generator to produce steam and direct it into the mixture. The steam inlet should reach below the surface of the liquid in the distillation flask.

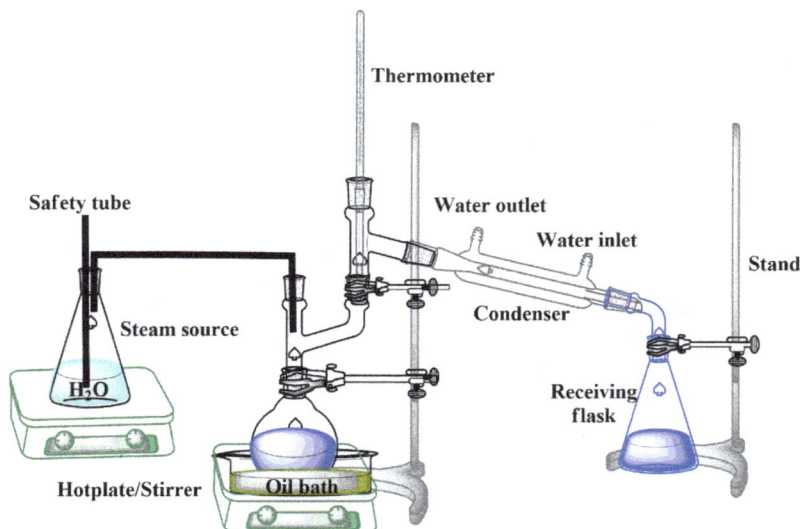

Figure 19.9: Schematic diagram for steam distillation setup.

- The flask may be heated gently with a burner to facilitate vaporization. If steam is supplied from a lab steam line, it is important to install a water trap in the line to remove any condensed water that could otherwise accumulate in the flask.
- Continue the steam flow until no more water-insoluble organic material is observed in the condensate, indicating the end of distillation.

19.6 Evaporation

Laboratory-scale evaporation (Figure 19.10) is a process of removing a solvent (typically a volatile liquid) from a solution by converting it into vapor, usually with the help of heat. In a chemistry lab, this technique is used to concentrate solutions or isolate dissolved compounds. Evaporation is done as follows:
- The solution is placed in the flask or beaker.
- Heat is applied to increase the temperature of the solvent.
- As the solvent reaches its boiling point, it vaporizes.
- If a condenser is used, the vapor travels through it and condenses back into a liquid.
- The remaining substance in the flask becomes more concentrated or is recovered as a solid/residue after complete evaporation.

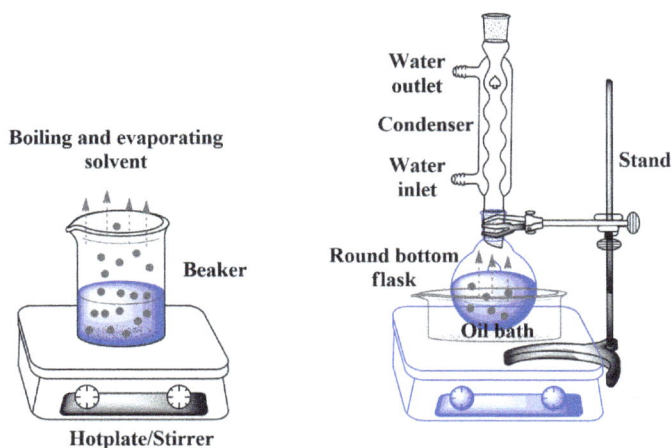

Figure 19.10: Simple evaporation setup.

19.7 Sublimation

Sublimation is the process by which certain solids transition directly into the vapor phase without passing through the liquid state. This occurs because the vapor pressure of solids increases with temperature, allowing some solids to evaporate directly under suitable conditions. A simplified representation of the sublimation setup is illustrated in Figure 19.11.

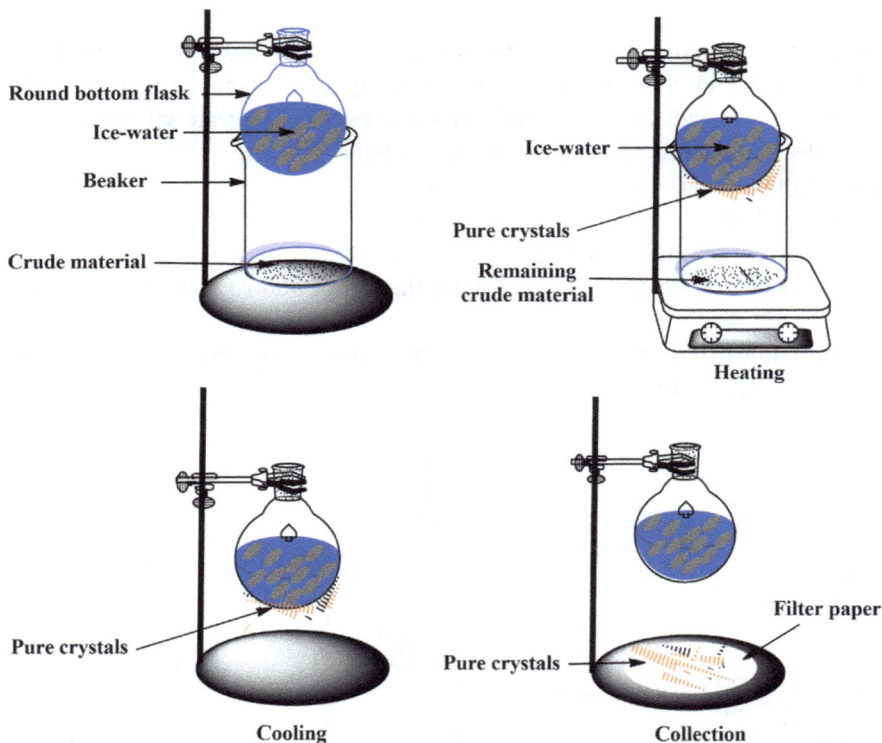

Figure 19.11: Simple sublimation setup.

This technique is commonly employed to purify solids when the desired compound has a significantly higher vapor pressure than the associated impurities.

19.7.1 Advantages of Sublimation as a Purification Method

– No solvent is required, making the process environmentally friendly.
– It is generally faster than crystallization.
– Effective in removing more volatile impurities from the target compound.

– Useful for separating nonvolatile or less volatile substances from more volatile components.

19.7.2 Limitation

Sublimation is less selective than crystallization, especially when the vapor pressures of the components are close in value, which may result in incomplete separation.

19.7.3 Typical Sublimation Procedure

– Sample Preparation. Place about 1 g of dry crude solid (e.g., naphthalene) in a clean, dry beaker.
– Sublimation Setup. Place a small round-bottomed flask containing ice over the beaker. Ensure the flask fits snugly but does not touch the sample (see Figure 19.11).
– Heating. Gently heat the beaker to provide a steady and uniform heat supply. The solid vaporizes and the vapors rise to contact the cold surface of the flask, where they condense and crystallize.
– Cooling. After sufficient sublimation, stop heating and allow the apparatus to cool down.
– Collection. Carefully collect the purified crystals from the surface of the flask using a dry, clean watch glass or filter paper.

19.8 Freeze-Drying

Freeze-drying or lyophilization (Figure 19.12) is a dehydration technique carried out under low temperature and vacuum. Water in the sample is first frozen and then removed as vapor through sublimation, bypassing the liquid phase. This process keeps the product frozen throughout, minimizing thermal degradation and preserving stability. Because of its ability to protect heat-sensitive materials and extend shelf life, freeze-drying is widely used for long-term storage. The four basic steps of this process are presented in Figure 19.13 and summarized in Table 19.6.

19.9 Column Chromatography

Column chromatography (Figure 19.14) is a technique for isolating and purifying components in a mixture according to their affinity. The column has the stationary phase, which is typically a solid such as silica gel or alumina, and the mixture to be separated is placed on top of the column, with a solvent (the mobile phase) used to trans-

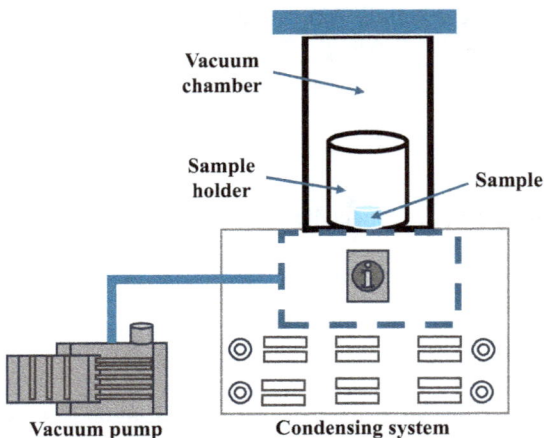

Figure 19.12: Schematic diagram for freeze-drying technique.

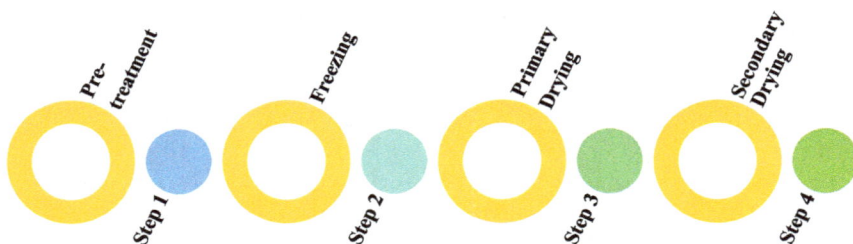

Figure 19.13: The four basic steps of the freeze-drying process.

Table 19.6: Basic steps of the freeze-drying (lyophilization) process.

Step	Description
Pretreatment	Preparation of the product before freezing. May include concentration, formulation with stabilizers, or adjustment of pH to enhance drying efficiency.
Freezing	The product is cooled below its eutectic or glass transition temperature so that all free water becomes solid ice.
Primary drying (sublimation)	Under vacuum, heat is applied to remove ice by sublimation. This is the longest phase and must be carefully controlled to prevent product collapse.
Secondary drying (desorption)	Removes unfrozen water molecules bound to the product matrix by increasing temperature under continued vacuum. Final moisture is typically <1–2%.

port the components through it. As the solvent flows down, the components move at various speeds based on how strongly they interact with the stationary phase, resulting in separation.

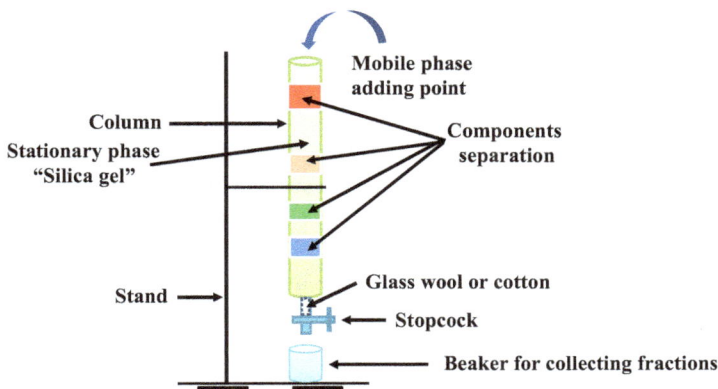

Figure 19.14: Schematic diagram for column chromatography setup.

19.9.1 Liquid Column Techniques

Column chromatography is one of the most widely used and versatile separation techniques in chemistry. The following guidelines can help improve its effectiveness:
- Use a column with a minimum length-to-diameter ratio of 20:1 for efficient separation.
- Estimate the correct amount of sorbent needed. A stationary phase (sorbent) to sample ratio of 20:1 to 100:1 (i.e., 20–100 g of sorbent per gram of sample) is frequently recommended for silica-based column chromatography to ensure effective separation. For more challenging or partition-based separations, chemists often advise increasing this ratio to 100:1 or more, sometimes up to 500:1, to improve resolution.
- When using a dry-packed column, add solvent immediately after packing to prevent the sorbent from drying out. Allow any heat produced during solvent addition to dissipate and let the column return to room temperature. Use of a cooling jacket can speed up the cooling process.
- For higher resolution, use slurry packing. Pour the sorbent suspended in an appropriate solvent into the column in small portions while gently tapping the column. This helps eliminate air bubbles and results in a tightly and uniformly packed column.
- Introduce the sample uniformly to avoid disturbing the packed bed. This can be done by mixing the sample with some sorbent and adding it as a slurry, or by applying the sample onto a filter disk placed on top of the column bed.

19.9.2 Sorbents for Column Chromatography

In column chromatography, the sorbent (stationary phase) is the key material responsible for the differential retention and separation of sample components. Unlike thin-layer chromatography (TLC), where binders are required to keep the sorbent adhered to a plate, no binder is necessary in column chromatography because the sorbent is physically retained by packing it into a vertical glass or plastic column.

The two most critical properties of a sorbent in column chromatography are:
- Particle Size. The average size of sorbent particles significantly affects both resolution and flow rate. Smaller particles provide more surface area for interactions between the analyte and the stationary phase, leading to better separation. However, small particles also cause higher resistance to solvent flow, requiring either gravity (hydrostatic pressure) or a low-pressure pump to move the mobile phase through the column.
- Particle Size Distribution. A narrow size distribution (where most particles are the same size) leads to more uniform packing of the column. This reduces channeling (uneven flow paths) and leads to sharper, better-resolved bands. In contrast, a wide size distribution can result in irregular packing, poor flow control, and reduced separation efficiency.

19.9.3 Solvent Mixtures for Use in Column Chromatography

The mobile phase in column chromatography is often a solvent mixture chosen to optimize the elution of components based on their polarity and affinity for the sorbent:
- Unlimited Combinations. There is no single best solvent for all separations. Due to the large number of available solvents and the ability to mix them in any ratio, the possibilities for creating solvent systems are virtually endless. This allows chemists to tailor the elution strength (also called eluent polarity) to specific separation challenges. Single solvents may be too strong (causing all compounds to elute together) or too weak (causing no elution). Solvent mixtures also allow fine-tuning of the eluting power, especially in gradient elution, where polarity is gradually increased during the run.
- Solvent System Selection. Solvents are often chosen based on the "like dissolves like" principle, polar solvents for polar analytes, and nonpolar solvents for nonpolar analytes. Mixtures are commonly formulated by experimentation, starting with known systems (e.g., hexane/ethyl acetate and dichloromethane/methanol) and adjusted based on TLC behavior or test column runs. Examples of common solvent pairs are presented in Table 19.7.

Table 19.7: Common solvent pairs in column chromatography.

Nonpolar to polar systems	Typical use
Hexane/ethyl acetate	General organic separations
Dichloromethane/methanol	Polar compound separations
Toluene/acetone	Aromatic and moderately polar samples
Chloroform/methanol	Polar lipids or glycosides

19.10 Ion-Exchange Chromatography

I-EC is a type of liquid chromatography used to separate mixtures of ions. This method takes advantage of charged molecules' attraction to oppositely charged sites on a stationary phase. Synthetic polymeric resins, frequently, polystyrene cross-linked with divinylbenzene (DVB), are commonly used in modern I-EC. These resins are produced into beads with either cationic or anionic exchange sites on their surfaces. Acidic functional groups like carboxylic acid ($-COOH$) or sulfonic acid ($-SO_3H$) are bonded to the polymer in cation-exchange resins. These groups can release a proton, leaving behind a negatively charged site that attracts positively charged ions (cations). Anion-exchange resins include basic groups such as primary and secondary amines ($-NH_2$, $-R_2N$). These groups form positively charged quaternary ammonium ions, allowing the resin to attract and bind negatively charged ions (anions). Both cationic and anionic exchange groups can be classed as strong or weak based on their degree of ionization at various pH levels. Table 19.8 depicts typical ion-exchange resin bottle labels and interpretations, whereas Table 19.9 provides an example resin label interpretation.

Table 19.8: Typical ion-exchange resin bottle labels and interpretations.

Label term	Example	Meaning/interpretation
Resin type	Strong cation, weak anion	Indicates the strength and type of exchanger "Strong" implies full ionization at all pH.
Functional group	$-SO_3^-$ (sulfonic acid), $-NH_3^+$ (amine)	Specifies the charged group responsible for ion exchange.
Matrix/base material	Styrene–divinylbenzene	Refers to the polymer backbone supporting the functional groups.
Ionic form	H^+, Na^+, Cl^-, OH^-	The form in which the resin is supplied. Must be converted to the appropriate form if needed.
Crosslinking (%)	8% DVB	Degree of crosslinking with divinylbenzene; affects porosity and swelling.

Table 19.8 (continued)

Label term	Example	Meaning/interpretation
Mesh size	16–50 mesh	Particle size range (smaller mesh = finer resin); affects flow and resolution.
Moisture content	45–55%	Indicates hydration level; useful for determining resin capacity and mass.
Total exchange capacity	1.8 meq/mL (wet)	Maximum amount of ions the resin can exchange per volume or weight.
Brand/trade name	Dowex 50WX8, Amberlite IR120	Commercial name; includes resin type and grade.

Table 19.9: Label interpretation of Amberlite IR120, H⁺ form, 8% cross-linked, 16–50 mesh, strong acid cation exchange resin.

Component	Interpretation
Amberlite IR120	Trade name (strong cation exchanger).
H⁺ form	Supplied in hydrogen form.
8% cross-linked	Moderate crosslinking, controls porosity and mechanical strength.
16–50 mesh	Moderate particle size suitable for typical column use.
Strong acid cation	Has sulfonic acid groups, ionized over a broad pH range.

Ion-exchange resins can be used in glass columns similar to the one shown in Figure 19.14 or in ion-exchange high-performance liquid chromatography systems equipped with ion-exchange packed columns, as illustrated in Figure 19.15.

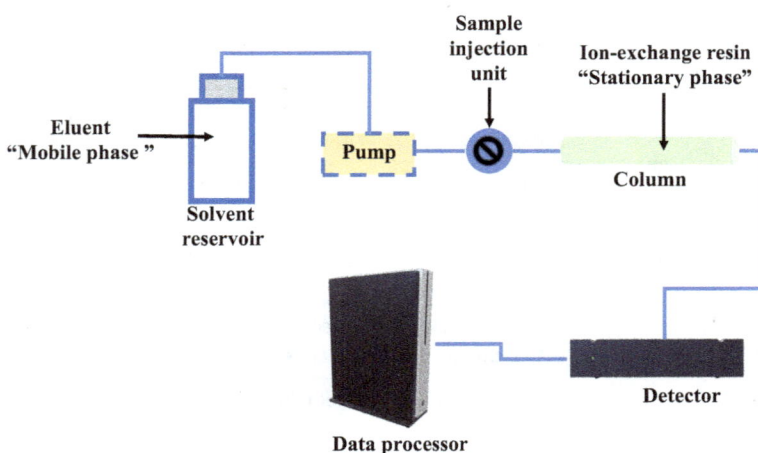

Figure 19.15: Ion-exchange high-performance liquid chromatography schematic diagram.

The following steps outline the conditioning procedure for ion-exchange resins in glass columns:
- Pretreatment Sequence. Fresh resin should be sequentially washed with 2 M HCl → water rinse → 2 M NaOH → water rinse, until eluate is neutral and salt-free. This removes contaminants and regenerates the exchange sites. "Analytical grade" resins typically come pretreated.
- Hydration. Soak the resin in water for ≥2 h. Highly cross-linked resins (e.g., 10% DVB) swell less and equilibrate faster than low-cross-linked ones (e.g., 2% DVB). Hydration, in general, fully swells resin to its operational form.
- Column controlled packing
- This prevents channeling and ensures uniform flow, and is done as follows:
  - Place a glass-wool plug at the bottom and half-fill with water.
  - Add the hydrated resin using a funnel without letting it dry.
  - Back-flush with water to remove air bubbles and let the resin settle.
  - Measure and adjust flow rate. Never allow the resin to dry to avoid channeling.
- Sample Loading. Use a separatory funnel for controlled sample application once the column is ready.

19.11 Gel Electrophoresis

Gel electrophoresis is an effective method for separating and purifying macromolecules, particularly proteins and nucleic acids. It utilizes an electric field to separate molecules in a gel matrix based on their size and charge. Table 19.10 offers an overview of electrophoretic applications and safety issues.

19.11.1 Principle of Gel Electrophoresis

Electrophoresis relies on the migration of charged particles in an electric field. Molecules move toward the electrode with opposite charge; negatively charged molecules (e.g., DNA) migrate toward the positive electrode (anode). The gel matrix acts as a molecular sieve, impeding the movement of larger molecules more than smaller ones. Key factors influencing migration:
- Size of the molecule (smaller molecules migrate faster)
- Charge (higher charge increases migration speed)
- Gel concentration (higher % gels have smaller pores)
- Buffer composition (affects ionic strength and pH)

19.11.2 Types of Gels

Gel type	Application	Typical sample
Agarose	Separation of nucleic acids	DNA, RNA
Polyacrylamide (PAGE)	High-resolution separation of proteins or small nucleic acids	Proteins, small DNA/RNA
SDS-PAGE	Denaturing gel for proteins	Linearized, negatively charged proteins

19.11.3 Buffer Systems

Buffers maintain pH and ionic strength to ensure consistent migration. Common buffers include:
- Tris-acetate-EDTA (TAE) for agarose gel electrophoresis of DNA
- Tris-borate-EDTA (TBE) for DNA requiring higher resolution
- Laemmli buffer in SDS-PAGE for protein analysis

19.11.4 Methodology

A typical gel electrophoresis (Figure 19.16) procedure involves:
- Preparation of Gel. Agarose or polyacrylamide dissolved in buffer and poured into a casting tray.
- Loading Samples. Mixed with loading dye and pipetted into wells.
- Running the Gel. Electric current is applied; molecules migrate based on size and charge.
- Staining and Visualization.
  - DNA: Stained with ethidium bromide, SYBR Green, or GelRed, and visualized under UV light.
  - Proteins: Stained with Coomassie Brilliant Blue or silver stain.

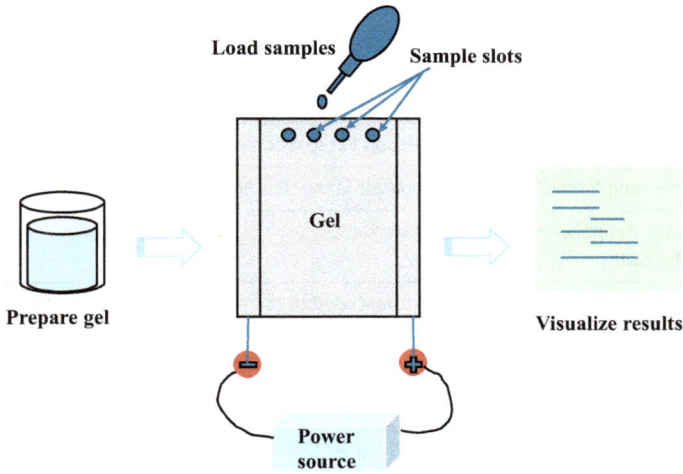

Figure 19.16: Gel electrophoresis schematic diagram.

Table 19.10: Applications and safety considerations in electrophoresis.

Category	Details
Applications	
Molecular biology	Analyzing restriction enzyme digestion, PCR products, and plasmid integrity
Genetics	DNA fingerprinting, genotyping, and mutation detection
Biochemistry	Protein purification, enzyme assays, and subunit analysis
Clinical diagnostics	Detecting hemoglobin variants, and immunoglobulin profiles
Safety and handling	
Chemical hazards	*** Ethidium bromide is widely used but the most hazardous, requiring special waste handling. SYBR Safe is a safer, commercially available alternative developed for improved lab safety. GelRed is considered even safer due to its inability to cross cell membranes and is nonmutagenic in the Ames test.
	Acrylamide (monomer) is neurotoxic; wear gloves and prepare gels in a fume hood.
UV exposure	UV transilluminators can damage eyes and skin; always use protective shields and eyewear.
Electrical safety	Electrophoresis units carry high voltage; ensure proper insulation and grounding to prevent accidents.

19.12 Questions and Answers

Questions	Answers
1. What is the main principle of distillation?	Separation based on differences in boiling points.
2. What is the role of a separating funnel?	To separate immiscible liquid layers.
3. Which technique separates components based on solubility and crystal growth?	Recrystallization.
4. What is an eluent in chromatography?	The solvent used to carry the sample through the stationary phase.
5. Which method separates compounds based on mass and volatility?	Gas chromatography (GC).
6. How does centrifugation separate mixtures?	By spinning samples at high speeds to separate by density.
7. What is the basis of separation in electrophoresis?	Charge and size of molecules under an electric field.
8. What is decantation?	Pouring off a liquid to separate it from a settled solid.
9. Which technique uses a resin to exchange ions in a solution?	Ion-exchange chromatography.
10. Why sublimation is used for separation?	To purify solids that can directly convert from solid to vapor.

Chapter 20
Physical Property Determination Techniques

20.1 Density

The density of any given substance is calculated by dividing the mass of the substance by the volume that it occupies. Density is expressed in grams per cubic centimeter (g/cm^3) or grams per milliliter (g/mL):

$$Density = \frac{Mass\ of\ the\ substance}{Volume\ of\ the\ substance}$$

20.1.1 Common Methods for Determining Density

20.1.1.1 Density Bottle Method
The density bottle and pycnometer methods are common laboratory procedures for determining the density of liquids and finely powdered particles. Although there are only small changes between the two procedures (Table 20.1), both use a calibrated glass vessel of known volume (Figure 20.1) to estimate a substance's mass, which is then used to calculate its density.

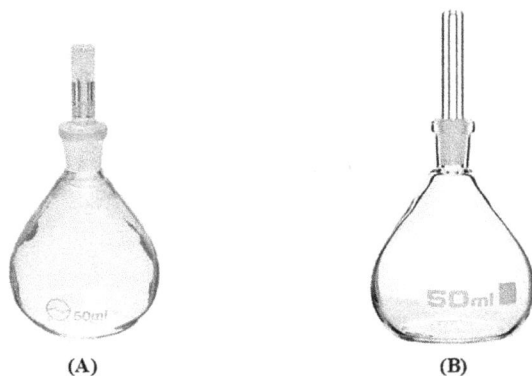

(A) (B)

Figure 20.1: Density bottle (A) and pycnometer (B).

The procedures for determining the density of liquids are outlined as follows:
- Ensure the density bottle, commonly 25 or 50 mL, is clean and dry before use.
- Weigh the empty bottle.
- Record the mass of the dry, empty bottle (M_1) using an analytical balance.
- Fill the bottle with the liquid.

https://doi.org/10.1515/9783112218105-020

Table 20.1: Key differences between pycnometer and density bottle methods.

Feature	Pycnometer method	Density bottle method
Apparatus	Uses a pycnometer, a small, precision-calibrated glass vessel with a ground-glass stopper having a fine capillary opening.	Uses a density bottle (also called a specific gravity bottle), usually larger and with a simple ground-glass stopper.
Volume accuracy	High accuracy, volume is precisely calibrated (often to 0.1 mL or better).	Less precise than a pycnometer; typically used for general-purpose density measurements.
Applications	Preferred for high-precision density measurements in analytical chemistry.	Common in teaching labs and routine industrial or field measurements.
Typical volume	Small, usually 10–50 mL.	Slightly larger, 25–100 mL or more.
Closure	Stopper with a narrow capillary bore allows excess liquid to escape, ensuring accurate volume.	Stopper may not have a capillary bore, requires careful handling to avoid trapped air or excess fluid.
Accuracy and reproducibility	More accurate and reproducible for precise work.	Good for routine use, but slightly less accurate due to stopper and volume variability.

- Carefully fill the bottle with the liquid, ensuring no air bubbles remain. Insert the stopper if present.
- Weigh the filled bottle.
- Record the mass of the bottle filled with liquid (M_2).
- Calculate the mass of the liquid:

$$\text{Mass of the liquid} = M_2 - M_1$$

- Determine the bottle volume.
- Use the volume V engraved or calibrated on the bottle.
- Calculate density:

$$\text{Density } (\rho) = \frac{M_2 - M_1}{v}$$

On the other hand, the procedure for determining the density of finely powdered solids is outlined as follows:
- Weigh the empty bottle (M_1).
- Add dry solid and weigh again (M_2).
- Add water to fill the bottle and weigh (M_3).
- Clean the bottle, fill with water only, and weigh (M_4).
- Calculate the density of the solid:

$$\text{Density } (\rho) = \frac{M_2 - M_1}{(M_4 - M_1) - (M_3 - M_2)}$$

20.1.1.2 Westphal Balance Method

The Westphal balance method (also known as the Mohr–Westphal balance) is a traditional method for determining the density or specific gravity (SG) of liquids (Figure 20.2). This method, based on Archimedes' principle, calculates the buoyant force acting on a glass float suspended in a liquid.

The approach assumes that the upward buoyant force exerted on a body submerged in a fluid equals the weight of the displaced fluid. By balancing this force with calibrated riders (small weights) on a precision arm, the liquid's SG (relative density) may be determined straight from the scale.

The apparatus required for determining the SG of a liquid using the Westphal balance method includes a sensitive Westphal balance, which features a calibrated scale for precise measurements. A glass float (also known as a plummet) is used; this is a standard-volume body suspended by a wire that is immersed in the liquid. A temperature-controlled vessel is essential to maintain a consistent temperature, typically 20 or 25 °C, ensuring accurate and reproducible readings. Additionally, calibrated riders, which are small precision weights, are used to balance the float on the beam and determine the liquid's SG. The procedure is outlined as follows:

- Fill the sample beaker with the liquid whose density is to be measured.
- Suspend the glass float in the liquid using the attached wire.
- Adjust the riders on the calibrated balance arm until the beam is level.
- Read the SG directly from the scale and rider positions.
- If required, convert to absolute density using:

Density (g/cm^3) = (specific gravity) × (density of water at measurement temperature)

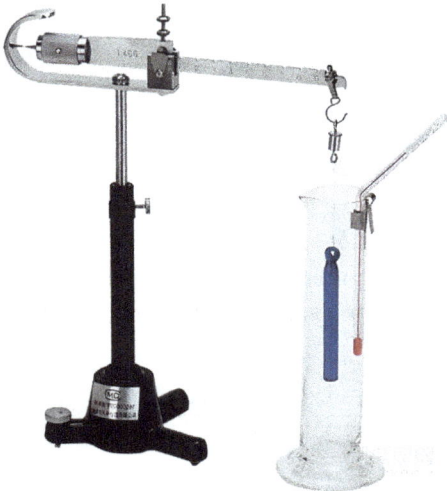

Figure 20.2: The Westphal balance. Image credit: https://www.lazada.com.ph/products/mohr-westphal-balance-liquid-gravity-i176875804.html.

20.1.1.3 Float Method

The float method (also known as the floating or flotation method) is a simple approach for estimating a solid's density by examining whether it sinks, floats, or remains suspended in a liquid of known density (Figure 20.3). This approach is especially beneficial for irregularly shaped objects and is based on the buoyancy concept proposed by Archimedes. When a solid is put in a liquid, solids sink when they have a higher density than liquids, and float when they have a lower density. If it suspends (does not sink or float), the solid and liquid densities are equivalent. This may be checked by altering the liquid's density. At the point where the solid stays suspended, the liquid's density equals that of the solid.

The procedure for this method is outlined as follows:

- Select the sample. Choose a small, clean, dry, solid sample.
- Prepare the liquid mixture. Use two miscible liquids with known densities (e.g., water and alcohol, or water and glycerin).
- Adjust liquid density. Gradually mix the liquids while testing whether the solid sinks, floats, or stays suspended.
- Determine matching point. When the solid remains suspended in the liquid mixture, the density of the liquid equals that of the solid.
- Measure liquid density. Use a hydrometer, pycnometer, or other method to determine the exact density of the liquid at the matching point.

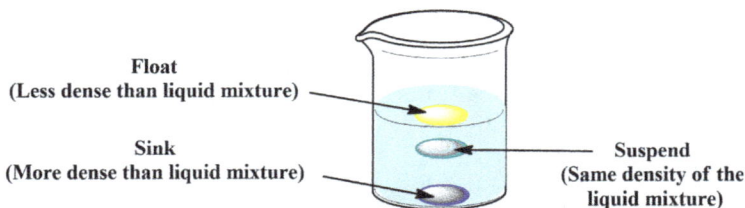

Figure 20.3: Schematic diagram for density determination by float method.

20.2 Specific Gravity

SG is the ratio of the density of a substance (solid) to the density of a reference substance, typically water for solids. It indicates whether a substance will sink or float in water (SG > 1 → sinks, SG < 1 → floats):

$$\text{Specific gravity (SG)} = \frac{\text{Density of the substance (solid)}}{\text{Density of water}}$$

20.2.1 Laboratory Techniques to Determine SG

There are several methods used in laboratories to measure SG, depending on the type and state of the material.

20.2.1.1 Hydrometer Method for SG of Liquids

The hydrometer technique is based on Archimedes' principle, which states that a floating item displaces an equivalent volume of liquid as its weight. The hydrometer descends to a depth determined by the liquid's density (or specific gravity). The density of the liquid determines how high the hydrometer floats. The less thick the liquid, the further it sinks. Hydrometers come precalibrated to read SG directly. The hydrometer (Figure 20.4) range varies with the predicted SG. The measurement requires a graduated cylinder or hydrometer jar, a thermometer, and a liquid sample. The technique includes the following steps:

- Clean the hydrometer and measuring jar with distilled water and dry them.
- Fill the hydrometer jar about ¾ full of the liquid sample.
- Measure the temperature of the liquid (typically standardized at 20 or 25 °C).
- Gently lower the hydrometer into the liquid. Avoid splashing or allowing it to touch the sides.
- Let the hydrometer stabilize and float freely.
- Read the scale at the bottom of the meniscus (unless otherwise indicated by the manufacturer).
- Record the SG value and the temperature.

Hydrometers are manufactured with different ranges to suit various applications. The SG range of a hydrometer depends on the type of liquid being measured. A hydrometer that reads midscale in the expected SG range is advisable for the best accuracy. Table 20.2 summarizes the typical hydrometer SG ranges by applications.

20.2.1.2 Pycnometer Method for SG of Liquids

A pycnometer is a compact, calibrated glass container typically equipped with a ground glass stopper featuring a fine capillary opening, which allows air or excess liquid to escape and ensures accurate volume control. The pycnometer method relies on precise mass measurements of a known volume. By weighing the pycnometer when empty (W_{empty}), then when filled with the test sample (W_{sample}), and finally when filled with a reference liquid, commonly water (W_{water}), the SG of the sample can be determined using the following approach:

$$\text{Specific gravity (SG)} = \frac{\text{Mass of sample}}{\text{Mass of equal volume of water}}$$

Figure 20.4: Specific gravity hydrometer. Image credit: https://thermcoproducts.com/products/specific-gravity-baume-hydrometers/.

Table 20.2: Typical hydrometer SG ranges by applications.

Application	SG range	Hydrometer type
Water-based solutions	0.990–1.050	General-purpose hydrometer
Alcohol solutions	0.790–1.000	Alcoholmeter or spirit hydrometer
Battery acid (lead-acid)	1.100–1.300	Battery hydrometer
Milk	1.025–1.035	Lactometer
Brines and salt solutions	1.000–1.250	Salinometer
Heavy oils and acids	1.000–2.000	Heavy liquid hydrometer
Light hydrocarbons	0.600–0.950	Petroleum hydrometer

Or, more formally:

$$SG = \frac{W_{sample} - W_{empty}}{W_{water} - W_{empty}}$$

where W_{sample} is the mass of pycnometer + sample, W_{water} is the mass of pycnometer + water, and W_{empty} is the mass of empty pycnometer.

The procedure involves the following steps:
–   Clean and dry the pycnometer thoroughly.
–   Weigh the empty pycnometer using an analytical balance. Record as W_{empty}.
–   Fill with the liquid sample at the desired temperature (usually 20 or 25 °C).
–   Insert the stopper carefully, ensuring the capillary allows excess liquid/air to escape.

- Wipe and dry the outside, then weigh the filled pycnometer. Record as W_{sample}.
- Empty and rinse the pycnometer.
- Refill with distilled water at the same temperature, insert the stopper, and weigh again. Record as W_{water}.
- Calculate SG using the formula above.

20.2.1.3 Archimedes' Method (Water Displacement Method) for SG of Solids

This method is best for irregularly shaped solids (nonporous and insoluble in water). The procedure involves the following steps:

- Weigh the solid in air (W_1).
- Suspend the solid in water using a thread or wire and weigh it while submerged (W_2).
- Calculate SG using this formula:

$$SG = \frac{W_1}{W_1 - W_2}$$

$W_1 - W_2$ gives the apparent loss in weight, which equals the weight of water displaced = volume of the solid in cm^3.

Example:
- Weight in air = 50.0 g
- Weight in water = 30.0 g
- SG = 50.0/(50.0 − 30.0) = 2.50

20.2.1.4 Pycnometer Method for SG of Fine Solids or Powders

This method is used to determine the SG of fine, dry, solids or powders by comparing the mass of a known volume of the material to the mass of an equal volume of water. The procedure involves weighing the pycnometer in various states and calculating the volume displaced by the powder using water as the displacement fluid. The procedure involves the following steps:

- Clean and dry the pycnometer thoroughly. Weigh it empty and record the mass as W_1.
- Add a known amount of dry, fine solid (about one-third to one-half of the pycnometer volume). Weigh again and record as W_2.
- Add distilled water to the pycnometer containing the solid. Fill completely, ensuring no air bubbles remain. Insert the stopper, wipe the outside, and weigh. Record this mass as W_3.
- Clean, refill, and weigh the pycnometer with water only (at the same temperature as in step 3). Record this as W_4.
- Use this formula for calculation:

$$SG_{solid} = \frac{W_2 - W_1}{(W_4 - W_1) - (W_3 - W_2)}$$

Example:
- *Mass of empty pycnometer (W₁) = 25.000 g*
- Mass of pycnometer + dry powder (W_2) = 30.000 g
- Mass of pycnometer + powder + water (W_3) = 48.000 g
- Mass of pycnometer + water only (W_4) = 50.000 g

$$SG_{solid} = \frac{30 - 25}{(50 - 25) - (48 - 30)} = 0.714$$

20.2.1.5 Volume Displacement in Graduated Cylinder Method

Best for regular or slightly irregular solids that do not dissolve in water. This method is based on Archimedes' principle, which states that a body submerged in a fluid displaces a volume of fluid equal to its own volume. By measuring the volume of water displaced and the mass of the solid, the density and SG can be calculated:

$$Density = \frac{Mass\ of\ solid\ (g)}{Volume\ displaced\ (cm^3)}$$

$$Specific\ gravity\ (SG) = \frac{Density\ of\ solid}{Density\ of\ water\ (usually\ 1.000\ g/cm^3)}$$

The procedure involves the following steps:
- Partially fill a graduated cylinder with water and record the volume (V_1)
- Gently place the solid into the cylinder and record new volume (V_2)
- Measure solid's mass using a balance (M).
- Volume of solid = $V_2 - V_1$
- Calculate density = $M/(V_2 - V_1)$
- Calculate SG = Density/1.000 g/cm^3 (since water density ≈1 g/cm^3)

Example:
- Mass of solid = 12.50 g
- Initial water volume = 25.0 mL
- Final water volume = 30.2 mL
- Volume displaced = 30.2 − 25.0 = 5.2 mL
- Density = 12.50/5.2 = *2.40 g/cm^3*
- SG = 2.40/1.000 = *2.40*

20.3 Melting Point

The temperature at which a crystalline solid transforms from a solid to a liquid under atmospheric pressure is known as its melting point. Chemical purity is shown by the narrow melting temperatures of pure organic molecules, which are typically within 1 °C. Melting point analysis is the recommended technique for determining the purity of solid compounds since impurities raise the range while decreasing the melting point.

The temperature range between the first softening or collapse of the crystals and the full transformation of the material into a liquid is known as the melting point range. Melting point measurement is a useful and instructive technique for identification and purity evaluation since the majority of organic compounds melt between 50 and 300 °C.

Mixtures of these compounds typically melt over a larger range and at a lower temperature, even if several compounds may have the same melting point. Melting point depression is a phenomenon that is very helpful for verifying a compound's identification.

For example, if a pure reference material and an unknown sample are chemically equivalent, combining them yields the same melting point. The ingredients are different if the combination has a reduced or expanded melting range. A mixture's melting point may occasionally be as much as 50 °C lower than the pure compound's.

General guidelines for interpreting melting point data include:
- A melting range of 0.1–0.3 °C typically indicates high purity.
- A 1 °C range is acceptable for most analytical and laboratory purposes.
- Commercial-grade materials often have a melting range of 2–3 °C.
- A wide range (e.g., >5 °C) strongly suggests the presence of impurities or degradation products.

Melting points are usually measured using capillary melting point apparatuses or modern digital devices, both of which are essential tools in organic synthesis, compound characterization, and quality control workflows.

20.3.1 Capillary Tube Method

The capillary technique is commonly used to determine melting points. A tiny amount of the finely ground material is placed into a capillary tube with thin walls, which is then attached to a thermometer's stem (Figure 20.5). To precisely observe the melting point, this assembly is submerged in a hot oil bath that is gradually heated.

The capillary should be placed in the bath early and closely monitored as the temperature gets closer to the desired range for compounds that decompose a few degrees below their melting point. For compounds that sublime, both ends of the capillary tube should be sealed to prevent material loss. The following is the detailed process:

- Prepare a capillary tube:
  - Use a commercially available melting point capillary tube or fabricate one by drawing out 12 mm soft-glass tubing over a flame.
- Load the sample into the capillary:
  - Scrape a small amount of the finely powdered compound into a pile on clean paper or glass.
  - Insert the open end of the capillary tube into the pile to collect the sample.
  - Tap the closed end of the capillary on a hard surface or gently tap the tube with a file to pack the powder tightly at the bottom.
  - Ensure the sample height is about 3–4 mm.
- Attach to thermometer and immerse:
  - Secure the capillary tube alongside the thermometer using a small rubber band or capillary holder.
  - Immerse the assembly into a hot-oil bath (e.g., silicone or mineral oil) with the compound level submerged below the oil surface.
- Initial heating (rapid phase):
  - Heat the oil bath rapidly until the temperature reaches approximately 5 °C below the expected melting point.
  - Stir the bath continuously to maintain uniform temperature distribution.
- Final heating (slow phase):
  - Reduce the heating rate to approximately 1 °C/min.
  - Continue stirring and observe the sample closely.
- Observe and record:
  - Note the temperature at which the first sign of melting (softening or collapse of crystals) appears.
  - Record the melting point range ending when the compound is completely liquefied.
- Dispose of used capillary:
  - Once the determination is complete, discard the used capillary tube in the proper glass waste container.

20.3.2 Electric Melting Point Apparatus

A melting point apparatus (Figure 20.6) is a specialized scientific equipment for determining the temperature at which a solid substance becomes liquid. This equipment is essential in many industrial and scientific applications where the melting point must be understood. The apparatus works on the assumption that a pure chemical experiences a modest temperature change from solid to a liquid form. The melting point is the temperature at which this change occurs.

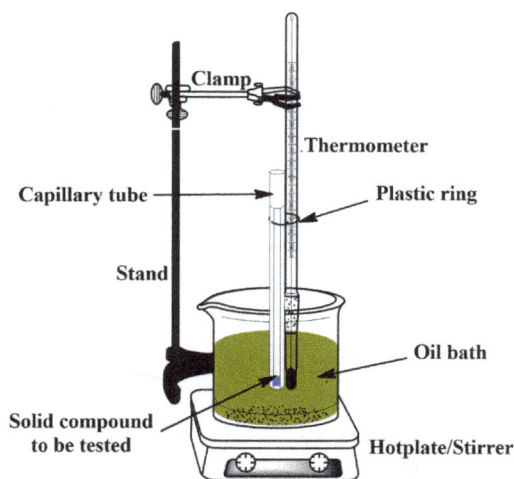

Figure 20.5: Melting point measurement by capillary tube method.

The process works as follows:
- Sample Preparation. A little sample of the substance whose melting point to be measured is placed in a small capillary tube.
- Insertion of Capillary Tube. The capillary tube containing the sample is inserted into the melting point device. The tube is frequently positioned such that the sample may be seen through a magnifying lens or a digital camera.
- Heating. The device includes a heating element, which is often an electrical coil or a hot plate. The heating rate may be adjusted to provide a progressive increase in temperature. This gradual heating enables exact measurement of the melting point.
- Observation and Monitoring. As the apparatus heats up, the substance begins to warm. At a certain temperature, it starts to melt, shifting from the solid phase to the liquid phase. This transition is noticeable through visible changes, such as the appearance of a liquid layer, alterations in color, or the disappearance of crystalline features.
- Recording the Melting Point. The temperature at which the substance fully changes from solid to liquid is recorded as its melting point. Pure substances typically exhibit a sharp melting point, meaning the transition occurs within a very narrow temperature range. In contrast, mixtures or impure samples generally show a lower and broader melting range.
- Cooling and Cleaning. Once the melting point is determined, the apparatus is allowed to cool. Any residual sample in the capillary tube should then be removed, and the tube cleaned in preparation for further use.
- Repetition. To improve reliability, the procedure can be repeated several times with fresh samples of the substance.

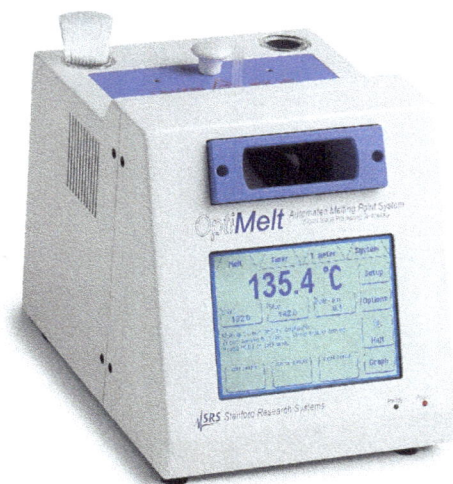

Figure 20.6: Electric melting point apparatus. Image credit: https://www.thinksrs.com/prod ucts/ mpa100.html.

20.4 Boiling Point

The boiling point of a liquid is the temperature at which the pressure of the saturated vapor is equal to the pressure of the atmosphere under which the liquid boils. The boiling point, as a physical property, is useful for the identification and assessment of the purity of organic compounds. Boiling point data is especially useful for distinguishing between liquid compounds with similar structures and for evaluating the success of purification steps, such as distillation. Normally, boiling points are determined at standard atmospheric pressure: 760 mmHg (torr) or 1 atm. The boiling point of a liquid is sensitive to atmospheric pressure and varies with it. As the pressure decreases, the boiling point will drop; at approximately normal pressure, it will drop about 0.5 °C for each 10 mm drop in pressure. At much lower pressures, close to 10 mmHg, the temperature will drop about 10 °C when the pressure is halved.

There are several methods for determining boiling points. These are summarized as follows.

20.4.1 Capillary Tube Method

The capillary tube method is a simple and fast microscale technique that uses a capillary tube placed inverted in the liquid sample. The boiling point is determined based on the formation and cessation of bubbles as the sample is heated and cooled. The boiling point of a liquid in microquantities can be determined using the capillary tube method, as illustrated in Figure 20.7. The procedure is outlined below:

- Place two to three drops of the liquid sample in a small, clean, test tube.
- Seal one end of a capillary tube by heating and melting the glass.
- Invert the sealed capillary and place it into the test tube containing the liquid (sealed end up).
- Attach the test tube to the thermometer using a rubber band.
- Immerse the setup in a hot oil bath, heated slowly.
- Heat gently until a continuous stream of bubbles is observed coming from the capillary.
- Remove the heat and allow the liquid to cool.
- Record the temperature at which the bubbling stops; this is the boiling point.

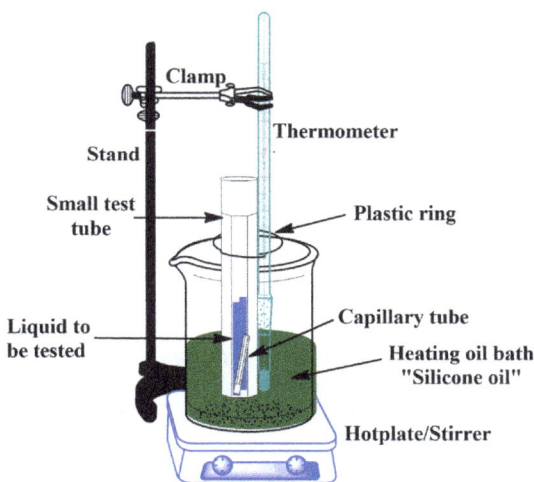

Figure 20.7: Boiling point determination by capillary tube method.

20.4.2 Thiele Tube Method

The Thiele tube (Figure 20.8) is a specialized glass device designed with a side arm that facilitates uniform heating through convection currents, enabling more precise boiling point measurements than conventional oil baths. For a comparative overview, refer to Table 20.3. This method is also suitable for determining melting points.

For performing this experiment, the following items are required:
- Thiele tube
- Capillary tube (sealed at one end)
- Thin-walled boiling point tube (for holding the sample)
- Thermometer
- Mineral or silicone oil
- Heat source (such as a microburner or small flame)

The procedure is straightforward and can be carried out as follows:
- Place two to three drops of the liquid into the boiling point tube.
- Invert a sealed capillary tube into the sample (sealed end up).
- Attach the boiling point tube to the thermometer.
- Clamp the thermometer so the sample is submerged in the oil inside the Thiele tube.
- Heat the side arm of the Thiele tube gently with a flame. The design creates convection currents for even heating.
- Observe bubble formation from the capillary.
- When bubbling is continuous, remove the heat.
- Record the temperature when bubbling stops as the boiling point.

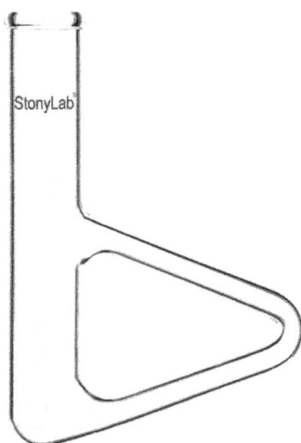

Figure 20.8: Thiele tube. Image credit: https://www.a.ubuy.com. kw/en/product/ICOO0D7DU-stonylab-thiele-melting-point-tube-bo rosilicate-glass-measuring-point-lab-melting-point-tube-triangle-shape-thiele-tube-150-mm.

Table 20.3: Comparison of capillary tube with Thiele tube.

Feature	Capillary tube method	Thiele tube method
Sample volume	Very small (~0.1 mL)	Small (~0.1–0.2 mL)
Heating control	Manual oil bath, variable stirring	Natural convection via side arm
Accuracy	Moderate	Higher due to even heating
Equipment requirement	Simple, general labware	Requires a Thiele tube
Temperature observation	Using attached thermometer	Using thermometer immersed in oil
Applications	Quick assessments, teaching labs	Precise determination, research labs

20.5 Viscosity

Viscosity measures a fluid's resistance to flow. It displays the internal friction that happens when adjacent layers of fluid pass past each other. Simply, viscosity refers to how "thick" or "sticky" a liquid is. Viscosity is an important physical property that may be measured using a number of established techniques. The following are three often utilized approaches in academic and industrial laboratories.

20.5.1 Ostwald Viscometer

A classical glass capillary device (Figure 20.9) that measures the time it takes for a liquid to flow between two marks under gravity. It is widely used for Newtonian fluids (Table 20.4) and dilute solutions. This procedure for measuring viscosity involve the following steps:
– Clean the viscometer properly to remove any residues or contaminants.
– Fill the reservoir bulb with distilled water using a pipette.
– Allow the viscometer to attain thermal equilibrium in a constant temperature water bath (typically 25 °C).
– To raise water beyond the upper time mark in the capillary arm, use mild suction (e.g., pipette or rubber tube). This is frequently accomplished by flipping the viscometer for more control.
– Measure the flow time. Using a stopwatch, check how long it takes for the meniscus to pass between the two calibration markings. Repeat the measurement at least three times to ensure accuracy and determine the average time.
– Drain and thoroughly dry the viscometer.
– Fill the viscometer with the test liquid using a pipette, using the same volume and process as for water.
– Repeat the timing process and keep the liquid at a consistent temperature (e.g., 25 °C).
– To calculate the sample's relative viscosity, use the formula below:

$$\eta_{rel} \frac{t_{sample}}{t_{water}}$$

where η_{rel} is the relative viscosity, t_{sample} is the average flow time for the test liquid, and t_{water} is the average flow time for water (standard).

The absolute viscosity of a liquid (η) can be determined by comparing its flow time and density to those of a reference liquid (usually water) using the formula:

$$\eta = \eta_0 \frac{t.\rho}{t_0.\rho_0}$$

where η (Greek letter eta) is the viscosity of the test liquid, η_0 is the viscosity of the reference liquid (e.g., water at 25 °C: 0.890 cP), t is the flow time of the test liquid, t_0 is the flow time of the reference liquid (water), ρ is the density of the test liquid, and ρ_0 is the density of the reference liquid.

If the test liquid and reference liquid have very similar densities, the equation can often be simplified by omitting the density ratio:

$$\eta = \eta_0 \frac{t}{t_0}$$

Example
- *Flow time of sample = 125 s*
- *Flow time of water = 100 s*
- *Density of sample = 1.05 g/cm^3*
- *Density of water = 0.998 g/cm^3*
- *Viscosity of water at 25 °C = 0.890 cP*
- *$\eta = 0.890 \cdot 125 \cdot 1.05 / 100 \cdot 0.998 = 1.17cP$*

Table 20.4: Newtonian fluids.

Fluid type	Definition	Common examples
Newtonian fluids	Fluids with constant viscosity regardless of shear rate	– Water – Ethanol – Benzene – Glycerol – Mineral oil

Figure 20.9: Ostwald viscometer. Image credit: https://www.eiscolabs.com/products/ch0386.

20.5.2 Ubbelohde Viscometer

The Ubbelohde viscometer (Figure 20.10) operates on the same basic principle as the Ostwald viscometer, measuring the time it takes a liquid to flow through a capillary under gravity. However, it differs in that it maintains a constant driving pressure due to its design, which features a closed upper arm open to the atmosphere through a side arm. This allows for more accurate viscosity determination, particularly for dilute solutions or when comparing liquids with different densities.

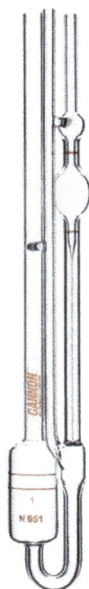

Figure 20.10: Ubbelohde viscometer. Image credit: https://cannoninstrument.com/manual-glass-viscometers/ubbelohde-viscometer.html.

To carry out this viscosity experiment, the following items are required
- Ubbelohde viscometer (suspended-level type)
- Thermostatic water bath (typically set at 25 °C)
- Rubber tubing or suction bulb
- Stopwatch
- Pipette
- Test liquid and reference liquid (e.g., distilled water)

The procedure is straightforward and can be carried out as follows:
- Clean and dry the Ubbelohde viscometer thoroughly before use.
- Introduce a measured volume (typically 10–15 mL) of the reference liquid (e.g., water) into the viscometer using a pipette.
- Place the viscometer in a constant temperature water bath and allow sufficient time for thermal equilibration at 25 °C.
- Apply gentle suction to the side arm (not the open reservoir tube) to draw the liquid above the upper timing mark.
- Start the stopwatch as the meniscus passes the upper mark and stop it when it reaches the lower mark. Repeat for consistency.
- Drain and dry the viscometer.
- Repeat the process with the test liquid under identical temperature conditions.
- Record the flow times and calculate viscosity. The kinematic viscosity (ν) is determined by:

$$v = C \cdot t$$

where v is the kinematic viscosity in mm^2/s (or cSt), C is the viscometer constant (depends on instrument and provided by the manufacturer), and t is the flow time in seconds.

To get dynamic viscosity (μ or η)

$$\eta = v \cdot \rho$$

where μ (sometimes "η" is used) is the dynamic viscosity (mPa $\cdot$ s or cP) and ρ is the density of the liquid (g/cm^3).

20.5.3 Saybolt Viscometer

The Saybolt viscometer (Figure 20.11) is an empirical device for determining the viscosity of petroleum-based liquids such as fuels, lubricants, and other refined petroleum products. It measures how long it takes for a certain volume of fluid to flow through a defined aperture at a given temperature. In a typical Saybolt viscometer, a specified quantity of liquid is heated. When the temperature reaches a certain point (usually 100 or 210 °F), the hole is opened. The time taken (in seconds) for 60 mL of liquid to pass through the orifice and into a receiving flask is recorded. The result is reported as Saybolt Universal Seconds (SUS) for low- to medium-viscosity oils or Saybolt Furol Seconds (SFS) for more viscous oils (using a larger orifice and higher temperature).

20.5.4 Brookfield Viscometer

The Brookfield viscometer (Figure 20.12) is a common rotational viscometer that calculates fluid apparent viscosity by measuring the torque required to rotate a spindle at a constant speed while submerged in the sample. It is particularly useful for investigating non-Newtonian fluids (Table 20.5), whose viscosity changes with shear rate. The general steps for utilizing the viscometer are:
- Select an appropriate spindle based on the sample viscosity range.
- Place the sample in a clean container, ensuring it covers the spindle's immersion mark.
- Set the temperature, usually with a circulating water bath if required.
- Attach the spindle and lower it slowly into the sample to avoid air bubbles.
- Start the instrument at a selected Revolution Per Minute (RPM) (shear rate).
- Allow the equilibrium, then record the viscosity value displayed.
- Repeat at various shear rates for non-Newtonian fluids to observe flow behavior.

Figure 20.11: Saybolt viscometer. Image credit: https://controls-group.com/product/saybolt-visc ometers/.

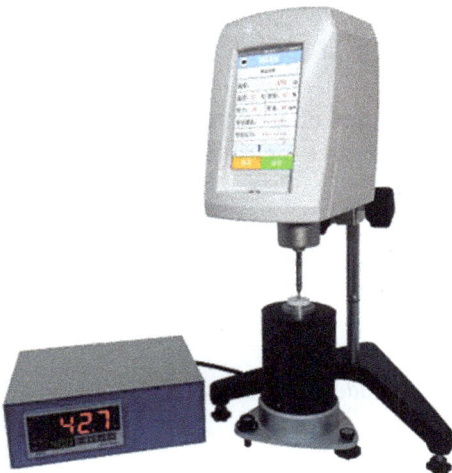

Figure 20.12: Brookfield viscometer. Image credit: https://www.kaycanlab.com/product/brookfield- viscometer.

Table 20.5: Types of non-Newtonian fluids with examples.

Type of non-Newtonian fluid	Flow behavior	Common examples
Pseudoplastic (shear-thinning)	Viscosity decreases with increasing shear rate	Paints, ketchup, blood, yogurt, and polymer solutions
Dilatant (shear-thickening)	Viscosity increases with increasing shear rate	Cornstarch in water (oobleck), wet sand, and some slurries
Bingham plastic	Requires a yield stress to flow, then behaves like a Newtonian fluid	Toothpaste, mayonnaise, mud, and printing ink
Thixotropic	Viscosity decreases over time under constant shear	Paint, clay suspensions, and ketchup
Rheopectic	Viscosity increases over time under constant shear	Some printer inks, gypsum pastes, and cream formulations

20.6 Hydrogen Ion Concentration (pH)

20.6.1 Indicators

Hydronium ion (or hydrogen ion) concentration can be estimated using acid–base indicators, complex organic compounds (Figure 20.13) that act as weak acids or bases. These indicators exhibit different colors in their molecular and ionic forms, with color changes occurring rapidly within a pH range of about 2 units. The observed color reflects the solution's pH. To use an indicator:
- Select a transition range that includes the required pH (see Table 20.6 for alternatives and preparations, and Table 20.7 for acid–base indicators for various titration methods).
- Add a few drops into the test solution.
- Compare the produced color to the indicator's recognized range.
- Use this comparison to estimate and record the pH.

Titration example:
Titrating acetic acid (CH₃COOH) with sodium hydroxide (NaOH):
- Acetic acid = weak acid
- NaOH = strong base
- Equivalence point = 8.7
- Choose an indicator whose transition range includes 8.7. Phenolphthalein (8.2–10.0) is the best choice.

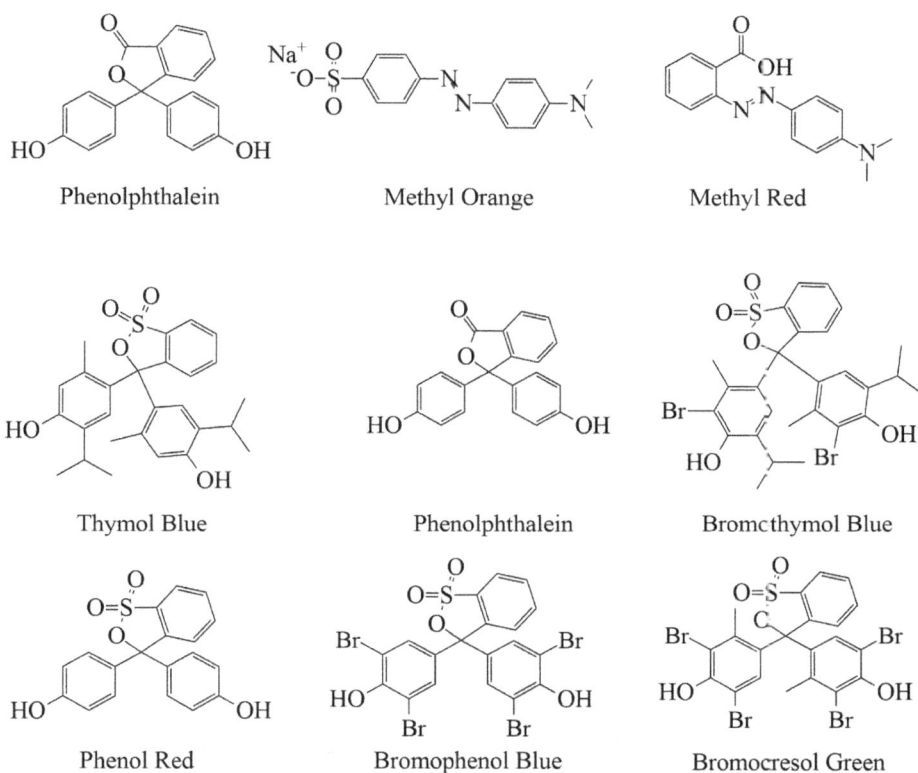

Phenolphthalein   Methyl Orange   Methyl Red

Thymol Blue   Phenolphthalein   Bromcthymol Blue

Phenol Red   Bromophenol Blue   Bromocresol Green

Figure 20.13: Chemical structures of indicators.

Table 20.6: Common acid–base indicators, properties, preparation, and applications.

Indicator	Preparation (0.1% w/v)	Acid color	Base color	pH range	Typical use
Methyl orange	Dissolve in water or ethanol	Red	Yellow	3.1–4.4	Strong acid versus weak base titrations
Methyl red	Dissolve in ethanol	Red	Yellow	4.4–6.2	Strong acid versus weak base
Bromothymol blue	Dissolve in ethanol	Yellow	Blue	6.0–7.6	Strong acid versus strong base
Phenolphthalein	Dissolve in ethanol	Colorless	Pink	8.2–10.0	Weak acid versus strong base

Table 20.6 (continued)

Indicator	Preparation (0.1% w/v)	Acid color	Base color	pH range	Typical use
Thymol blue	Dissolve in ethanol	Red → Yellow (1.2–2.8) Yellow → Blue (8.0–9.6)	Blue	Acid/base indicator in wide pH range	Two distinct transitions (diprotic acid)
Bromocresol green	Dissolve in ethanol	Yellow	Blue	3.8–5.4	Biochemical buffers and titrations
Bromophenol blue	Dissolve in ethanol	Yellow	Purple	3.0–4.6	Gel electrophoresis tracking dye
Litmus	Extract in water, use on paper	Red	Blue	4.5–8.3	Simple pH testing
Phenol red	Dissolve in ethanol	Yellow	Red	6.8–8.4	Enzyme assays, aquaria, and pools
Red cabbage extract	Boil chopped cabbage in water and filter before use	Red	Green–Blue	2–11	Educational pH tests

Table 20.7: Selected acid–base indicators for titration types.

Titration type	Equivalence point pH	Recommended indicator	Indicator pH range	Notes
Strong acid + strong base	7	Bromothymol blue	6–8	Sharp color change at neutral pH
Weak acid + strong base	8–9	Phenolphthalein	8–10	Transition near endpoint
Strong acid + weak base	4–6	Methyl orange	3–5	Suited for lower pH
Weak acid + weak base	6–8 (gradual)	Use pH meter		

So when the pH of the solution reaches 8.7, phenolphthalein will sharply change from colorless to pink, indicating the endpoint of the titration.

20.6.2 pH Test Paper

pH test paper (Figure 20.14) is a quick and inexpensive tool used to estimate the acidity or alkalinity of a solution. It consists of paper strips impregnated with one or more acid–base indicators that change color in response to the hydrogen ion concentration of a solution. Comparison of pH paper types and their characteristics is shown in Table 20.8.

Figure 20.14: pH test paper. Image credit: https://www.grainger.ca/en/product/PH-PAPER/p/WWG3UUT4.

The pH test paper works on the principle that acid–base indicators exhibit a distinct color change over specific pH ranges. When the treated paper comes into contact with a liquid sample, the color of the indicator changes based on the pH of the solution. It is used as follows:
- Dip or touch the test paper to a small drop of the solution to be tested (avoid immersing it).
- Observe the color change on the strip immediately (within 5–10 s).
- Compare the resulting color to a standard color chart provided with the strips.
- Estimate the pH by matching the observed color to the chart.

Table 20.8: Comparison of pH paper types and their characteristics.

Type	pH range	Primary use	Advantages	Limitations	Typical applications
Litmus paper	No specific range; indicates acid/base	Quick acid/base detection	– Fast and easy to use – Inexpensive – No calibration needed	– No exact pH value – Not suitable for weak acids/bases	Educational use, quick classroom or lab demos
Universal pH paper	Broad range (typically 1–14)	General pH testing across full scale	– Covers wide pH range – Easy color comparison	– ±1 pH unit accuracy – Affected by sample color/turbidity	Soil/water testing, food, textile, and cosmetic industries
Narrow-range pH strips	Limited range (e.g., 4.0–5.5 or 7.0–8.0)	Precise measurement within specific pH range	– Higher resolution (0.2–0.5 pH units) – More accurate in defined intervals	– Only useful within specified range – Still subjective color interpretation	Buffer prep, biological systems, titration endpoints, and sensitive environmental tests

20.6.3 pH Meter

A pH meter (Figure 20.15) uses hydrogen ion concentration (pH) to detect whether a liquid is acidic or alkaline. It typically comprises a glass electrode and a reference electrode coupled to a digital meter that shows the pH level.

20.6.3.1 Procedure for pH measurement
– Warm Up. Switch to STANDBY and let the instrument warm up for around 30 min.
– Electrode Preparation. Remove the electrode from its storage solution. Rinse with distilled water and gently dry with a lint-free cloth.
– Standardization. Calibrate the meter using a buffer solution that is close to the predicted sample pH (e.g., pH 4.00, 7.00, or 10.00).

20.6.3.2 Measurement
– Place the beaker with the test solution under the electrode.
– Lower the electrode into the solution and adjust the temperature compensator to match the solution temperature.
– Switch the selector to pH and record the reading.

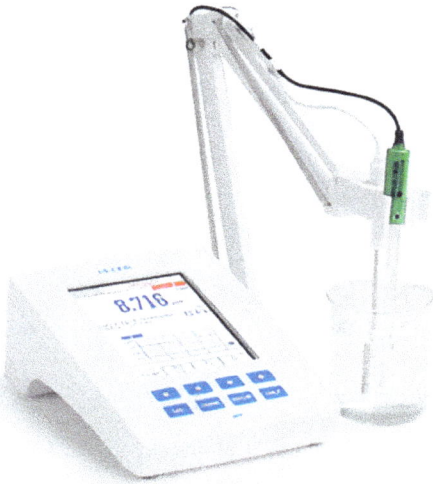

Figure 20.15: pH meter. Image credit: https://www.ddbiolab.com/product/0D-24-41?language=en.

20.6.3.3 Postmeasurement
– Switch to STANDBY.
– Rinse the electrode with distilled water.
– Store it in distilled water or a recommended storage solution.

Note: Never allow electrodes to dry out. Always keep the pH meter connected to power when in regular use to prevent humidity effects and extend component life. When not in use for long periods, disconnect the device and store it properly.

20.6.3.4 Calibration and Buffer Selection
– Daily calibration is essential using certified buffer solutions.
– Choose a buffer near the sample's expected pH for optimal accuracy:
    – pH 4.00 for acidic solutions.
    – pH 7.00 for neutral.
    – pH 9.00–10.00 for alkaline samples.

Ensure buffer and sample temperatures are similar (within ±10 °C) to reduce measurement errors.

20.7 Refractive Index

The refractive index (denoted as n) is a fundamental physical property that describes how light propagates through a medium. It is defined as the ratio of the speed of light in vacuum to the speed of light in the medium:

$$n = \frac{c}{v}$$

where n is the refractive index, c is the speed of light in vacuum (3.00×10^8 m/s), and v is the speed of light in the substance.

When light passes from one medium to another (e.g., air to glass or air to diamond), it bends or refracts. The refractive index (see refractive indices of common materials in Table 20.9) indicates how much the light will bend:
- Higher "n" → Slower light, greater bending.
- Lower "n" → Faster light, less bending.

Example 1: Water
Refractive index of water: $n = 1.333$
Speed of light in vacuum: $c = 3.00 \times 10^8$ m/s
Find v, the speed of light in water:
$v = c/n = 3.00 \times 10^8/1.333 = 2.25 \times 10^8$ m/s

Example 2: Glass (crown glass)
Refractive index $n = 1.52$
Speed of light in vacuum $c = 3.00 \times 10^8$ m/s
Find v, the speed of light in glass: $v = c/n = 3.00 \times 10^8/1.52 = 1.97 \times 10^8$ m/s

Example 3: Diamond
Refractive index $n = 2.417$
Speed of light in vacuum $c = 3.00 \times 10^8$ m/s
Find the speed of light in diamond: $v = c/n = 3.00 \times 10^8/2.417 = 1.24 \times 10^8$ m/s

Example 4: Ethanol
Refractive index $n = 1.361$
Speed of light in vacuum $c = 3.00 \times 10^8$ m/s
Find the speed of light in ethanol: $v = c/n = 3.00 \times 10^8/1.361 = 2.204 \times 10^8$ m/s

Table 20.9: Refractive indices of common materials (at 20–25 °C, $\lambda = 589$ nm).

Material	Refractive index (n)	Notes
Air (STP)	1.00029	Standard temperature and pressure
Water (25 °C)	1.333	Pure distilled water
Ethanol	1.361	100% anhydrous ethanol
Glycerol	1.473	Pure, at 20 °C
Glass (crown or optical glass)	1.52	Common optical crown glass
Benzene	1.501	Pure liquid benzene
Olive oil	1.47	May vary with composition
Quartz (crystalline)	1.544–1.553	Birefringent: ordinary versus extraordinary
Sapphire (Al_2O_3)	1.76	Isotropic approximation
Diamond	2.417	Very high optical density

20.7.1 Laboratory Determination

The common instruments used for determining refractive index are:
- The Abbe refractometer (Figure 20.16) is the most commonly used in chemistry labs
- Immersion refractometer
- Digital refractometer
- Pulfrich refractometer (high precision)

20.7.2 Typical Procedure with Abbe Refractometer

- Calibrate with distilled water or a known standard.
- Place a drop of the sample on the prism surface.
- Close the cover to spread the sample evenly.
- Use illumination and adjustment knobs to bring the boundary line into view.
- Read the refractive index directly from the scale, usually at 20 °C using the D-line of sodium (589 nm).

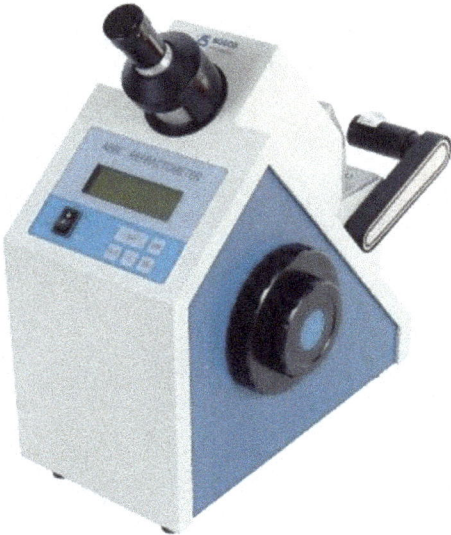

Figure 20.16: Abbe refractometer. Image credit: https://www.boeco.com/boeco-abbe-refractometer/boeco-digital-abbe-refractometer&sk=136.

20.8 Optical Rotation

Optical rotation describes the ability of some chemicals, known as optically active compounds, to rotate the plane-polarized light as it travels through them. This effect arises when light interacts with chiral compounds with no mirror symmetry. Rotation can be clockwise (dextrorotatory "+") or counterclockwise (levorotatory "–") depending on solution concentration, light path length, and chemical composition. The relationship can be described by the following formula:

$$[\alpha] = \frac{\alpha}{l.c}$$

where [a] is the specific rotation (a characteristic of the substance), a is the observed rotation in degrees (in degrees "°"), l is the path length of the sample tube or cell (in decimeters "dm"), and c is the concentration of the solution (in g/mL).

Example
A 10 cm (0.1 dm) polarimeter tube is filled with a solution of glucose. The observed rotation is +5.20°, and the solution concentration is 0.1 g/mL. Calculate the specific rotation:

$$[a] = a/l \cdot c = 5.20/0.1 \times 0.1 = 5.20/0.01 = +520°$$

Optical rotation is measured using an instrument called a polarimeter, and the technique is known as polarimetry. The polarimeter (Figure 20.17) consists of:
– Light source (e.g., sodium lamp at 589 nm)
– Polarizer: produces plane-polarized light
– Sample tube: holds the optically active substance
– Analyzer: measures the angle of rotation

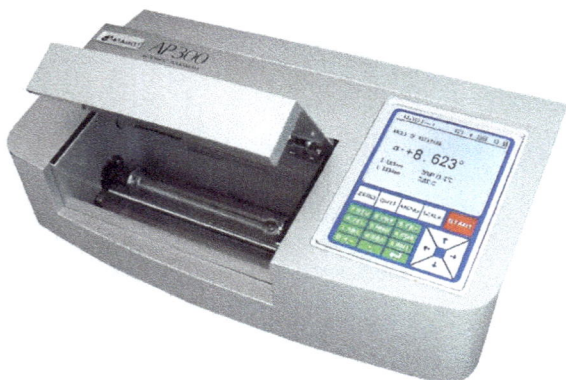

Figure 20.17: Polarimeter. Image credit: https://www.buch-holm.com/products/polarimeters/polarimeter-ap-100-type-ap-10-0-6227906.

The steps for operating a polarimeter are as follows:
- Calibrate the polarimeter using solvent (e.g., water or ethanol).
- Fill the sample tube with the test solution (ensure no air bubbles).
- Place the tube in the polarimeter and record the angle of rotation.
- Use the observed rotation and the formula to calculate specific rotation.
- Compare with literature values for identification.

The key parameters affecting optical rotation, along with examples of optically active compounds and polarimetry applications, are summarized in Tables 20.10, 20.11, and 20.12.

Table 20.10: Parameters affecting optical rotation.

Parameter	Effect on optical rotation	Explanation
Concentration	Rotation increases proportionally with concentration	More chiral molecules increase interaction with polarized light
Path length	Longer path increases observed rotation	Light travels through more of the sample, increasing cumulative rotation
Wavelength of light	Rotation varies with wavelength (optical rotatory dispersion)	Measured at standard 589 nm (sodium D-line), unless stated otherwise
Temperature	Rotation may increase or decrease	Changes in temperature affect solute solubility and molecular orientation
Solvent	May increase or decrease rotation	Solvent–solute interactions influence optical activity
Chemical nature	Each compound has a unique specific rotation $[\alpha]$	Depends on stereochemistry and chiral centers in the molecular structure

Table 20.11: Specific rotations of optically active compounds.

Compound	Specific rotation $[\alpha]$	Optical activity	Notes
D-Glucose	+52.7°	Dextrorotatory	Naturally occurring sugar; chiral
Lactic acid	+14.8°/ −14.8°	Depends on isomer	L-(+)-Lactic acid in muscle and dairy; D-(−)-form synthetic
(R)-Limonene	+115.5°	Dextrorotatory	Found in orange peel oil
(S)-Limonene	−115.5°	Levorotatory	Found in lemon oil
Tartaric acid	+12° (for L(+)-form)	Dextrorotatory	Naturally present in grapes
L-Alanine	+14.5°	Dextrorotatory	Essential amino acid; optically active L-isomer
(−)-Nicotine	−167.4°	Levorotatory	Natural alkaloid from tobacco; optically active

Table 20.12: Polarimetry applications.

Field	Use of polarimetry
Pharmaceuticals	Identification and purity of chiral drugs
Food industry	Sugar analysis (sucrose concentration in juices or syrups)
Essential oils	Authenticity and quality control
Organic chemistry	Stereoisomer differentiation and enantiomeric excess
Biochemistry	Study of amino acids, carbohydrates, and proteins

20.9 Surface Tension

Surface tension is a physical property of liquids caused by the cohesive forces between molecules at the liquid–air interface. Molecules in the bulk of a liquid are surrounded by neighboring molecules and subjected to cohesive forces from all sides, resulting in a net force of zero. However, because there are no continuous molecules above them, molecules on the surface experience a net inward force. This imbalance stresses the surface, causing it to behave like a stretched elastic membrane and driving the liquid to decrease its surface area. Surface tension is defined quantitatively as the force per unit length that acts on a liquid's surface to withstand external forces while decreasing surface area. Mathematically, it is expressed as:

$$\text{Surface tension}(\gamma) = \frac{\text{Force } (F)}{\text{Length } (L)}$$

Surface tension units, influencing factors, and measurement techniques are summarized in Table 20.13.

Table 20.13: Surface tension units, examples, influencing factors, and measurement techniques.

Aspect	Details
Units	– SI: N/m (Newtons per meter) – CGS: dyn/cm – 1 N/m = 1,000 dyn/cm (or = 1,000 mN/m) – 1 mN/m = 0.001 N/m – 1 mN/m = 1 dyn/cm
Common examples	– Water: ~72.8 mN/m = 0.0728 N/m = 72.8 dyn/cm – Ethanol: ~22.3 mN/m = 0.0223 N/m = 22.3 dyn/cm – Benzene: ~28.9 mN/m = 0.0289 N/m = 28.9 dyn/cm – Mercury: ~485 mN/m = 0.485 N/m = 485 dyn/cm
Factors affecting	– Temperature: ↑Temp → ↓Surface tension – Impurities/surfactants: ↓Surface tension – Intermolecular forces: ↑IMF → ↑Tension

Table 20.13 (continued)

Aspect	Details
Measurement methods	– Capillary rise method: Uses Jurin's law to relate rise height to surface tension. – Drop weight/volume method: Uses Tate's law. – Du Noüy ring method: Measures force to detach a ring from liquid surface. – Wilhelmy plate method: Measures force on a thin plate. – Pendant drop method: Uses shape of hanging drop for calculation. – Bubble pressure method: Based on pressure in air bubble through liquid.

The surface tension or the interfacial tension of liquids is measured with analytical instrument called tensiometer (Figure 20.18).

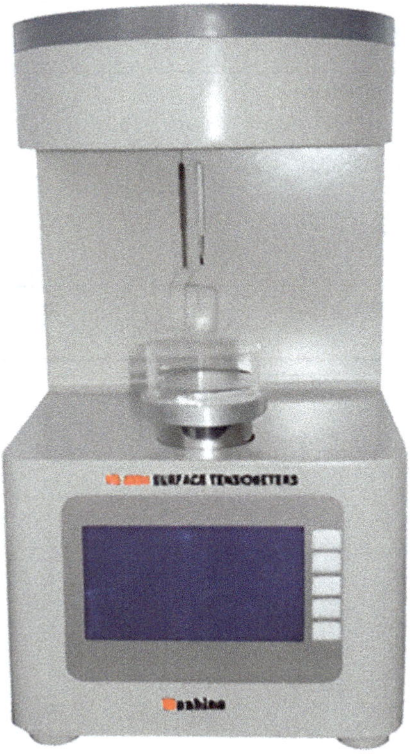

Figure 20.18: Tensiometer. Image credit: https://we shinelectric.en.made-in-china.com/product/ YmXRTNKSnWVb/China-ASTM-d971-Surface-Tension-Measure-Transformer-Oil-Tension-Meter-Surface-Tensi ometer.html.

20.10 Questions and Answers

Questions	Answers
1. What is the primary use of a pycnometer?	To determine the density of liquids and fine powders.
2. Which method uses a calibrated bottle to measure liquid density?	Density bottle method.
3. What physical property does a Westphal balance measure?	Density of liquids.
4. What principle does the float method rely on?	Archimedes' principle of buoyancy.
5. Which technique is used to determine boiling points of liquids?	Distillation or simple boiling point apparatus.
6. What property is measured using a melting point apparatus?	Melting point of solids.
7. How is refractive index commonly determined in the lab?	Using a refractometer.
8. What physical property is measured using a viscometer?	Viscosity of liquids.
9. Which technique is used to determine the optical rotation of compounds?	Polarimetry.
10. What instrument is used to measure the surface tension of a liquid?	Tensiometer.

Chapter 21
Laboratory Glassware: Usage, Cleaning, Drying, Measuring, and Transfer Techniques

21.1 Glassware Items

Glassware used in chemical labs is typically constructed of various types of glass, while specific items may be manufactured of polymers for specialized use. Laboratory-grade glassware is normally made of borosilicate glass, which is extremely resistant to heat stress and chemical damage. Pyrex, Quickfit, SVL, and MBL are well-known makers of durable borosilicate glassware used in academic and industrial labs. There are several varieties of glass used in laboratory and industrial applications, each with its unique composition and properties:

- Soda Lime Glass (Sometimes Called Soda Glass). It consists of silicon dioxide (SiO_2), sodium carbonate (Na_2CO_3), and calcium carbonate ($CaCO_3$). It is inexpensive but has less heat and is chemical-resistant. Frequently used in disposable glasses.
- Lead Glass or Crystal. Silica is created by combining lead oxide. It has a high refractive index and clarity, but because of its limited chemical resistance, it is more commonly utilized in decorative glassware than in scientific applications.
- Potash-Lime Glass. It is made up of silica, potassium carbonate (K_2CO_3), and calcium carbonate. It has a greater melting point than soda-lime glass and is more chemically stable.
- Borosilicate Glass (Pyrex). Its main components are silica, boric oxide (B_2O_3), sodium oxide, and aluminum oxide. This combination is highly heat and chemical-resistant, making it ideal for most scientific applications.
- Special Glasses. Some glass formulations may incorporate barium oxide, cerium oxide, or other chemicals that alter optical or mechanical properties for specific scientific uses.

A selection of common laboratory glassware items is listed in Table 21.1, including beakers, flasks, pipettes, burettes, and condensers, among others.

https://doi.org/10.1515/9783112218105-021

Table 21.1: Selected laboratory glassware.

Name	Purpose	Picture
Reduction adapter	Useful for the connection of dissimilar-sized ground glass joints	
Receiver adapter plain bend	Useful for the connection of condenser to receiving flask	
Three-way adapter or still head adapter	Used in distillation assemblies for connecting flasks to condenser	
Beakers	Useful as a reaction container or to hold liquid or solid samples	
Reagent bottle	Used for mixing and storing liquids, reagents, and analytical standards	
Density bottles	Used in density measurement. These bottles have exact volume engraved on each bottle	
Dropping bottle	Used to transfer small amounts or drops of certain reagents	

Table 21.1 (continued)

Name	Purpose	Picture
Burette	Graduated glass tube with a tap at one end for delivering known volumes of a liquid, especially in titrations	
Condenser	Suitable for either reflux or distillation	
Graduated cylinder	Used for measuring and transferring liquids	
Desiccator	Used for drying solids and preserving moisture-sensitive items	
Crystallizing dish	Used as heating and cooling baths	
Mortar and pestle	Used for crushing solids into powders for easier handling	

Table 21.1 (continued)

Name	Purpose	Picture
Filtration system	Used for filtration	
Conical flask	Used for collecting and transferring liquids	
Büchner flask	Used for vacuum filtration	
Round-bottomed flask	Used in as distilling flasks and receiving flasks for the distillate and also to contain chemical reactions especially for reflux setups and laboratory-scale synthesis	
Three-necked round-bottomed flask	Used to connect three components for complex distillation or reaction	
Volumetric flask	Used for precise dilutions and preparation of standard solutions	

Table 21.1 (continued)

Name	Purpose	Picture
Stem funnel	Used for gravity filtration	
Sintered glass disc Büchner funnel	Used for suction filtration	
Fractionating column	Used in fractional distillation	
Dropping funnel	Used for adding reagents and liquids drop-wise	
Pressure equalizing funnel	Used for adding liquids into vessels under vacuum	

Table 21.1 (continued)

Name	Purpose	Picture
Separatory funnel	Used for liquid–liquid extraction	
Joint clips	Used to prevent a joint from separating during a reaction process	
Graduated pipette	Used to accurately measure and transfer a volume of liquid from one container to another	
Pipette filler	Used to release air, draw liquid into the pipette, and accurately release liquid	
Test tube rack	Used to hold upright multiple test tubes at the same time	

Table 21.1 (continued)

Name	Purpose	Picture
Clamp and ring stand	Used to support other pieces of equipment and glassware such as burettes, columns, and flasks	
Stirrer bars or magnetic stir bars	Used to stir liquids in any container	
Stirring glass rods	Used for mixing liquids, or solids and liquids	
Stoppers	Used to seal containers	
Stopcock	Used to control the flow of a liquid	
Thermometer	Used to measure the boiling point of liquids during chemistry experiments	
Thermometer adapter	Allows the use of a standard laboratory thermometer in any chemistry reaction	

Table 21.1 (continued)

Name	Purpose	Picture
Centrifuge tube	Used to contain liquids during centrifugation	
Test tube	Used to handle chemicals, especially for qualitative experiments	
Tweezer	Used for grasping objects too small to be easily handled with the human fingers	
Watch glass	Used to evaporate liquids and cover beakers during sample preparation	
Weighing scoop	Used on weighing scales	
Scoopula	Used to transfer solids	
Spatula	Used to mix, spread, and lift solids	

21.2 Cleaning Laboratory Glassware

21.2.1 General Rules

- Clean your glassware immediately after use, as it is much easier to remove debris before it dries and hardens.
- Be careful when cleaning glassware, especially heavy flasks and long, thin columns.

– Rinse the glassware carefully with water to remove any soap or detergent residues to prevent any possible contamination.

21.2.2 Cleaning Volumetric Glassware

– Always rinse volumetric glassware equipment three times with distilled water after you have emptied and drained it. This prevents solutions from drying on the glassware, causing difficulty in cleaning.
– Dry volumetric glassware at room temperature, never in a hot oven. Expansion and contraction may change the calibration.
– The glass surfaces should be wetted evenly. Spotting is caused by grease and dirt. Grease can be removed by rinsing and scrubbing with a hot detergent solution followed by adequate distilled water rinses. Dirt can be removed by filling or rinsing with a dichromate cleaning solution. Allow standing for several hours, if necessary. Follow with multiple distilled water rinses.

21.2.2.1 Pipettes
Pipettes can be cleaned with a hot detergent solution or a cleaning solution. Pull the bulb into enough liquid to fill it to about one-third of its capacity. Holding it almost horizontally, carefully rotate the pipette to cover all the inner surfaces. Drain upside down and rinse thoroughly with distilled water. Check for water leaks and repeat the cleaning cycle as often as needed.

21.2.2.2 Burettes
Clean the tube thoroughly with detergent and a long brush. If water breaks do not disappear after washing, clamp the burette upside down by lowering its end into a glass of washing solution. Connect the hose from the burette tip to the vacuum line. Gently draw the cleaning solution into the burette before reaching the stopcock.

Leave it on for 10–15 min, then rinse it well with distilled water. After some use, the lubricating oil on the stopcock tends to harden; small grease particles can break off and flow to the edges of the burettes and clog them. Degreasing can be done with a thin, flexible wire. Finally, wash off any remaining particles with water.

21.2.3 Cleaning Glassware Soiled with Stubborn Films and Residues

When you cannot completely clean glassware by scrubbing it with a detergent solution, more drastic cleaning methods must be used.

21.2.3.1 Dichromate-Sulfuric Acid Cleaning Solution

Use great caution when handling and preparing this cleaning solution. Avoid skin and clothing contact.

After dissolving 92 g of sodium dichromate ($Na_2Cr_2O_7 \cdot 2H_2O$) in 458 mL of water, carefully add 800 mL of concentrated H_2SO_4 while stirring. The flask's contents will get very hot and solidify into a red mass. Add just enough sulfuric acid to bring the bulk into the solution. Before moving the solution to a different glass bottle, let it cool.

Before using this solution, thoroughly clean the glassware with soap and giving it a gentle water rinse and then add a tiny bit of the chromate solution to the glassware, making sure the solution runs down the whole glass surface. Refill the stock bottle with the solution. Once the glass surface looks clean, empty the glassware and rinse with tap water and then distilled water. You can keep using this solution until it takes on the green color of the chromium(III) ion. It should be disposed of after this.

21.2.3.2 Diluted Nitric Acid Cleaning Solution

Residues inside bottles and flasks can be removed with dilute nitric acid, followed by multiple rinses with distilled water.

21.2.3.3 Aqua Regia Cleaning Solution

The aqua regia is composed of one-part concentrated HNO_3 and three parts concentrated HCl. Although this cleaning solution is quite powerful, it is also very caustic and unsafe. Use the hood with extreme caution.

21.2.3.4 Alcoholic Potassium Hydroxide or Sodium Hydroxide Cleaning Solution

Add about 1 L of 95% ethanol to 120 mL of H_2O containing 120 g NaOH or 105 g KOH.

"This is a very good cleaning solution. Avoid prolonged contact with ground-glass joints on inter-joint glassware because the solution will etch glassware and damage will result. This solution is excellent for removing carbonaceous materials."

21.2.3.5 Trisodium Phosphate Cleaning Solution

Add 57 g Na_3PO_4 and 25.5 g sodium oleate to 470 mL H_2O. "This solution is good for removing carbon residue. Soak glassware for a short time in the solution and then brush vigorously to remove the incrustations."

21.2.3.6 Nochromix Cleaning Solution

This is a commercial oxidizer solution, but it contains no metallic ions. The powder is dissolved in concentrated sulfuric acid, yielding a clear solution. The solution turns orange as the oxidizer is used up. Use with care.

21.2.4 Ultrasonic Cleaning

Ultrasonic cleaning equipment emits high-frequency sound waves with frequencies ranging from 20 to 400 kHz, which are greater than what the human ear can perceive. Through the quick rupture of microscopic cavitation bubbles in the cleaning solution, these sound waves produce localized high-energy impacts. This method efficiently removes contaminants from surfaces with gaps, thin channels, and irregular forms.

Ultrasonic cleaners are perfect for precision equipment that is difficult to clean by hand such as manometers, narrow-bore pipettes, laboratory glassware, and optical components. Turn the machine on to start the cleaning cycle after thoroughly immersing the items in the appropriate cleaning solution, which is frequently a mixture of water and mild detergent or solvent. The process is nonabrasive, thorough, and time-efficient, making it ideal for delicate or complex equipment. After cleaning, items should be thoroughly rinsed with distilled water to remove any detergent residues and then dried appropriately to avoid contamination.

21.3 Laboratory Glassware Drying

After the glassware is thoroughly cleaned and washed, it must be dried.

21.3.1 Drainage Boards and Drainage Racks

Drainage boards and drainage racks (Figure 21.1) are used for drainage and drying glassware of various sizes and shapes. The brackets have pins and wedges anchored in an inclined position to ensure drainage. Some drain plates are equipped with hollow wedges and a hot air fan to speed up the drying process. Place the glassware securely on the rack. Do not allow parts to touch each other and cause accidental breakage.

21.3.2 Drying Ovens

Drying ovens (Figure 21.2) are designed to dry glassware at high speed. They have different sizes and power ratings; some of them are equipped with timers.

21.3.3 Quick Drying

Drying the inside of flasks or similar vessels can be done by lightly heating them on a Bunsen flame and then gently passing a jet of compressed air through a glass tube leading to the bottom of the flask until they are dry.

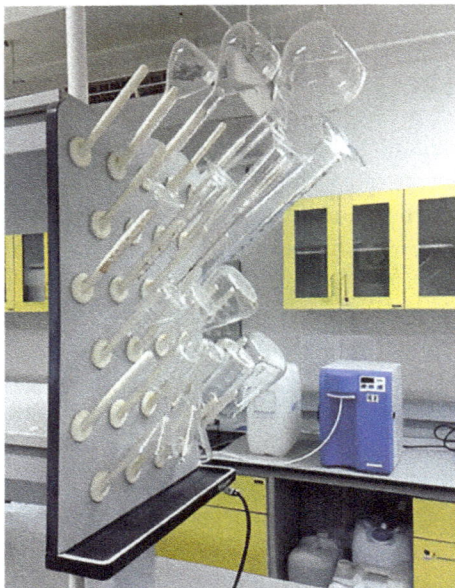

Figure 21.1: Glassware drainage racks. Image credit: https://en.wikipedia.org/wiki/Laboratory_ drying_ rack.

Figure 21.2: Laboratory drying ovens. Image credit: Image credit: https://www.testmak.com/ Laboratory-Oven-120-Liters-Capacity.

21.3.4 Rinsing Wet Glassware with Acetone

Water wet glassware can be dried more quickly by rinsing it with several small portions of acetone and discarding the acetone after each rinse. Then place them in a safe oven or gently pull air through the glassware by connecting a pipette with rubber tubing to an aspirator and inserting the pipette into the glassware.

21.4 Measuring and Transferring Liquids by Selected Glassware

21.4.1 Pipettes

21.4.1.1 Volumetric Pipettes
- Purpose
  Deliver a fixed, precise volume of liquid (e.g., 10.00 mL).
- Design
  Bulb in the center with a narrow stem; calibrated for a single volume.
- Use
  - Rinse with the solution to be used.
  - Use a pipette bulb or aspirator (never a mouth pipette).
  - Fill to the calibration mark.
  - Allow the liquid to drain by gravity; do not blow out the final drop.

21.4.1.2 Graduated (Mohr and Serological) Pipettes
- Purpose
  Deliver variable volumes within a range.
- Design
  Marked with graduation lines, Mohr pipettes are calibrated from the tip upward, while serological pipettes are often calibrated to be "blown out" at the end.
- Use
  Similar to volumetric pipettes but requires careful reading at both start and end volumes.

21.4.1.3 Micropipettes
- Purpose
  Accurately measure and transfer small volumes (typically 0.1–1,000 μL).
- Types
  Fixed volume or adjustable volume.
- Use
  - Attach a disposable tip.
  - Set the desired volume.
  - Press plunger to first stop, immerse tip, release slowly to aspirate.
  - Dispense by pressing to the second stop.

21.4.2 Burettes

- Purpose
  Deliver variable, measurable volumes of liquid, especially in titration.

- Design
  Long, narrow tube with volume graduations and a stopcock at the bottom.
- Use
  - Rinse with the titrant.
  - Fill above the zero mark and remove air bubbles.
  - Record initial volume.
  - Deliver titrant slowly into a reaction flask until the endpoint is reached.
  - Record final volume and calculate the volume dispensed.

21.4.3 Graduated Cylinders

- Purpose
  Measure approximate volumes of liquids.
- Design
  Cylindrical container with graduation marks; available in sizes from 10 mL to 2 L.
- Use
  - Place on a flat surface.
  - Pour liquid slowly.
  - Read the bottom of the meniscus at eye level.

21.4.4 Volumetric Flasks

- Purpose
  Prepare precise standard solutions of known volume.
- Design
  Bulb-shaped bottom with a long narrow neck and a single calibration mark.
- Use
  - Add solute, then partially fill with solvent.
  - Swirl until the solute dissolves.
  - Add solvent carefully up to the mark.
  - Stopper and invert several times to ensure homogeneity.

21.5 Questions and Answers

Questions	Answers
1. What glassware is commonly used to measure exact volumes of liquids?	Volumetric flask.
2. Which piece of glassware is used for titrations?	Burette.
3. What is the primary use of a graduated cylinder?	Measuring approximate liquid volumes.
4. Which glassware is used to prepare solutions of a fixed volume?	Volumetric flask.
5. What is the purpose of a pipette?	To transfer specific volumes of liquid accurately
6. Which glassware is used to contain reactions or heat liquids?	Beaker or Erlenmeyer flask.
7. What is a round-bottomed flask commonly used for?	Heating or distillation processes.
8. What is the function of a glass stirring rod?	To mix or stir solutions.
9. What is the purpose of a separatory funnel?	To separate immiscible liquid layers.
10. Which glassware is commonly used in distillation setups?	Round-bottomed flask, condenser, receiving flask.

Chapter 22
Grades of Chemical Purity and Preparation of Common Laboratory Solutions

22.1 Introduction

In research labs, industrial settings, and educational institutions, reliable and repeatable findings depend on the proper selection, handling, and solution preparation of chemicals. Efficiency and safety in the laboratory depend on an understanding of the nature, categorization, and purity of chemicals. In the chemical sciences, it is the basis of every experimental endeavor. In almost all chemical processes, the preparation of the solution is an essential step. All chemists must be able to create solutions with exact concentration and composition, whether they are using titrations in analytical processes or reaction mixes in synthetic operations. Chemical stoichiometry, concentration units, solubility principles, and the proper use of lab equipment must all be thoroughly understood. This chapter covers the many types and grades of laboratory chemicals as well as the principles of solution preparation and the related practical approaches. The common calculations used in solution chemistry are emphasized.

22.2 Grades of Purity of Chemicals

The quality and purity of chemicals used in laboratories and industry play a crucial role in ensuring the accuracy, safety, and reproducibility of products. Chemicals have various grades of purity (Figure 22.1); each grade is tailored to specific applications, from routine academic labs to high-precision analytical techniques and pharmaceutical manufacturing. These grades reflect the extent to which a chemical is free from impurities and meet established standards set by organizations such as the American Chemical Society and the United States Pharmacopeia (USP). Understanding these classifications helps chemists, researchers, and technicians select the most appropriate chemicals for their intended use. Table 22.1 below summarizes the most common grades of chemical purity, along with their typical applications.

22.3 Laboratory Solutions

Laboratory solutions are homogeneous mixtures made to supply exact solute concentrations in solvents, most often water. They are required for carrying out chemical reactions, calibrating equipment, performing titrations, and assuring repeatability in

https://doi.org/10.1515/9783112218105-022

Table 22.1: Grades of purity of chemicals.

Grade	Description	Typical uses
ACS grade	Meets or exceeds American Chemical Society specifications for purity.	Analytical chemistry and research labs
Reagent grade (RG)	Very high purity; suitable for most analytical work.	Qualitative and quantitative analysis
USP/NF grade	Complies with U.S. Pharmacopeia/National Formulary standards.	Pharmaceutical manufacturing and compounding
BP/EP grade	Meets British or European Pharmacopoeia standards.	Drug formulation and clinical use in Europe
CP grade	Chemically pure; lower than reagent grade.	Routine lab procedures and general synthesis
Lab grade	Moderate purity; not suitable for analytical work.	Educational labs and general science experiments
Technical grade	Lowest purity; may contain impurities.	Industrial processes, cleaning, and manufacturing
HPLC grade	High purity with low UV absorbance; filtered for chromatography.	High-performance liquid chromatography (HPLC)
Spectrophotometric grade	High optical purity with minimal absorbance in UV/vis range.	UV-vis and fluorescence spectroscopy
LC-MS grade	Ultra-high purity for use in LC-MS systems.	Trace analysis and bioanalytical research

experimental methods. The precision and consistency of laboratory solutions are dependent on adequate preparation processes and trustworthy chemical calculations.

22.3.1 Chemical Calculations for Preparation of Laboratory Solutions

Mass Percent (w/w%). Grams of solute per 100 g of solution.
Example: 20 g NaCl + 100 g water = 120 g solution → (20/120) × 100 = 16.67%
[20 g of NaCl in 100 g of H_2O → mass of solution = 20 g (NaCl) + 100 g (H_2O) = 120 g → mass percent = (20 g ÷ 120 g) × 100 = 16.67% w/w]

Volume Percent (v/v%). mL of solute per 100 mL of solution.
Example: 10 mL ethanol + 90 mL water = 100 mL → (10/100) × 100 = 10%
[10 mL ethyl alcohol + 90 mL water = 100 mL solution → volume percent = (10 mL ÷ 100 mL) × 100 = 10% v/v]

Image credit:
https://www.sigmaaldrich.com/SA/en/product
/mm/100658?utm_source=google&utm_medi
um=cpc&utm_campaign=22162084963&utm_
content=&gad_source=1&gad_campaignid=22
162087891&gbraid=0AAAAAD8kLORpchxn7FjZ
JkPjahLMjBgu78qgclid=CjwKCAjwyb3DB8iEIw
AqZLeSG-
MQVnXAb8R_tRECcVkD2nX_CuTON_PgrCvDL
ue4-5-TwA8d8K7Lb6Ceq4QAvD_BwE.

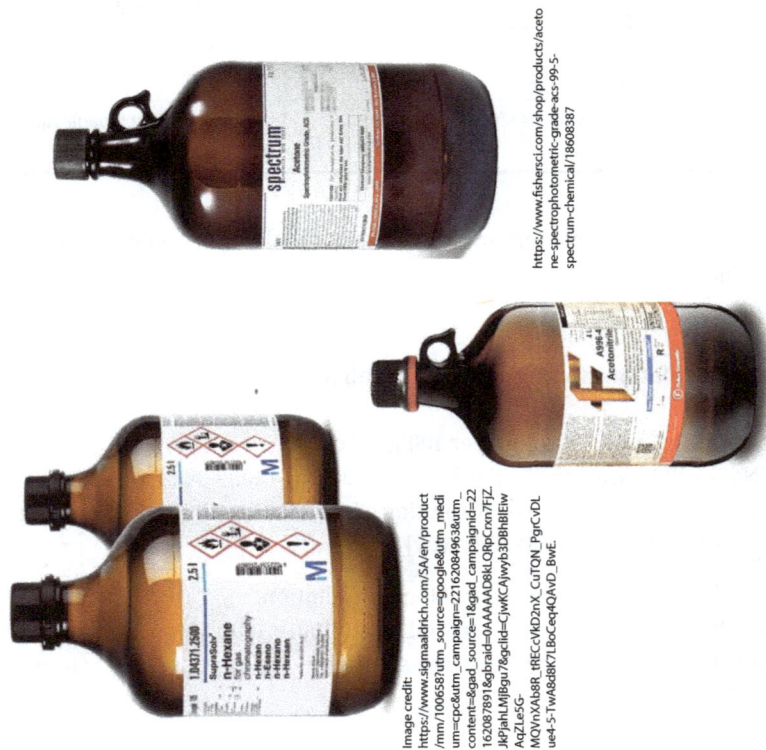

Image credit: https://www.fishersci.ca/shop/products/acetonitrile-
optima-fisher-chemical-6/p-215675.

https://www.fishersci.com/shop/products/aceto
ne-spectrophotometric-grade-acs-99-5-
spectrum-chemical/18608387

Figure 22.1: Various grades of purity of chemicals.

Molality (m). Moles of solute per 1,000 g (1 kg) of solvent.

Example: 58.44 g NaCl in 1,000 g water = 1.0 m

[58.44 g NaCl (1 mol) in 1,000 g of water = 1.0 m *and* 29.22 g NaCl (0.5 mol) in 1,000 g water = 0.5 m]

Molarity (M). Moles of solute per liter (1,000 mL) of solution.

Example: 58.44 g NaCl in total volume 1.0 L = 1.0 M

[58.44 g NaCl (1 mol) dissolved and made up to 1,000 mL total volume = 1.0 M]

Normality (N). Equivalents of solute per liter of solution.

Example: To prepare 3 L of 0.1 N H_2SO_4:

Normality = 0.1 N, volume = 3 L, eq. wt = 49 g = 0.1 × 49 × 3 = 14.7 g [H_2SO_4 has n = 2 eq./mol → equivalent weight = molar mass ÷ n = 98 ÷ 2 = 49 g/equiv.]

****Water is the most common solvent used in solution preparation. Types include:*
Distilled Water. Used for general lab work.
Deionized Water. Free from ions, suitable for analytical and biological work.
Ultrapure Water. For HPLC and molecular biology applications.

22.3.2 Preparation of Common Acids, Common Bases, and Standard Laboratory Solutions

Tables 22.2, 22.3, and 22.4 present the calculations required for preparing commonly used acids, bases, and standard laboratory solutions. A schematic diagram illustrating the solution preparation process is provided in Figure 22.2.

Table 22.2: Preparation of common acids (1.0 L solutions).

Acid	Desired molarity (M)	Concentrated stock	Volume to use	Notes
HCl	1.0 M	~37% w/w HCl, d = 1.19 g/mL, and ~2.08 M	82.78 mL	Add acid to ~500 mL water, mix, and dilute to 1 L.
H_2SO_4	1.0 M	~98% w/w, d = 1.84 g/mL, and ~8.4 M	54.35 mL	Add slowly to water (not vice versa), stir, and dilute to 1 L.
HNO_3	1.0 M	~70% w/w, d = 1.42 g/mL, and ~16 M	62.5 mL	Toxic vapors, work in fume hood.
CH_3COOH	1.0 M	Glacial acetic acid, ≥99–100% w/w, d = 1.05 g/mL, and ~17.4 M	57.5 mL	Flammable; mix with water carefully.
H_3PO_4	1.0 M	~85% w/w, d = 1.70 g/mL, and ~14.7 M	68.0 mL	Viscous; mix slowly.

Table 22.3: Preparation of common bases (1.0 L solutions).

Base	Desired molarity (M)	Substance	Mass or volume to use	Notes
NaOH	0.1 M	Solid NaOH pellets (MW = 40.00 g/mol)	4.00 g	Dissolve in ~700 mL water and then dilute. Exothermic.
NaOH	1.0 M	Same as above	40.00 g	Standardize with KHP (primary standard).
KOH	0.1 M	Solid KOH (MW = 56.11 g/mol)	5.61 g	Hygroscopic; store in airtight container.
NH_4OH	~1 M	~28% $NH_3(aq)$, $d \approx 0.90$ g/mL (~14.5 M)	69 mL	Use in fume hood due to ammonia fumes.
$Ca(OH)_2$	Saturated	Solid $Ca(OH)_2$	Add excess to water, stir, and filter	Used in qualitative analysis; solubility ~0.02 M.

Table 22.4: Preparation of standard laboratory solutions (1.0 L).

Solution	Concentration	Substance	Mass/ volume required	Preparation notes
Na_2CO_3	0.05 N	Anhydrous Na_2CO_3 (MW = 105.99 g/mol and $n = 2$)	2.65 g	Dry at 270 °C, dissolve; use to standardize HCl.
NaOH	0.1 N	Solid NaOH	4.00 g	Secondary standard; standardize with KHP.
HCl	0.1 N	~37% HCl (~12 M)	8.3 mL	Use standard Na_2CO_3 for titration standardization.
$KMnO_4$	0.02 N	Solid $KMnO_4$ (MW = 158.04 g/mol, $n = 5$)	0.632 g	Dissolve in warm water; standardize with oxalic acid.
Iodine (I_2)	0.05 N	I_2 + KI	6.35 g I_2 + 10 g KI	Dissolve in ~500 mL water; store in amber bottle.
Sodium thiosulfate	0.1 N	$Na_2S_2O_3 \cdot 5H_2O$ (MW = 248.18 g/mol, $n = 1$)	24.82 g	Use boiled distilled water; store in dark bottle.
Oxalic acid	0.05 N	$H_2C_2O_4 \cdot 2H_2O$ (MW = 126.07 g/mol, $n = 2$)	3.15 g	Primary standard; used for $KMnO_4$ standardization.
EDTA (disodium salt)	0.01 M	$Na_2EDTA \cdot 2H_2O$ (MW = 372.24 g/mol)	3.72 g	Adjust pH to ~10 with buffer (NH_4OH/NH_4Cl).

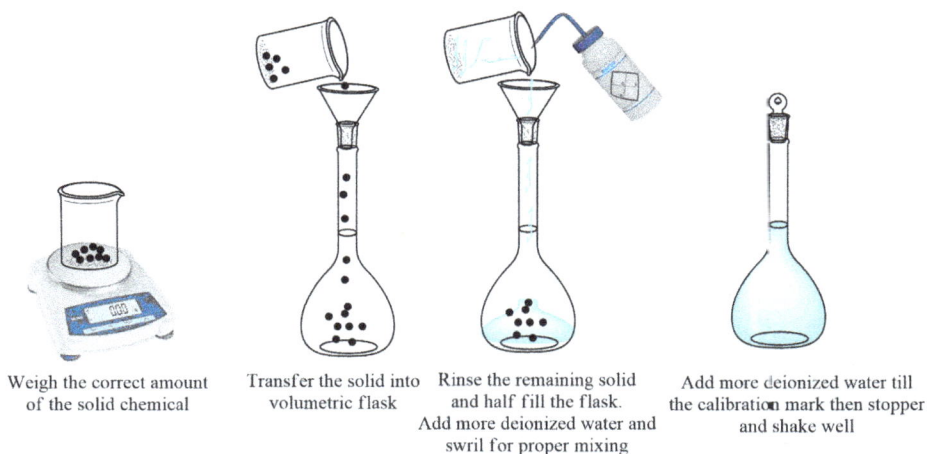

Weigh the correct amount of the solid chemical

Transfer the solid into volumetric flask

Rinse the remaining solid and half fill the flask. Add more deionized water and swril for proper mixing

Add more deionized water till the calibration mark then stopper and shake well

Figure 22.2: A schematic diagram illustrating the solution preparation process.

22.4 Questions and Answers

Questions	Answers
1. How to prepare 1 L of 1 M H_2SO_4 from 98% concentrated sulfuric acid?	Concentration of stock acid = 98% H_2SO_4. Density of 98% H_2SO_4 = 1.84 g/mL = 1,840 g/L. Purity of H_2SO_4 is in that liter = 98% of 1,840 g = 0.98 × 1,840 = 1,803.2 g. Molar mass of H_2SO_4 = 98 g/mol. Moles = 1,803.2 g /98 g/mol = 18.4 mol = 18.4 M. Dilution formula: 1 M/18.4 M × 1,000 mL = 54.35 mL.
2. How to prepare 1 L of 1 M HCl from 37% concentrated hydrochloric acid?	Concentration of stock acid = 37% HCl. Density of 37% HCl = 1.19 g/mL = 1,190 g/L. Purity of HCl is in that liter = 37% of 1,190 = 0.37 × 1,190 = 440.3 g. Molar mass of HCl = 36.46 g/mol. Moles = 440.3 g/36.46 g/mol = 12.08 mol = 12.08 M. Dilution formula: 1 M/12.08 M × 1,000 mL = 82.78 mL.
3. How many grams of KCl are needed to prepare 500 mL of 0.2 M solution?	Moles = 0.2 mol/L × 0.5 L = 0.1 mol Mass = 0.1 mol × 74.55 g/mol = 7.46 g
4. Calculate molality of solution with 10 g NaOH in 500 g water.	Moles = 10 g ÷ 40 g/mol = 0.25 mol. Molality = 0.25 mol ÷ 0.5 kg = 0.5 m.

(continued)

Questions	Answers
5. How much 12 M HCl is needed to prepare 250 mL of 1 M HCl?	$C_1V_1 = C_2V_2 \rightarrow V_1 = 1 \times 250/12 = 20.83$ mL.
6. Determine normality of solution made by dissolving 49 g H_2SO_4 in 1 L solution.	Moles = 49 g ÷ 98 g/mol = 0.5 mol; normality = 0.5 mol × 2 (H^+ per mol) = 1 N.
7. Calculate % w/v of solution with 25 g glucose in 250 mL solution.	% w/v = (25 g ÷ 250 mL) × 100 = 10%.
8. pH of 0.01 M HCl solution?	pH = $-\log[H^+]$ = $-\log(0.01)$ = 2.
9. How many milliliters of 2 M NaOH required to prepare 100 mL of 0.1 M NaOH?	$V_1 = 0.1 \times 100/2 = 5$ mL.
10. How to prepare 100 mL of 0.01 M EDTA solution?	Weigh 0.01 mol × 372.24 g/mol = 0.372 g disodium EDTA, dissolve and dilute to 100 mL.

Chapter 23
Automation, Sensors, and Digital Tools in Laboratories

23.1 Introduction

Automation, sensor technologies, and digital equipment have revolutionized the chemical laboratory by increasing productivity, accuracy, and safety. Because they provide researchers and students more control over processes and real-time data processing, these technologies are increasingly being used in experimental workflows. This chapter discusses the many automation systems, sensors, and digital instruments often used in chemical laboratories, as well as their benefits, and most recent innovations.

23.2 Automation in the Laboratory

Laboratory automation refers to the use of machines, robots, and software systems to perform repetitive or complex tasks in the lab, often replacing manual labor. Automation can significantly improve the precision and speed of experiments while reducing human error.

23.2.1 Types of Laboratory Automation

Automation technologies (Table 23.1) are increasingly transforming laboratory operations by enhancing precision, reproducibility, and throughput. These systems minimize human error, reduce manual labor, and improve data integrity. From sample preparation to data acquisition, various types of automation (Figure 23.1) play critical roles in analytical and synthetic chemistry workflows.

Table 23.1: Common types of laboratory automation and their applications.

Automation type	Description	Typical applications
Sample preparation systems	Automated units for sample weighing, dilution, filtration, and mixing	Pretreatment of samples prior to instrumental analysis (e.g., ICP-MS and HPLC)
Automated titrators	Instruments that perform titrations with automated endpoint detection and reagent addition	Acid–base, redox, complexometric, and Karl Fischer titrations

https://doi.org/10.1515/9783112218105-023

Table 23.1 (continued)

Automation type	Description	Typical applications
Robotic liquid handling systems	Robotic arms or pipetting stations that accurately dispense liquids	High-throughput screening, qPCR, ELISA, and enzyme and cell assays
Automated synthesis platforms	Integrated systems for planning and executing chemical syntheses	Organic synthesis, medicinal chemistry, and polymer synthesis
Data acquisition and control systems	Software–hardware platforms for real-time monitoring and data logging	Chromatography (e.g., GC and HPLC), spectroscopy, and electrochemical analysis

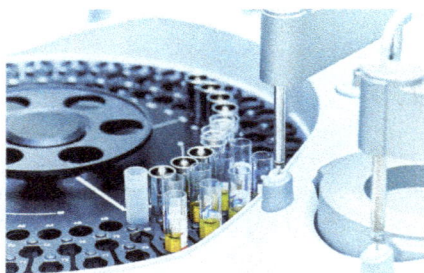

Image credit: https://genemod.net/blog/lab-automation.

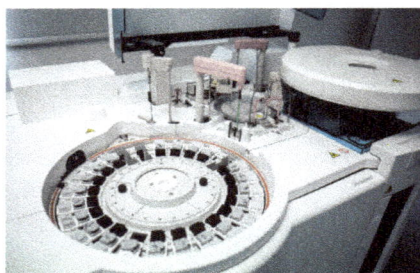

Image credit: https://anaheimautomation.com/lab-automation-and-diagnostics.

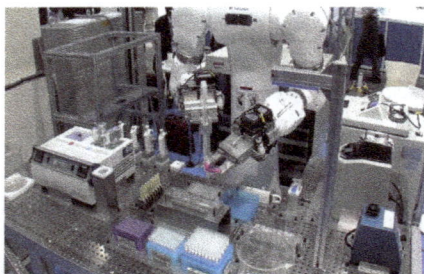

Image credit: https://www.pharmiweb.com/press-release/2024-02-21/lab-automation-market-pioneers-the-future-poised-to-surpass-us-55-billion-by-2033-insights-and-g.

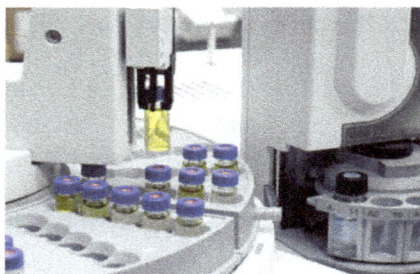

Image credit: https://www.electrocraft.com/motors-for/lab-automation/.

Figure 23.1: Various types of laboratory automation.

23.2.2 Benefits of Automation

The use of automation in chemical laboratories provides various advantages that boost operating efficiency and data reliability. By automating routine and challenging operations, laboratories may boost production, minimize unpredictability, and maximize resource use. Figure 23.2 demonstrates some of the key advantages of implementing automation technology in lab environments.

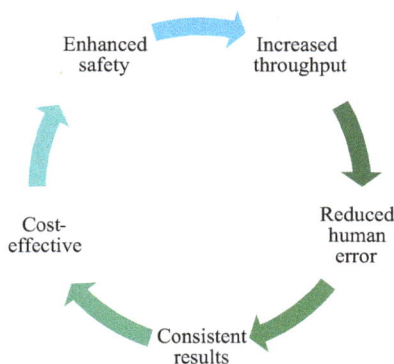

Figure 23.2: Key benefits associated with implementing automation systems in laboratory settings.

23.3 Sensors in the Laboratory

23.3.1 Types of Laboratory Sensors

Sensors (Table 23.2) are essential tools for real-time data gathering, monitoring, and control in laboratory research. They assess physical, chemical, and environmental characteristics and give an immediate response. Sensors offer several advantages (Figure 23.3). Among them, they provide excellent sensitivity and accuracy, which is especially important in delicate or complicated studies. Furthermore, integrating sensors with automated systems enables closed-loop control, in which the system automatically adjusts conditions depending on sensor inputs.

Table 23.2: Common sensor types and their applications in chemistry laboratories.

Sensor types	Functions	Typical applications
pH sensors	Measure the hydrogen ion concentration to determine acidity or alkalinity	Acid–base titrations, biological assays, and environmental testing
Temperature sensors	Monitor and control temperature in reactions and instruments	Calorimetry, incubators, and temperature-sensitive reactions
Pressure sensors	Measure the pressure of gases or liquids in closed systems	Gas chromatography (GC), vacuum distillations, and reaction vessels
Conductivity sensors	Measure the electrical conductivity, indicating ion concentration	Water purity analysis, ionic strength monitoring, and electroplating

Table 23.2 (continued)

Sensor types	Functions	Typical applications
Spectroscopic sensors	Detect and quantify light absorption/ emission at specific wavelengths	UV-Vis, IR, fluorescence spectroscopy, and colorimetric assays
Gas sensors	Detect and measure concentrations of specific gases	Air quality monitoring, fermentation, and glove box atmosphere control
Mass flow sensors	Measure the mass flow rate of gases or liquids	Gas and liquid chromatography and controlled reagent delivery

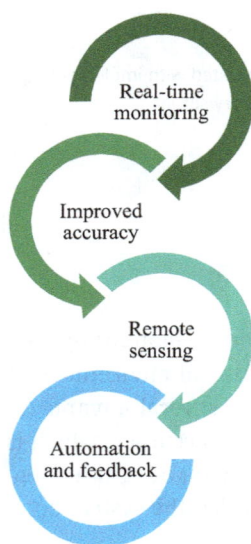

Figure 23.3: Advantages of sensors.

23.3.2 Examples of Sensor Integration

– Automated Titrations. A combination of robotic liquid handling and pH sensors allows titrations to be performed without manual intervention.
– In Situ Monitoring. Sensors in reaction vessels can track temperature, pressure, and concentration in real time, adjusting conditions as needed to optimize reactions.
– Gas Detection. Integrated systems can monitor gas concentrations in reaction chambers and adjust ventilation or shut down systems if dangerous levels are detected.

23.4 Digital Tools in the Laboratory

23.4.1 Laboratory Information Management Systems (LIMS)

LIMS are software systems used to manage samples, track data, and integrate laboratory equipment and sensors into a single database. Key functions include:
- Sample tracking (from collection to analysis)
- Data integration from multiple sources (e.g., instruments and sensors)
- Report generation and compliance documentation
- Equipment maintenance tracking

23.4.2 Data Analysis and Visualization Tools

Advanced software tools can help analyze large datasets generated by laboratory experiments, facilitating interpretation and decision-making:
- Statistical Analysis Tools. Softwares such as R, MATLAB, and Origin provide advanced statistical functions for data processing.
- Graphing and Visualization. Tools such as GraphPad Prism or SigmaPlot can create publication-ready graphs and plots.
- Cheminformatics Software. Programs such as ChemDraw, Chem3D, and Gaussian assist in molecular modeling, structure visualization, and simulation.

23.4.3 Digital Control and Automation Software

Digital controllers are used to manage various laboratory instruments, ensuring accurate and synchronized operation. Some applications include:
- Automated Reactors. Control and monitor temperature, pressure, and stirring in real time.
- Spectrometers. Software interfaces allow remote control and analysis of UV-Vis, NMR, or FTIR spectrometers.
- Robotics. Software allows for integration of robotic arms for liquid handling and sample preparation, offering a fully automated workflow.

23.4.4 Cloud Computing and Remote Access

Cloud-based platforms enable laboratory data to be stored and accessed remotely. This allows for:
- Real-time access to data from anywhere in the world

- Collaborative work with remote teams
- Backup and storage of experimental data in secure environments

23.5 Safety Considerations and Best Practices

23.5.1 Automation and Safety

- System Integrity. Regular maintenance and calibration are crucial to ensure the automation system operates as intended.
- Redundancy. Critical systems (e.g., temperature control and gas flow) should have fail-safe mechanisms to prevent accidents.
- Laboratory Supervision. While automation can reduce the need for human intervention, lab personnel need to monitor systems, especially during initial runs.

23.5.2 Sensor Calibration and Accuracy

- Routine calibration of sensors is necessary to ensure accurate readings.
- Environmental factors (temperature and humidity) can affect sensor performance, so consistent conditions are required.
- Preventive Maintenance. Sensors should be checked regularly for wear and replaced if necessary to avoid drift in measurements.

23.5.3 Data Security and Integrity

With the increased use of digital tools and cloud storage, protecting data becomes paramount. Best practices include:
- Regular backups of critical data
- Secure passwords and encryption for cloud-based data storage
- Proper access control and authorization for sensitive data

23.6 Emerging Trends in Laboratory Automation and Digital Tools

- Artificial intelligence (AI) and machine learning (ML). AI-driven systems can now predict experimental outcomes, optimize reaction conditions, and automate the analysis of complex datasets.
- Lab-on-a-Chip. Miniaturized lab systems that combine multiple analytical functions (e.g., sample preparation and analysis) into a single, compact device.

– Internet of Things (IoT). IoT-enabled lab equipment can communicate with each other, sharing data and providing real-time insights into experimental conditions.

23.7 Questions and Answers

Questions	Answers
1. What is laboratory automation?	The use of technology to perform lab tasks with minimal human intervention.
2. What is the main function of a laboratory robot?	To handle repetitive tasks like sample preparation and analysis.
3. What digital tool is used to store and manage experimental data?	Laboratory Information Management System (LIMS).
4. What is the function of a digital burette?	To accurately and digitally dispense liquids during titrations.
5. What kind of sensor can be used to measure liquid level in a container?	Ultrasonic, capacitive, or float sensors.
6. What is the role of a digital balance in automation?	To provide precise and automated mass measurements.
7. What is an autosampler?	A device that automatically loads samples into analytical instruments.
8. What does IoT stand for in laboratory environments?	Internet of Things.
9. How does a conductivity sensor work?	It measures a solution's ability to conduct electricity.
10. Why are digital tools important in modern laboratories?	They increase accuracy, efficiency, data integrity, and reproducibility.

Essential Keywords

Absorption. The process by which one substance is taken up into the interior of another substance, such as a gas absorbed into a liquid.

Accident. An incident arising from carrying out work that results in personal injury.

Acid–base titration. A method used to determine the concentration of an acid or base by neutralizing it with a known volume of the opposite.

Acids. Substances that release hydrogen ions (H^+) in water and have a pH less than 7.

Activated carbon. A form of carbon processed to have a large surface area that adsorbs impurities, often used in filtration.

Administrative controls. Policies, training, procedures, and scheduling practices that reduce the risk of exposure to hazards in the workplace.

Adsorption. The process by which atoms, ions, or molecules adhere to a surface.

Aeration. The process of introducing air into a liquid or solid, typically to increase oxygen levels or facilitate chemical reactions.

Airborne contaminants. Hazardous substances, such as dust, fumes, or gases, suspended in the air.

Airborne nanoparticles. Nanoscale particles suspended in air that may pose inhalation risks during handling or processing.

Analytical chemistry. The branch of chemistry that deals with the identification and quantification of substances.

Analytical instruments Tools and equipment used to perform analysis, such as spectrometers, balances, and chromatographs.

Analytical balance. A high-precision balance used to measure mass with very fine accuracy, typically to the milligram.

Anhydrous. Describes a substance that contains no water, commonly used when preparing moisture-sensitive solutions.

Artificial intelligence (AI). The use of computer algorithms to mimic human decision-making and learning, enabling predictive analysis, and process optimization in laboratories.

Asbestos. A group of minerals with fiber-like crystals used in insulation and other materials, known for its carcinogenic properties.

Atomic absorption spectroscopy. A technique used to measure the concentration of elements in a sample by analyzing light absorption.

https://doi.org/10.1515/9783112218105-024

Atomic mass. The mass of an atom, usually expressed in atomic mass units (amu), based on the number of protons and neutrons.

Autoclave. A pressurized vessel used for sterilization or hydrothermal reactions by applying steam at high temperature and pressure.

Auto-ignition temperature. The lowest temperature at which a vapor will ignite spontaneously when mixed with air.

Automation. The use of control systems and technology to perform laboratory tasks with minimal human intervention, improving efficiency and consistency.

Azeotrope. A mixture of two or more liquids that boil at a constant temperature and cannot be separated by simple distillation.

Barometer. An instrument used to measure atmospheric pressure.

Baseline. The reference point in an experiment or measurement, often used in spectroscopic analysis to assess changes in data.

Beaker. A cylindrical glass container used for mixing, heating, and holding liquids or solids in the laboratory.

Biodegradation. The breakdown of organic substances by microorganisms, often used to describe the decomposition of waste.

Biohazardous waste. Waste materials contaminated with biological agents that pose a risk to human health or the environment.

Boiling point. The temperature at which the pressure of the saturated vapor of the liquid is equal to the pressure of the atmosphere under which the liquid boils.

Bourdon gauge. A mechanical pressure gauge that measures gauge pressure using a curved, flexible, tube that straightens as pressure increases.

Buffer solution. A solution that resists changes in pH when small amounts of acid or base are added, typically made from a weak acid and its conjugate base.

Bunsen burner. A common laboratory device used to produce a controlled flame for heating substances.

Calibration. The process of adjusting the output of an instrument to match a known standard or reference.

Capillary action. The ability of a liquid to flow in narrow spaces without external forces, such as liquid rising in a thin tube.

Centrifugation. A technique that uses centrifugal force to separate components of different densities, commonly used for separating solids from liquids or different phases of a mixture.

Chemical compatibility. The property that determines whether different chemicals can be safely stored or mixed without hazardous reactions.

Chemical inventory. A detailed list of chemicals stored in a laboratory or facility, used to track quantities and manage waste effectively.

Chemical reactions. Processes in which substances are transformed into new products through the breaking and forming of bonds.

Chemical safety. The practice of ensuring safe handling, storage, and disposal of chemicals to prevent accidents and exposures.

Chemical spill. An unintentional release of a chemical substance, which can pose a risk to health or the environment.

Chemical spill kit. A set of tools and materials (e.g., absorbents, neutralizers, and PPE) prepared in advance for safely cleaning up chemical spills.

Chemical waste. Discarded chemicals that are no longer needed and require safe disposal due to their potential hazard.

Chromatography. A technique used to separate mixtures into their individual components based on their movement through a medium.

Cleaning solution. A chemical agent (e.g., dilute acid or detergent) used to remove residues from glassware surfaces to ensure cleanliness and prevent contamination.

Cooling bath. A mixture of substances (e.g., ice–water, ice–salt, or dry ice–solvent) used to achieve and maintain low temperatures for reactions or sample preservation.

Column chromatography. A type of chromatography where a mixture is passed through a column filled with adsorbent material to separate its components.

Combustible material. A substance that can catch fire and burn at relatively high temperatures when exposed to a heat source.

Concentration. The amount of solute dissolved in a given amount of solvent, often expressed as molarity or molality.

Condensation. The process by which vapor transforms into liquid, often occurring in condensers during distillation or reflux.

Condenser. A laboratory device used to cool vapors and condense them back into liquid form, commonly used in distillation.

Containment. Measures taken to prevent hazardous materials from spreading, such as using fume hoods or sealed containers.

Corrosion. The deterioration of metals due to chemical reactions with their environment, often involving oxidation.

Corrosive chemicals. Chemicals that result in an immediate, acute erosive effect on body tissue.

Cryogenic liquids (cryogens). Liquids with a boiling point less than −73 °C.

Crystallization. The process by which a solid forms from a liquid or gas and develops a structured arrangement of atoms or molecules.

Dangerous occurrence. Occurrence arising from work activities in a chemical laboratory that results in a hazardous situation.

Decantation. The process of carefully pouring off a liquid from a settled solid or from another immiscible liquid layer without disturbing it.

Deionized water. Water purified by ion exchange processes, commonly used for rinsing and preparing analytical solutions.

Density. The mass of a substance per unit volume (e.g., g/mL); an important parameter in identifying and characterizing substances.

Desiccator. A sealed container with a drying agent (desiccant) used to keep glassware or samples free from moisture.

Distillation. A technique used to separate mixtures based on differences in boiling points.

Dry ice. Solid carbon dioxide (CO_2) that sublimates at −78.5 °C; commonly used in cooling baths for low-temperature reactions.

Drying oven. An apparatus used to dry laboratory glassware at controlled temperatures after washing and rinsing.

Emergency drill. A planned exercise that simulates an emergency scenario to test the effectiveness of response procedures and improve preparedness.

Emergency exit. A clearly marked, unobstructed route used for the safe and quick evacuation of occupants during an emergency.

Emergency plan. A documented strategy that outlines actions to be taken before, during, and after an emergency to minimize risk and ensure safety.

Emergency shower. A safety fixture used to rinse hazardous chemicals from the body in the event of a spill or splash.

End point. The point in a titration when the reaction is complete, often indicated by a color change.

Engineering controls. Safety mechanisms like HEPA filters, local exhaust ventilation, and sealed enclosures designed to limit exposure to nanomaterials.

Equivalence point. The point at which, the standard solution is chemically equivalent to the substance being titrated.

Evacuation The organized movement of personnel away from a hazardous area to a safer location in response to an emergency.

Extraction. A process used to separate a substance from a mixture by selectively dissolving it in a solvent.

Fire extinguisher. A portable device that discharges an extinguishing agent to control or extinguish small fires; must be used according to fire class.

Fire triangle. The three essential elements for fire – fuel, heat, and oxygen – removal of any of which will prevent or extinguish the fire.

First aid. The initial medical assistance given to an injured or ill person before professional medical help is available.

Flammable liquid. A liquid that can easily ignite and burn, typically having a flash point below 37.8 °C (100 °F).

Flame test. A method used to identify the presence of metal ions by observing the color of their flame when heated.

Flash point. The lowest temperature at which a liquid emits enough vapor to ignite in air, important for safety classification.

Filtration. The process of removing material from a substrate in which it is suspended.

Flashback. The rapid combustion of heavy vapors of organic compounds that collect in areas distant from their source and when burning, lead the flame back to their source to cause a large fire or explosion.

Flash point. The lowest temperature at which compounds in an open vessel gives off sufficient vapors to produce a momentary flash of fire. This happened when a flame, a spark, an incandescent wire, or another source of ignition is brought near the surface of the liquid.

Fractional distillation. The separation and purification of a mixture of two or more liquids, present in appreciable amounts, into various fractions.

Freeze-drying. This process, also known as lyophilization, involves the drying of the product under low temperature and vacuum.

Fume hood. Equipment designed to control chemists and laboratory technicians exposure to hazardous chemicals.

Gas chromatography. A technique used to separate and analyze volatile substances by passing them through a column with a gas as the mobile phase.

Gauge pressure. Pressure measured relative to atmospheric pressure. It reads zero at atmospheric pressure and does not account for absolute vacuum.

Glove box. A sealed container with built-in gloves that allows safe handling of highly toxic, reactive, or air-sensitive substances.

Good laboratory practices. A compilation of procedures and practices designed to promote the quality and validity of all laboratory studies.

Good manufacturing practices. A procedure that regulates the manufacturing and associated quality control of products.

Gravity filtration. A process in which the filtrate passes through the filter medium under the forces of gravity and capillary attraction between the liquid and the funnel stem.

Gravimetric analysis. A method of quantitative analysis where the amount of a substance is determined by measuring its mass.

Green chemistry. The design of chemical products and processes that minimize the use and generation of hazardous substances.

Hazard. Anything that can cause harm.

Hazard classes. Categories of hazards defined by WHMIS, including physical hazards, health hazards, and environmental hazards.

Hazard communication. The practice of informing personnel about chemical hazards through labels, signs, and safety data sheets (SDS).

Hazard identification. The process of recognizing and classifying hazards based on chemical properties and health effects.

Hazard symbol. A pictogram used on labels to quickly communicate the type of hazard a chemical poses, such as flammability or toxicity.

HAZMAT (hazardous materials). Substances that pose a risk to health, safety, or the environment, requiring special handling and disposal procedures.

Health risks control measure. A hierarchical approach combining varieties of both engineering and operational/procedural control measures.

Heating mantle. A laboratory device that provides uniform heating to round-bottomed flasks using an insulated, flexible heating element.

Hierarchy of controls. A framework that ranks hazard control strategies from most to least effective: elimination, substitution, engineering controls, administrative controls, and PPE.

Hydrometer. A glass container, weighted at the bottom, having a slender stem calibrated to a standard.

Ignition source. Any item or condition that can initiate a fire, such as open flames, hot surfaces, sparks, or electrical equipment.

Ignition temperature. Lowest temperature at which the vapors over the surface of the liquid ignite.

Incident. A work-related occurrence during which injury, illness, or fatality happened or could have happened.

Infrared spectroscopy. A technique used to identify molecular structures by measuring the absorption of infrared light at specific wavelengths.

International Organization for Standardization. An international standard-setting body composed of representatives from various national standards organizations.

Internet of Things (IoT). A network of interconnected devices embedded with sensors and software, enabling communication and data exchange in smart laboratories.

Ion exchange. A reversible chemical reaction in which ions from a solution are exchanged with ions from a solid material, like a resin.

ISO 9001. A standard provides a model for quality assurance in the design, production, and supply of products or services.

ISO 9002. A model for quality assurance in production and installation, not however, for research and development.

ISO 9003. A model for quality assurance when only final inspection and testing are required.

ISO 9004. A model dealing with guidelines for developing quality management.

Labeling. The practice of clearly marking waste containers with contents, hazard information, and dates to ensure safe handling and compliance.

Laboratory fire alarm. An alert system that signals the presence of fire and prompts immediate evacuation of laboratory personnel.

Laboratory safety. The practices and procedures put in place to minimize risks and ensure the safe handling of chemicals and equipment in a lab.

Lab Information Management System (LIMS). A software system used to manage samples, data, and workflow in laboratory environments.

LC50. The lethal concentration in air of a substance that produces death in 50% of the exposed test population within a specified time.

LD50. The lethal dose required to produce death in 50% of the exposed test population within a specified time.

Liquid chromatography. A separation technique which uses two phases in contact with each other; the stationary phase can be an immiscible liquid or solid, but the mobile phase must be a liquid.

Machine learning. A subset of AI where algorithms improve automatically through experience, used in pattern recognition and predictive modeling in laboratories.

Magnetic stirrer. A device that uses a rotating magnetic field to spin a stir bar inside a liquid, ensuring homogeneous mixing during heating or reaction.

Mass spectrometry (MS). A technique that measures the mass-to-charge ratio of ions to identify and quantify molecules based on their mass.

Melting point. The melting point of a crystalline solid is the temperature at which the solid substance begins to change into a liquid.

Meniscus. The curved surface of a liquid in a container; measurements are read at the bottom of the meniscus at eye level for accuracy.

Mercury barometer. A barometer with a column of mercury whose height varies according to the atmospheric pressure.

Molality. Known also as molal concentration is the number of gram molecular masses of solute per 1,000 g of solvent.

Molarity. Known also as molar solution is the number of gram molecular masses of solute per liter or 1,000 mL of solution.

Nanomaterial. A material with at least one dimension in the nanoscale range (1–100 nm), exhibiting unique physical and chemical properties.

Nanoparticle. A particle with dimensions in the nanometer range that may display different reactivity and toxicity compared to bulk materials.

Normality. Normal solution has a specific number of equivalent masses of the acid or base dissolved in the solution per liter.

NMR (nuclear magnetic resonance). A technique used to determine the structure of organic compounds by analyzing the magnetic properties of atomic nuclei.

Oil bath. A container filled with oil used to heat samples at temperatures higher than boiling water, offering stable and uniform heating.

Personal protective equipment (PPE). Specialized clothing and gear (e.g., gloves, lab coats, and respirators) used to reduce exposure to hazardous nanomaterials.

pH. A scale used to measure the acidity or alkalinity of a solution, ranging from 0 (acidic) to 14 (basic).

pH sensor. A sensor that measures the hydrogen ion concentration in a solution, indicating its acidity or alkalinity.

Pipette. A laboratory instrument used to measure and transfer small volumes of liquids accurately.

PPM (parts per million). A unit of concentration used for very dilute solutions; 1 ppm = 1 mg of solute per liter of solution.

Precipitation. The formation of a solid from a solution when the concentration of a dissolved substance exceeds its solubility.

Pressure. The force exerted per unit area, often measured in atmospheres, pascals, or mmHg.

Proper handling. The correct use and care of glassware to avoid contamination, breakage, or injury during laboratory operations.

Pure substance. A material made of only one type of particle, either an element or a compound.

Pycnometer. A calibrated-volume ground, glass vessel fitted with a closure and a thermometer.

Pyrophoric chemicals. Are chemicals that may spontaneously ignite upon exposure to air.

Quality assurance (QA). A system of planned and systematic actions to ensure that laboratory processes meet established standards.

Quality control (QC). The operational techniques and activities used to monitor and maintain the quality of laboratory procedures and results.

Reagent. A substance used in a chemical reaction to detect, measure, or produce other substances.

Reagent grade. Chemicals that meet the purity requirements for laboratory reagents but may vary slightly depending on supplier standards.

Recrystallization. A purification technique where an impure solid is dissolved in a hot solvent and then crystallized upon cooling.

Recycling. The process of reclaiming usable materials from chemical waste to reduce environmental impact and resource consumption.

Redox reactions. Reactions involving the transfer of electrons between species, resulting in the reduction of one and oxidation of another.

Reducing agents. Are chemicals that are good sources of hydride and thus react vigorously with other chemicals or materials.

Reflux. A technique where a reaction mixture is continuously boiled and condensed to prevent loss of volatile components while allowing extended heating.

Refractive index. A measure of how much light is bent, or refracted, when entering a substance; used to identify and characterize liquids.

Rinsing. The process of washing glassware with water or solvent to remove residual chemicals, often followed by drying.

Risk. A likelihood that the hazard will cause actual harm.

Risk assessment. The process of identifying potential hazards and evaluating the likelihood and severity of their impact during an emergency.

Rotary evaporator. Laboratory equipment that uses vacuum to lower the boiling point of solvents, allowing their removal by evaporation under reduced pressure.

Safety. A protection or be away from danger, risk, or injury.

Safety culture. An attitude, rather than a set of rules or procedures.

Sand bath. A heating medium made from dry sand placed in a metal container, providing even heat distribution for glassware.

Schlenk line. A dual manifold system used for air-sensitive chemistry, allowing manipulation under inert gas and vacuum.

Sealed tube reaction. A chemical reaction carried out in a sealed tube capable of withstanding elevated pressure and temperature.

Secondary standard. A solution whose concentration is determined by comparison with a primary standard.

Segregation. The practice of separating incompatible wastes to prevent dangerous reactions and ensure safe storage.

Sensor. A device that detects a physical property (e.g., temperature, pressure, and pH) and converts it into a readable signal.

Solubility. The ability of a substance to dissolve in a solvent to form a homogeneous mixture at a given temperature and pressure.

Solute. The substance dissolved in a solvent to form a solution.

Solution. A homogeneous mixture of two or more substances where one is dissolved in the other.

Solvent. A substance that dissolves a solute, forming a solution, typically in larger quantities than the solute.

Solvency. The substance to be purified should be sparingly soluble in the solvent at room temperature yet should be very soluble in the solvent at its boiling point.

Specific gravity. The mass of a substance divided by the mass of an equal volume of water.

Spectrophotometry. A technique that measures the intensity of light absorbed by a substance as a function of wavelength.

Spill response. The procedures followed to safely contain and clean up nanomaterial spills, preventing further exposure or environmental contamination.

Standardization. The process of determining the exact concentration of a solution, often by titration against a primary standard.

Standard operating procedure (SOP). A detailed, written set of instructions designed to achieve uniformity in the performance of a specific function.

Standard solution. A reagent of known composition used in a titration.

Steam distillation. Separating and purifying organic compounds by volatilization.

Stock solution. A concentrated solution that is diluted to a lower concentration for actual use in experiments.

Stoichiometry. The calculation of reactants and products in chemical reactions based on the conservation of mass and the mole concept.

Sublimation. A phenomenon in which solids can go from the solid to the vapor state without passing through the liquid state.

Supplier label. The label provided by the chemical supplier that includes product identifiers, hazard pictograms, precautionary statements, and first-aid measures.

Surfactant. A substance that reduces the surface tension of a liquid, commonly used in detergents and emulsifiers.

Technical grade. Chemicals of lower purity used in industrial or noncritical laboratory applications where exact purity is not required.

Temperature sensor. A device that measures temperature, often used for monitoring reactions or environmental conditions in laboratories.

Thin-layer chromatography. A simple, rapid, and inexpensive method for analyzing a wide variety of materials ranging from inorganic ions to high-molecular-weight biological compounds.

Titration. A laboratory technique used to determine the concentration of a substance by reacting it with a solution of known concentration.

Torr. A unit of pressure equal to 1/760 of atmospheric pressure; commonly used in vacuum measurements.

Toxicity. The degree to which a substance can cause harm to living organisms, particularly by poisoning or injury.

UV–visible spectroscopy. A technique that measures the absorption of ultraviolet or visible light by a sample to analyze its chemical composition.

Vacuum distillation. A distillation process conducted under reduced pressure, allowing the separation of components at lower temperatures.

Vacuum filtration. A faster filtration method using reduced pressure to draw liquid through a filter medium, typically used for drying crystals.

Vacuum pump. A device used to remove air and other gases from a system to create a vacuum.

Vapor pressure. The pressure exerted by a vapor in equilibrium with its liquid or solid phase at a given temperature.

Vent. A controlled outlet in a pressure system to safely release gases or vapors.

Viscosity. A measure of a liquid's resistance to flow, often related to its thickness or stickiness.

Volatility. The volatility of a solvent determines the ease or difficulty of removing any residual solvent from the crystals which have formed.

Volumetric flask. A piece of glassware calibrated to contain a precise volume, commonly used in preparing standard solutions.

Volumetric pipette. A pipette that delivers a fixed volume of liquid with high precision.

Waste management. The proper collection, labeling, treatment, and disposal of nanomaterial waste to prevent unintended environmental or health impacts.

Water bath. A controlled heating apparatus filled with water, used for gentle and uniform heating of samples up to ~100 °C.

Wet chemistry. The branch of chemistry dealing with chemical reactions and analysis in solution, typically involving liquid samples.

Workplace label. A label applied by the workplace when a controlled product is transferred to a new container, providing necessary hazard information.

Abbreviations

Abbreviation	Full Form
AAR	Analytical reagent
AAS	Atomic absorption spectroscopy
ACS	American Chemical Society
AFM	Atomic force microscopy
AI	Artificial intelligence
ALCOA	Attributable, Legible, Contemporaneous, Original, and Accurate
ANSI	American National Standards Institute
RG	Reagent grade
BLEVE	Boiling liquid expanding vapor explosion
BP	Boiling point
BP	British Pharmacopoeia
BSL	Biosafety level
°C	Degrees Celsius
°F	Degrees Fahrenheit
CDC	Centers for Disease Control and Prevention
CE	Capillary electrophoresis
CLP	Classification, labeling, and packaging
CNS	Central nervous system
CEPA	Canadian Environmental Protection Act
COD	Chemical oxygen demand
COSHH	Control of substances hazardous to health
CP	Chemically pure
CSA	Canadian Standards Association
DMAIC	Define, Measure, Analyze, Improve, Control
DNA	Deoxyribonucleic acid
DPD	Dangerous preparations directive
DSD	Dangerous substances directive
EA	Environment Agency
EHS	Environment, Health, and Safety
ELN	Electronic lab notebook
EMA	European Medicines Agency
EP	European Pharmacopoeia
EPA	Environmental Protection Agency
EPRP	Emergency preparedness and response plan
ERP	Emergency response plan
EUH	European Union Hazard (for EUH codes)
FDA	Food and Drug Administration
FIFO	First-in, first-out
FM	Factory Mutual
FMEA	Failure mode and effects analysis
FPM	Feet per minute
FTIR	Fourier transform infrared spectroscopy
GC	Gas chromatography
GDPR	General Data Protection Regulation
GHS	Globally harmonized system

https://doi.org/10.1515/9783112218105-025

GLP	Good laboratory practice
GMP	Good manufacturing practice
GPC	Gel permeation chromatography
Gy	Gray
HA	Hazardous materials
HAZMAT	Hazardous materials
HAZOP	Hazard and operability
HCS	Hazard Communication Standard
HEPA	High efficiency particulate air
HPLC	High performance liquid chromatography
HVAC	Heating, ventilation, and air conditioning
IAEA	International Atomic Energy Agency
ICP-MS	Inductively coupled plasma mass spectrometry
IEC	International Electrotechnical Commission
I-EC	Ion-exchange chromatography
IR	Infrared spectroscopy
ISO	International Organization for Standardization
IU	International Unit
IUPAC	International Union of Pure and Applied Chemistry
JIT	Just-in-time
KF	Karl Fischer (titration)
LC	Liquid chromatography
LC-MS	Liquid chromatography–mass spectrometry
LC_{50}	Lethal concentration, 50%
LD_{50}	Lethal dose, 50%
LEV	Local exhaust ventilation
LFL	Lower flammable limit
LIMS	Laboratory Information Management System
LOD	Limit of detection
LOQ	Limit of quantification
LPG	Liquefied petroleum gas
LSI	Laboratory Safety Inspection
MEC	Minimum explosible concentration
ML	Machine learning
MP	Melting point
MAQ	Maximum allowable quantity
MS	Mass spectrometry
MSDS	Material safety data sheet (now SDS)
MT	Empty (for marking gas cylinders)
NF	National Formulary "standards"
NFPA	National Fire Protection Association
NIST	National Institute of Standards and Technology
NMR	Nuclear magnetic resonance
NOAEL	No observed adverse effect level
NRTL	Nationally Recognized Testing Laboratory
OECD	Organization for Economic Co-operation and Development
OHS	Occupational health and safety
OSHA	Occupational Safety and Health Administration
PEL	Permissible exposure limit

PIPEDA	Personal Information Protection and Electronic Documents Act
PPM	Parts per million
PPE	Personal protective equipment
PSI	Pounds per square inch
QA	Quality assurance
QC	Quality control
QMS	Quality management system
R&D	Research and development
RAMP	Recognize, Assess, Minimize, Prepare
RCRA	Resource Conservation and Recovery Act
Rf	Retardation factor (chromatography)
RNA	Ribonucleic acid
RSD	Relative standard deviation
RT	Retention time (or room temperature)
SAR	Supplied air respirators
SDS	Safety data sheet
SEM	Scanning electron microscopy
SG	Specific gravity
SOP	Standard operating procedure
SP	Stationary phase
STEL	Short-term exposure limit
Sv	Sievert
TC	To contain
TD	To deliver
TEM	Transmission electron microscopy
TGA	Thermogravimetric analysis
TLC	Thin-layer chromatography
TNT	Trinitrotoluene
UFL	Upper flammable limit
UL	Underwriters laboratories
UN	United Nations
USP	United States Pharmacopeia
UV	Ultraviolet
UV-Vis	Ultraviolet–visible spectroscopy
VOC	Volatile organic compound
WFD	Water Framework Directive
WHO	World Health Organization
WHMIS	Workplace Hazardous Materials Information System
XPS	X-ray photoelectron spectroscopy
XRD	X-ray diffraction
XRF	X-ray fluorescence

References, Resources, and Further Readings

Textbooks and Manuals

[1] Elzagheid, M. *Chemical Technicians: Good Laboratory Practice and Laboratory Information Management Systems*, Walter de Gruyter GmbH & Co KG, 1st edition, 2023. ISBN: 9783111191621.

[2] Elzagheid, M. *Chemical Laboratory: Safety and Techniques*, Walter de Gruyter GmbH & Co KG, 1st edition, 2022. ISBN: 9783110779110.

[3] Safety in Academic Chemistry Laboratories. *American Chemical Society 1155 Sixteenth Street*, 8th edition, NW Washington, DC 20036, 2017. ISBN: 97808412-37322.

[4] Furr, A. K. *CRC Handbook of Laboratory Safety*, 5th edition, New York: CRC Press, 2000. ISBN: 9780849325236.

[5] Shugar, G. J., Ballinger, J. T. *Chemical Technicians' Ready Reference Handbook*, 4th edition, New York, USA: McGraw-Hill, Inc, 1996. ISBN 0070571860.

[6] Campbell, B. N., Ali, M. M. *Organic Chemistry Experiments, Microscale and Semi-Microscale*, Pacific Grove, California, USA: Brooks/Cole publishing company, 1994. ISBN:978-0534176112.

[7] National Research Council. *Prudent Practices in the Laboratory: Handling and Management of Chemical Hazards*, National Academies Press, 2011. ISBN: 9780309138642.

[8] Hill, R. H., Finster, D. C. *Laboratory Safety for Chemistry Students*, 2nd edition, John Wiley & Sons, 2016, ISBN: 9781119027669.

[9] American Chemical Society Committee on Chemical Safety. *Safety in Academic Chemistry Laboratories*, 8th edition, American Chemical Society, 2017. ISBN: 9780841237322.

[10] Stricoff, R. S., Walters, D. B. *Handbook of Laboratory Health and Safety*, 2nd edition, John Wiley & Sons, 1995, ISBN: 9780471026280.

[11] Crowl, D. A., Louvar, J. F. *Chemical Process Safety: Fundamentals with Applications*, 4th edition, Pearson Education, 2019, ISBN: 9780134857770.

[12] Weinberg, S. *Good Laboratory Practice Regulations*, 4th edition, CRC Press, 2007, ISBN: 978-0849375835.

[13] World Health Organization. *Handbook: Good Laboratory Practice*, 2nd edition, World Health Organization, 2009. ISBN: 9789241547550.

[14] Skoog, D. A., Holler, F. J., Crouch, S. R. *Principles of Instrumental Analysis*, 7th edition, Cengage Learning, 2018, ISBN: 9781305577213.

[15] Harris, D., Lucky, C. *Quantitative Chemical Analysis*, 10th edition, W. H. Freeman, 2019, ISBN: 978-1319164300.

[16] Dean, J. R. *Extraction Techniques in Analytical Sciences*, John Wiley & Sons, 2009. ISBN: 9780470772850.

[17] Kitson, F. G., Larsen, B. S., McEwen, C. N. *Gas Chromatography and Mass Spectrometry: A Practical Guide*, Academic Press, 1996. ISBN: 9780124833852.

[18] Weiss, J., Weis, T. *Handbook of Ion Chromatography*, 3rd edition, Wiley-VCH, 2004, ISEN: 9783527287017.

[19] Miller, J. M. *Chromatography: Concepts and Contrasts*, 2nd edition, Wiley-Interscience, 2009, ISBN: 978-0470530252.

[20] Prichard, E., Barwick, V. *Quality Assurance in Analytical Chemistry*, Wiley-Interscience, 2007. ISBN: 9780470012031.

[21] Occupational Safety and Health Administration (OSHA). *Laboratory Safety Guidance*, U.S. Department of Labor, OSHA Publication 3404-11R, 2011.

[22] World Health Organization (WHO). *Laboratory Biosafety Manual*, 4th Edition, Geneva: WHO Press, 2020. ISBN: 9789240011311.

https://doi.org/10.1515/9783112218105-026

[23] OECD (1998). *OECD Principles on Good Laboratory Practice, OECD Series on Principles of Good Laboratory Practice and Compliance Monitoring, No. 1*, Paris: OECD Publishing, 1998. https://doi.org/10.1787/9789264078536-en.

[24] Rouessac, F., Rouessac, A. *Chemical Analysis: Modern Instrumentation Methods and Techniques*, 2nd Edition, Wiley, 2007, ISBN: 9780470859032.

[25] Elzagheid, M. *Polymers: Chemistry, Morphology, Characterization, Processing, Technology and Recycling*, Walter de Gruyter GmbH & Co KG, 2nd Edition, 2025. ISBN: 9783111585659.

[26] Mendham, J., Denney, R. C., Barnes, J. D., Thomas, M. J. K. *Vogel's Textbook of Quantitative Chemical Analysis*, 6th edition, Prentice Hall, 2000, ISBN: 9780582226289.

[27] U.S. Consumer Product Safety Commission & U.S. Department of Health and Human Services. *School Chemistry Laboratory Safety Guide*, 1st edition, U.S. Government Printing Office, 2006. ISBN: 9780160765871.

[28] Hill, R. H., Finster, D. C. *Laboratory Safety for Chemistry Students*, 1st edition, John Wiley & Sons, 2010, ISBN: 9780470344286.

[29] Pavia, D. L., Lampman, G. M., Kriz, G. S., Engel, R. G. *A Small-Scale Approach to Organic Laboratory Techniques*, 4th edition, Cengage Learning, 2015, ISBN: 9781305253926.

[30] Mohrig, J. R., Alberg, D. G., Hofmeister, G. E., Schatz, P. F. *Techniques in Organic Chemistry*, Freeman, W. H., 4th edition, 2014. ISBN: 9781464134227.

[31] Christian, G. D., O'Reilly, J. E. *Instrumental Analysis, Allyn and Bacon*, 2nd edition, 1986, ISBN: 9780205079054.

[32] Harris, D. C. *Quantitative Chemical Analysis*, Freeman, W. H., 9th edition, 2015. ISBN: 9781464135385.

[33] Shriner, R. L., Hermann, C. K. F., Morrill, T. C., Curtin, D. Y., Fuson, R. C. *The Systematic Identification of Organic Compounds*, Wiley, 8th edition, 2004. ISBN: 9780471215035.

[34] Lide, D. R. (Ed.) *CRC Handbook of Chemistry and Physics*, 85th edition, CRC Press, 2004. ISBN: 9780849304859.

[35] Braithwaite, A., Smith, F. J. *Chromatographic Methods*, 5th edition, Kluwer Academic Publishers, 1999, ISBN: 9780751401582.

[36] Harwood, L. M., Moody, C. J., Percy, J. M. *Experimental Organic Chemistry: Principles and Practice*, 2nd edition, Wiley-Blackwell, 2012, ISBN: 9781119965551.

[37] Skoog, D. A., West, D. M., Holler, F. J., Crouch, S. R. *Fundamentals of Analytical Chemistry*, 10th edition, Cengage Learning (Brooks/Cole), 2021, ISBN: 9780357450390.

[38] Mayo, D. W., Pike, R. M., Forbes, D. C. *Microscale Organic Laboratory: With Multistep and Multiscale Syntheses*, 5th edition, Wiley, 2010, ISBN: 978-0471215028.

Online Databases

[1] National Center for Biotechnology Information. PubChem Database, Available at: https://pubchem.ncbi.nlm.nih.gov.

[2] National Institute of Standards and Technology (NIST). Chemistry WebBook, Available at: https://webbook.nist.gov/chemistry.

[3] U.S. National Library of Medicine. TOXNET (archived), Available via NLM: https://www.nlm.nih.gov.

[4] Sigma-Aldrich. Safety Center and SDS Resources, Available at: https://www.sigmaaldrich.com.

[5] Fisher Scientific Canada. Available at: https://www.fishersci.ca/ca/en/home.html.

Internet Resources

- *Laboratory Chemical Safety and Procedures Manual*, Alberta, Canada: University of Lethbridge.
- Campus Safety – Safety Services, Faculty of Arts & Science. https://www.ualberta.ca/en/engineering/media-library/engg-ehs-lab-safety/general/labchemicalsafetymanual.pdf. Accessed on May 5, 2025.
- University of Alberta, Office of Environmental Health, and Safety. https://www.ualberta.ca/en/human-resources-health-safety-environment/environment-and-safety/index.html. Accessed on May 5, 2025.
- Laboratory Solution Preparation. https://www.flinnsci.ca/api/library/Download/18ce587821c24fb3b0ad7d878bd6a3d9. Accessed on May 5, 2025.
- TOWARDS A MATURE SAFETY CULTURE. https://www.icheme.org/media/10194/xvi-paper-49.pdf?utm_source=chatgpt.com. Accessed on May 5, 2025.
- Laboratory Safety Guidance (OSHA-3404-11R, 2011). Occupational Safety and Health Administration. https://www.osha.gov/sites/default/files/publications/OSHA3404laboratory-safety-guidance.pdf. Accessed on May 7, 2025.
- Laboratory Safety eBook – OSHA Lab Standard & PPE. https://www.osha.gov/sites/default/files/publications/OSHAfactsheet-laboratory-safety-osha-lab-standard.pdf. Accessed on May 7, 2025.
- Laboratory Safety: Chemical Hygiene Plan Fact Sheet (OSHA). https://www.osha.gov/sites/default/files/publications/OSHAfactsheet-laboratory-safety-chemical-hygiene-plan.pdf. Accessed on May 7, 2025.
- Laboratory Safety: Labeling & Transfer of Chemicals QuickFacts. https://www.osha.gov/sites/default/files/publications/OSHAquickfacts-lab-safety-labeling-chemical-transfer.pdf. Accessed on May 7, 2025.
- Lab Safety: Chemical Fume Hoods QuickFacts. https://www.osha.gov/sites/default/files/publications/OSHAquickfacts-lab-safety-chemical-fume-hoods.pdf. Accessed on May 8, 2025.
- Lab Safety: Centrifuges Quick Facts. https://www.osha.gov/sites/default/files/publications/OSHAquickfacts-lab-safety-centrifuges.pdf. Accessed on May 8, 2025.
- Lab Safety: Cryogens & Dry Ice Quick Facts. https://www.osha.gov/sites/default/files/publications/OSHAquickfacts-lab-safety-cryogens-dryice.pdf. Accessed on May 8, 2025.
- School Chemistry Laboratory Safety Guide. CDC/NIOSH. https://www.cdc.gov/niosh/docs/2007-107/pdfs/2007-107.pdf. Accessed on May 8, 2025.
- Laboratory Biosafety Manual, 4th Edition. WHO. https://www.who.int/publications/i/item/9789240011311. Accessed on May 11, 2025.
- Emergency Preparedness Resource Guide for Laboratories. CDC. https://reach.cdc.gov/sites/default/files/job-aids-resources/Emergency_Preparedness_Resource_Guide_for_Laboratories_508.pdf. Accessed on May 11, 2025.
- FDA GLP 101 – Nonclinical Lab Studies. FDA. https://www.fda.gov/media/165993/download. Accessed on May 12, 2025.
- US FDA – Lab EHS Procedures Manual (Volume III). FDA. https://www.fda.gov/media/74000/download. Accessed on May 12, 2025.
- OSHA's Laboratory Safety Publications Page. https://www.osha.gov/publications/bytopic/laboratory-safety. Accessed on May 12, 2025.
- OSHA Laboratories Standards Overview. https://www.osha.gov/laboratories/standards. Accessed on May 15, 2025.
- Lab Safety: Electrical Hazards QuickFacts. OSHA. https://www.osha.gov/sites/default/files/publications/OSHAquickfacts-lab-safety-electrical-hazards.pdf. Accessed on May 15, 2025.
- HSE – Chemical Safety Data Sheets (COSHH). UK HSE. https://www.hse.gov.uk/coshh/basics/datasheets.htm. Accessed on May 15, 2025.

- HSE – SDS Regulatory Overview. UK HSE. https://www.hse.gov.uk/chemical-classification/labelling-packaging/safety-data-sheets.htm. Accessed on May 15, 2025.
- Control of Substances Hazardous to Health (COSHH). UK HSE. https://www.hse.gov.uk/coshh/. Accessed on May 15, 2025.
- Compilation of SDS (Third Edition) Guidance. UK HSE. https://www.hse.gov.uk/pubns/books/l130.htm. Accessed on May 15, 2025.
- Advancing Safety: Understanding the 5 Levels of Safety Maturity (Part Two). https://www.avetta.com/blog/advancing-safety-understanding-the-5-levels-of-safety-maturity-part-two#:~:text=The%20levels%20of%20safety%20maturity,and%20strategic%20priority%20regarding%20safety. Accessed on May 15, 2025.
- Safety Culture Ladder. https://www.safetycenter.ch/en/certification/systeme-produkte/normen-standards/safety-culture-ladder#:~:text=The%20Safety%20Culture%20Ladder%20consists,based%20on%20the%20assessment%20criteria. Accessed on May 16, 2025.
- Safety culture: The ultimate goal. https://skybrary.aero/sites/default/files/bookshelf/1091.pdf. Accessed on May 16, 2025.
- Decoding the safety culture ladder (Part 1): Five levels of organisational maturity. https://cairnrisk.com/knowledge_bank/decoding-the-safety-culture-ladder-part-1-five-levels-of-organisational-maturity/. Accessed on May 16, 2025.
- Understanding your organization's HSE/Safety culture maturity. https://www.linkedin.com/pulse/understanding-your-organizations-hsesafety-culture-maturity-h-l-xihuf/. Accessed on May 17, 2025.
- The Keil Centre developed the Safety Culture Maturity® Model (SCM®M) to measure and facilitate discussion about safety culture. This enables a business to identify specific actions to improve its safety culture. https://keilcentre.co.uk/services/human-factors-ergonomics/safety-culture/scmm/. Accessed on May 17, 2025.
- What Are the 9 Principles of Prevention in Health & Safety? https://www.tsw.co.uk/blog/health-and-safety/principles-of-prevention/. Accessed on May 17, 2025.
- General Laboratory Safety Rules. https://ehs.okstate.edu/laboratory-safety/lab_safety_rules.html. Accessed on May 18, 2025.
- Protect Your Workplace Against Chemical Hazards. https://safetyculture.com/topics/chemical-hazards/. Accessed on May 19, 2025.
- Identifying Chemical Hazards https://www.ecoonline.com/blog/identifying-chemical-hazards/. Accessed on May 19,2025.
- Hazard statements. https://www.msds-europe.com/h-statements/. Accessed on May 25, 2025.
- Classification, Labelling & Packaging Regulation (CLP). https://aise.eu/priorities/product-stewardship/chemicals-management/clp/. Accessed on May 25, 2025.
- Chemical Hazard Class Definitions. https://ehs.stanford.edu/reference/chemical-hazard-class-definitions. Accessed on May 25, 2025.
- Extremely and Highly Toxic Chemicals. https://ehs.gatech.edu/extremely-and-highly-toxic-chemicals. Accessed on May 26, 2025.
- What is a Class 1 Flammable Liquid? https://samex-env.com/blog/what-is-a-class-1-flammable-liquid. Accessed on May 26, 2025.
- Flammable Liquids Classes & Categories. https://chemicalstrategies.com/flammable-liquids-classes-and-categories/. Accessed on May 26, 2025.
- Water-Reactive Chemicals. https://www.umt.edu/risk-management/safety-compliance/safety-fact-sheets/water-reactive-chemicals.php Accessed on May 26, 2025.
- Highly Reactive Chemicals. https://www.unr.edu/ehs/policies-manuals/chemical-hygiene-plan/chapter-5. Accessed on May 26, 2025.
- Corrosive Chemicals. https://www.brandeis.edu/environmental-health-safety/safety/labs/corrosives.html. Accessed on May 26, 2025.

- What is a Corrosive Substance? https://blog.storemasta.com.au/what-are-corrosive-substances. Accessed on May 26, 2025.
- Cryogenic Liquids Use. https://umdearborn.edu/environmental-health-and-safety/lab-safety/chemical-safety/cryogenic-liquids-use. Accessed on May 26, 2025.
- Cryogenic liquids: what they are, how they are obtained and what they are used for. https://cryospain.com/cryogenic-liquids-what-they-are. Accessed on May 26, 2025.
- Explosives. https://ehs.cornell.edu/research-safety/chemical-safety/laboratory-safety-manual/chapter-8-chemical-hazards/81-explosives. Accessed on May 26, 2025.
- Chemical Storage. https://ehs.wisc.edu/labs-research/chemical-safety/chemical-safety-guide/chemical-storage/. Accessed on May 27, 2025.
- https://www.safeworkaustralia.gov.au/safety-topic/hazards/chemicals/storing-hazardous-chemicals. Accessed on May 28, 2025.
- Chemical Inventory Management. https://ehs.princeton.edu/laboratory-research/chemical-safety/chemical_inventory_management. Accessed on May 28, 2025.
- Storing hazardous chemicals. https://www.safeworkaustralia.gov.au/safety-topic/hazards/chemicals/storing-hazardous-chemicals. Accessed on May 28, 2025.
- What is UL and FM and why is it important when manufacturing ductile iron pipe? https://www.mcwaneductile.com/blog/what-is-ul-and-fm-and-why-is-it-important-when-manufacturing-ductile-ironpipe/#:~:text=Preservation%20of%20life%20and%20property,standard%20for%20nearly%20a%20century. Accessed on June 4, 2025.
- Storing hazardous chemicals https://www.safeworkaustralia.gov.au/safety-topic/hazards/chemicals/storing-hazardous-chemicals. Accessed on June 4, 2025.
- Hazardous Chemical Waste. https://www.csusb.edu/ehs/campus-waste-and-disposal/hazardous-chemical-waste. Accessed on June 5, 2025.
- Chemical Fume Hoods. https://lsm.alfaisal.edu/documentations/chemical-fume-hoods/. Accessed on June 9, 2025.
- 5 types of fire extinguishers: A guide to using the right class. https://www.ifsecglobal.com/fire-extinguishers/choose-right-type-fire-extinguisher/. Accessed on June 10, 2025.
- Types of Fire Extinguishers and their uses. https://www.i2comply.com/health-safety/types-of-fire-extinguishers-and-their-uses/. Accessed on June 10, 2025.
- Fire and Explosion Hazard Management Guideline | Formerly IRP 18. https://www.energysafetycanada.com/Resource/Guidelines-Reports/FIRE-AND-EXPLOSION-HAZARD-MANAGEMENT-GUIDELINE. Accessed on June 10, 2025.
- Fire Triangle/Tetrahedron Information. https://fire-risk-assessment-network.com/blog/fire-triangle-tetrahedron/. Accessed on June 12, 2025.
- What Are the 3 Methods for Extinguishing A Fire? https://www.hardhattraining.com/what-are-the-3-methods-for-extinguishing-a-fire/?srsltid=AfmBOooqzAYpHDP0On6Wb2x8WIBhzAIkJMo-vMWTBsEssG6wZ_i4ocTp. Accessed on June 12, 2025.
- 5 types of fire extinguishers: A guide to using the right class. https://www.ifsecglobal.com/fire-extinguishers/choose-right-type-fire-extinguisher/. Accessed on June 12, 2025.
- Explosion Risk assessment. https://www.conformance.co.uk/explosion-risk-assessment#hazardous-substance-classification. Accessed on June 14, 2025.
- Laboratory Glassware Safety: Best Practices, Tips, and Essential Guidelines. https://www.labmanager.com/glassware-safety-in-the-lab-19779. Accessed on June 14, 2025.
- Heat Source Safety. https://www.nsta.org/blog/heat-source-safety. Accessed on June 14, 2025.
- PPE Maintenance. https://ehs.usc.edu/research/lab/manage/ppe2/ppe-maintenance/. Accessed on June 15, 2025.
- Chemicals and Materials: Spill Response – Chemicals. https://www.ccohs.ca/oshanswers/chemicals/spill-response-chemicals.html. Accessed on June 15, 2025.

–   Guide for Chemical Spill Response. https://www.acs.org/about/governance/committees/chemical-safety/publications-resources/guide-for-chemical-spill-response.html. Accessed on June 15, 2025.
–   Evaluating Hazards and Assessing Risks in the Laboratory. https://www.ncbi.nlm.nih.gov/books/NBK55880/. Accessed on June 15, 2025.
–   Types of Hazards and Risks in a Laboratory. https://www.labmanager.com/laboratory-types-of-hazards-and-risks-18238. Accessed on June 15, 2025.
–   Laboratory Emergencies. https://www.safety.fsu.edu/safety_manual/Laboratory%20Emergencies.pdf. Accessed on June 16, 2025.
–   Engineered nanoparticles: Health and safety considerations. https://www.canada.ca/en/employment-social-development/services/health-safety/reports/engineered-nanoparticles.html. Accessed on June 16, 2025.
–   Nanomaterials Type. https://www.sciencedirect.com/topics/materials-science/nanomaterials-type. Accessed on June 16, 2025.
–   Mercury barometer. https://www.britannica.com/technology/mercury-barometer. Accessed on June 18, 2025.
–   A barometer. https://byjus.com/physics/barometer/. Accessed on June 18, 2025.
–   ANEROID BAROMETER. https://maritimeacademia.wixsite.com/my-site-1/post/aneroid-barometer. Accessed on June 17, 2025.
–   pH meters. https://www.dwyeromega.com/en-us/resources/ph-meter?srsltid=AfmBOorX8pEHHWBHZNghzc7SPkJTX_b0m8zeUHdaFmzp_hdvCf6Z6a77. Accessed on June 17, 2025.
–   Lab Equipment Dryer/Oven/Washer Safety Guidelines.https://www.bu.edu/research/forms-policies/lab-equipment-dryerovenwasher-safety-guidelines/. Accessed on June 17, 2025.
–   What is Melting Point Apparatus. https://www.mrclab.com/what-is-melting-point-apparatus. Accessed on June 18, 2025.
–   Saybolt Viscometer. https://instrumentationtools.com/saybolt-viscometer/. Accessed on June 18, 2025.
–   How to Use Laboratory Vacuum Pumps. https://www.munroscientific.co.uk/how-to-use-laboratory-vacuum-pumps. Accessed on June 18, 2025.
–   IKA Rotary Evaporators. https://www.ika.com/en/Products-LabEq/Rotary-Evaporators-pg35/RV-10-digital-V-C-10004802/. Accessed on June 18, 2025.
–   Frit Porosity and Sizes. https://adamschittenden.com/technical/frits/frit-size. Accessed on June 23, 2025.
–   Recrystallization. https://people.chem.umass.edu/samal/267/owl/owlrecryst.pdf. Accessed on June 23, 2025.
–   Column Chromatography. https://www.sciencedirect.com/topics/agricultural-and-biological-sciences/column-chromatography. Accessed on June 26, 2025.
–   Ion Exchange Chromatography – Principle, Protocol, Applications, Examples. https://biologynotesonline.com/ion-exchange-chromatography/. Accessed on June 26, 2025.
–   HOW TO DETERMINE SPECIFIC GRAVITY OF SOIL? https://civilblog.org/2015/02/14/how-to-determine-specific-gravity-of-soil/. Accessed on June 28, 2025.
–   A Guide to Chemicals & Chemical Grades. https://westlabblog.wordpress.com/2018/03/26/a-guide-to-chemicals-and-chemical-grades/. Accessed on July 10, 2025.
–   Biological Risk Assessment Process. https://www.cdc.gov/safe-labs/php/biological-risk-assessment/process.html. Accessed on July 15, 2025.
–   Biosafety in Microbiological and Biomedical Laboratories. https://www.cdc.gov/labs/pdf/SF__19_308133-A_BMBL6_00-BOOK-WEB-final-3.pdf#page=557. Accessed on July 17, 2025.
–   Scintillation Counters: What It Is and What It's Used For. https://www.moravek.com/scintillation-counters-what-it-is-and-what-its-used-for/. Accessed on July 17, 2025.

- Dosimeters. https://dosimetry.web.cern.ch/dosimeters#:~:text=Dosimeters%20are%20devices% 20used%20to,energy%20deposited%20by%20ionising%20radiation. Accessed on July 19, 2025.
- Basic Radiation Safety for Lab Personnel. https://umanitoba.ca/environmental-health-and-safety/ sites/environmental-health-and-safety/files/2022-10/basic-lab-radiation-safety.pdf. Accessed on July 19, 2025.
- Environment, Health and Safety: Radiation Hazards. https://ehs.cornell.edu/research-safety/chemi cal-safety/laboratory-safety-manual/chapter-14-radiation-hazards. Accessed on July 22, 2025.
- Laboratory Emergency Preparedness. https://www.chemistry.utoronto.ca/chemistry-standard- operating-procedures-sops/laboratory-emergency-preparedness. Accessed on July 22, 2025.
- What are the key components of an effective Emergency Response Plan? https://sbnsoftware.com/ blog/what-are-the-key-components-of-an-effective-emergency-response-plan/. Accessed on July 25, 2025.
- Engineered nanoparticles: Health and safety considerations. https://www.canada.ca/en/employ ment-social-development/services/health-safety/reports/engineered-nanoparticles.html. Accessed on July 25, 2025.
- Different types of nanomaterials. https://wiki.anton-paar.com/ca-en/different-types-of- nanomaterials/. Accessed on July 28, 2025.
- A Guide to Good Laboratory Practice (GLP). https://safetyculture.com/topics/good-laboratory- practice-glp/. Accessed on July 28, 2025.
- VACUU·LAN® Lab Vacuum Systems. https://www.brandtech.com/products/laboratory-vacuum/va cuubrand-vacuulan-lab-vacuum-systems?gad_source=1&gad_campaignid=22712372527&gbraid= 0AAAAADNRdAGTOCLN4j8j-1XNKZTevoM51&gclid=Cj0KCQjwss3DBhC3ARIsALdgYxOg3pjv9kQXH m1eBJOo_CBw7ioNsd6m-1ga0f-eXYoOm6Ut8C2w2T8aAq6yEALw_wcB. Accessed on July 28, 2025.
- Pressure and Vacuum Systems. https://ehs.princeton.edu/laboratory-research/laboratory-safety/lab oratory-equipment-and-engineering/pressure-and-vacuum-systems. Accessed on July 28, 2025.
- Safety Guidelines for Working with Pressure and Vacuum Systems. https://www.labmanager.com/ pressure-and-vacuum-systems-20228. Accessed on July 28, 2025.
- Heating and Cooling Methods. https://chem.libretexts.org/Bookshelves/Organic_Chemistry/Or ganic_Chemistry_Lab_Techniques_(Nichols)/01%3A_General_Techniques/1.04% 3A_Heating_and_Cooling_Methods. Accessed on July 30, 2025.
- Temperature Control: Resources. https://www.chem.rochester.edu/notvoodoo/pages/tips.php? page=heating_and_cooling. Accessed on July 30, 2025.
- Common Laboratory Heating Techniques and Equipment. https://www.drawellanalytical.com/com mon-laboratory-heatingtechniques-and-equipment/. Accessed on August 1, 2025.
- Reactions in Sealed Pressure Vessels. https://www.auckland.ac.nz/assets/health-safety-wellbeing/ health-safety-topics/safety-in-labs/SMOU_reactions_sealed_pressure_vessels_v1.0.pdf. Accessed on August 1, 2025.
- Laboratory Analytical Techniques. https://www.intertek.com/analytical-laboratories/technologies/. Accessed on August 2, 2025.
- Gravimetric analysis. https://www.ebsco.com/research-starters/chemistry/gravimetric- analysis#drying/ignition. Accessed on August 2, 2025.
- Laboratory Methods of Sample Preparation. https://www.911metallurgist.com/blog/laboratory- methods-of-sample-preparation/Accessed on August 2, 2025.
- Sample Preparation. https://www.ndsu.edu/agriculture/sites/default/files/2024-09/3.%20SAMPLES% 20PREPARATION.pdf. Accessed on August 3, 2025.
- Volumetric Analysis. https://www.sciencedirect.com/topics/chemistry/volumetric-analysis. Accessed on August 3, 2025.
- INTRODUCTION TO VOLUMETRIC ANALYSIS. https://faculty.ksu.edu.sa/sites/default/files/unit_6_-_in troduction_to_volumetric_analysis_-_subjects_1_2.pdf. Accessed on August 3, 2025.

- Laboratory Techniques for Separation of Mixtures. https://pressbooks.bccampus.ca/chem1114langar acollege/chapter/1-3-laboratory-techniques-for-separation-of-mixtures/. Accessed on August 4, 2025.
- Classifying Separation Techniques. https://chem.libretexts.org/Bookshelves/Analytical_Chemistry/ Analytical_Chemistry_2.1_(Harvey)/07%3A_Obtaining_and_Preparing_Samples_for_Analysis/7.06% 3A_Classifying_Separation_Techniques.Accessed on August 4, 2025.
- Evaluation of basic physical properties. https://www.sciencedirect.com/science/article/abs/pii/ S0166526X06470033. Accessed on August 5, 2025.
- How to Clean Laboratory Glassware. https://www.labmanager.com/how-to-clean-laboratory-glassware-20325. Accessed on August 5, 2025.
- Volumetric Apparatus – Use & Calibration. https://lab-training.com/volumetric-apparatus-use-calibration/. Accessed on August 5, 2025.
- The Most Common Grades of Reagents and Chemicals. https://www.labmanager.com/master-everyday-lab-tasks-for-greater-efficiency-and-accuracy-32976. Accessed on August 5, 2025.
- Automation in the Life Science Research Laboratory. https://pmc.ncbi.nlm.nih.gov/articles/ PMC7691657/. Accessed on August 5, 2025.
- Types of Laboratory Automation Systems. https://highresbio.com/blog/automation-infrastructure/ laboratory-automation-types. Accessed on August 5, 2025.
- The digital lab manager: Automating research support. https://www.sciencedirect.com/science/arti cle/pii/S2472630324000177. Accessed on August 5, 2025.

Appendices

Appendix A: Common Laboratory Glassware and Equipment and Their Functions

1. General Glassware

Item	Function
Beaker	Holding, mixing, and heating liquids; approximate volume measurement
Erlenmeyer flask	Mixing and swirling liquids with a reduced risk of spillage
Volumetric flask	Preparing precise standard solutions
Graduated cylinder	Measuring liquid volumes more accurately than beakers
Test tube	Holding small amounts of substances for reactions or heating
Boiling tube	Larger than a test tube; used for vigorous boiling
Watch glass	Evaporation of liquids, weighing solids, and covering beakers
Reagent bottle	Storage of chemicals and solutions
Glass dropper/Pasteur pipette	Transferring small quantities of liquid dropwise
Glass stirring rod	Manually stirring solutions
Round-bottomed flask	Heating or refluxing reactions; used in distillation setups
Flat-bottomed flask	General purpose heating or mixing vessel
Separatory funnel	Liquid–liquid extraction
Burette	Precise delivery of titrants in volumetric analysis
Volumetric pipette	Measuring and delivering fixed volumes of liquid precisely
Graduated pipette	Measuring variable volumes with moderate accuracy
Condenser	Cooling vapor into liquid during distillation or reflux
Distillation flask	Separating mixtures based on boiling point differences
Filtering flask (Büchner flask)	Used with vacuum filtration setups

2. Heating and Cooling Equipment

Item	Function
Bunsen burner	Produces flame for heating and combustion
Hot plate	Electrically heats beakers and flasks
Heating mantle	Safe and even heating of round-bottom flasks
Crucible	Used for heating solids to high temperatures
Crucible tongs	Handling hot crucibles
Water bath	Gentle and uniform heating using water
Sand bath	Uniform heating of reaction vessels
Ice bath	Cooling reactions to slow or control kinetics
Refrigerated circulator	Provides controlled cooling for sensitive reactions

https://doi.org/10.1515/9783112218105-027

3. Measuring and Dispensing Equipment

Item	Function
Analytical balance	High-precision mass measurements
Top-loading balance	General-purpose mass measurements
Digital thermometer	Temperature monitoring
Barometer	Measures atmospheric pressure
Manometer	Measures gas pressure in reactions
pH meter	Measures pH of solutions electronically
Conductivity meter	Measures electrical conductivity of solutions
Spectrophotometer	Measures absorbance/transmittance of light by a sample
Colorimeter	Measures concentration using light absorption
Pipette filler (bulb or pump)	Aids in drawing liquids into pipettes safely
Micropipette	Precise dispensing of microliter volumes
Automatic dispenser	Dispensing fixed volumes repeatedly

4. Filtration and Separation Tools

Item	Function
Filter paper	Removes solids from liquids by filtration
Funnel	Aids in transferring liquids or supporting filter paper
Büchner funnel	Used with vacuum filtration setups
Hirsch funnel	For small-scale vacuum filtration
Centrifuge tube	Holds samples for centrifugation
Centrifuge	Separates mixtures by density via rapid spinning
Magnetic stirrer	Stirs solutions using a rotating magnetic field
Stir bar (magnetic)	Spins inside the solution with a magnetic stirrer
Desiccator	Removes moisture from samples or stores dry reagents
Drying oven	Removes moisture from samples and glassware

5. Safety Equipment

Item	Function
Fume hood	Ventilates toxic or flammable fumes during experiments
Safety goggles	Eye protection from splashes and debris
Face shield	Additional protection for the face and eyes
Lab coat	Protects clothing and skin from chemicals
Gloves (nitrile, latex, heat-resistant)	Hand protection against chemicals and heat
Fire extinguisher	For controlling small fires in the lab
Emergency shower	Washes chemicals off the body in case of spills
Eye wash station	Rinses chemicals out of eyes immediately
Spill kit	Contains materials for cleaning chemical spills
First-aid kit	Basic treatment for lab injuries

6. Organic Chemistry-Specific Equipment

Item	Function
Reflux condenser	Maintains constant boiling and condensation without the loss of solvent
Dean-Stark apparatus	Removes water from reaction mixtures (especially in esterification)
Drying tube	Prevents moisture from entering apparatus
Rotary evaporator	Removes solvents gently by evaporation under reduced pressure
Gas syringe	Measures gas volumes evolved in reactions
Gas washing bottle	Purifies gases by bubbling them through liquids
Oil bath	Provides controlled and even heating for high-temperature reactions

7. Specialized Glassware and Accessories

Item	Function
Claisen adapter	Allows the use of multiple connections in a distillation setup
Three-way adapter	Directs the flow to different parts of an apparatus
Thermometer adapter	Holds a thermometer in a jointed apparatus
Glass stopcock	Controls the flow of liquids or gases in glass setups
Inert gas line (Schlenk line)	For air-sensitive synthesis using nitrogen or argon
Vacuum desiccator	Removes moisture under vacuum conditions

Appendix B: Chemical Incompatibility List

Chemical 1	Chemical 2	Hazard/reaction
Acids (e.g., HCl and H_2SO_4)	Bases (e.g., NaOH and KOH)	Violent neutralization, heat generation
Acids	Cyanides	Toxic hydrogen cyanide (HCN) gas release
Acids	Sulfides	Toxic hydrogen sulfide (H_2S) gas formation
Acids	Sodium hypochlorite (bleach)	Toxic chlorine gas release
Acids	Sodium nitrite	Nitrogen oxides and toxic gas
Acetic acid	Chromic acid and nitric acid	Explosive reaction
Acetone	Concentrated nitric + sulfuric acids	Highly explosive mixture
Ammonia (NH_3)	Halogens (Cl_2, Br_2, and I_2)	Toxic halogenated amines
Ammonia	Silver, mercury compounds	Forms explosive fulminates
Ammonium nitrate	Acids, reducing agents, and organic matter	Explosion risk

(continued)

Chemical 1	Chemical 2	Hazard/reaction
Aniline	Nitric acid	Exothermic nitration and explosive potential
Benzoyl peroxide	Acids, bases, and metal salts	Violent decomposition
Bromine	Acetylene, ammonia, and butadiene	Explosive compounds
Calcium carbide	Water	Releases flammable acetylene gas
Chlorates	Sulfur, phosphorus, and organics	Explosive mixtures
Chlorine	Turpentine and ammonia	Fire or toxic gas
Chromium trioxide	Acetone, alcohol, and organics	Highly oxidizing and explosion potential
Copper (powder)	Acetylene	Explosive copper acetylide
Cyanides	Acids	HCN gas release (lethal)
Fluorine	Hydrogen, hydrocarbons	Explosive and highly toxic reactions
Hydrocarbons (e.g., hexane)	Concentrated oxidizing acids	Explosion and combustion
Hydrogen peroxide ($\geq 30\%$)	Copper, iron, and manganese salts	Catalytic decomposition and oxygen release
Hydrogen peroxide	Acetone, alcohols, and organics	Violent decomposition
Iodine	Acetylene and ammonia	Explosion and toxic vapors
Lithium, sodium, and potassium	Water and moist air	Violent reaction and hydrogen gas release
Mercuric oxide	Sulfur	Explosive mixture
Nitrates (e.g., KNO_3)	Organic matter and sulfur	Fire or explosion
Nitric acid	Alcohols, acetone, and organics	Highly exothermic or explosive reactions
Nitric acid	Urea	Rapid reaction and toxic gas release
Nitromethane	Amine bases	Explosive salts
Oxalic acid	Silver and mercury	Formation of explosive salts
Oxygen (compressed gas)	Oils, greases, and organics	Fire or explosion
Perchloric acid	Organic material and metals	Explosive perchlorates may form
Permanganates	Alcohols, glycerol, and organics	Exothermic or explosive oxidations
Phosphorus (white)	Air and oxidizers	Spontaneously flammable in air

(continued)

Chemical 1	Chemical 2	Hazard/reaction
Picric acid (dry)	Metals (lead and copper) and bases	Formation of highly sensitive metal picrates (explosives)
Potassium permanganate	Glycerol, ethanol, and organics	Delayed exothermic reaction leading to ignition
Silver nitrate	Ammonia and organics	Forms explosive silver compounds
Sodium	Water and alcohols	Violent reaction and hydrogen gas release
Sodium azide	Copper, lead, and heavy metals	Shock-sensitive and toxic compounds
Sodium hypochlorite	Ammonia and acids	Chlorine or chloramine gas release
Sulfuric acid	Water (added in reverse)	Splattering and violent heat release
Sulfuric acid	Potassium permanganate	Fire and explosion
Zinc	Acid + cyanide	Enhances HCN release

Appendix C: Ordering Reagents and Chemical Supply Sources

This appendix provides guidance on where and how to order laboratory chemicals responsibly and legally from reputable chemical suppliers.

Supplier name	Website	Region
Sigma-Aldrich (Merck)	www.sigmaaldrich.com	Global
Fisher Scientific	www.fishersci.com	Global
Alfa Aesar (Thermo Fisher Scientific)	www.alfa.com https://www.thermofisher.com/ca/en/home.html	Global
VWR International	www.vwr.com	Global
Loba Chemie	www.lobachemie.com	Asia and Europe
TCI Chemicals	www.tcichemicals.com	Global
Acros Organics	www.fishersci.com/shop/brands/AC	Global
BDH Chemicals (Avantor)	www.avantorsciences.com	Global
CDH Fine Chemicals	www.cdhfinechemical.com	Asia and Middle East

Appendix D: Common Laboratory Reagents and Their Uses

Reagent	Formula	Common use	Hazard
Hydrochloric acid	HCl	pH adjustment and titrations	Corrosive and fumes
Sulfuric acid	H_2SO_4	Dehydration and acid catalyst	Highly corrosive
Nitric acid	HNO_3	Oxidizer and nitration reactions	Corrosive and toxic fumes
Sodium hydroxide	$NaOH$	Base, saponification, and neutralization	Corrosive
Ammonium hydroxide	NH_4OH	Precipitation and buffer component	Corrosive and toxic vapors
Acetic acid (glacial)	CH_3COOH	Organic synthesis and pH control	Corrosive and flammable
Ethanol	C_2H_5OH	Solvent and disinfectant	Flammable
Acetone	$(CH_3)_2CO$	Solvent and cleaning glassware	Highly flammable
Sodium bicarbonate	$NaHCO_3$	Neutralizing acids and buffer component	Generally safe
Sodium thiosulfate	$Na_2S_2O_3$	Dechlorination and titration agent	Eye/skin irritant
Potassium permanganate	$KMnO_4$	Oxidizing agent and redox titrations	Strong oxidizer
Silver nitrate	$AgNO_3$	Precipitation of halides, microbial control	Light-sensitive and toxic
Phenolphthalein	$C_{20}H_{14}O_4$	Acid–base indicator	Irritant
Methylene blue	$C_{16}H_{18}ClN_3S$	Biological stain	Staining agent
Ferric chloride	$FeCl_3$	Etching metals and coagulants in water treatment	Corrosive

Appendix E: ANSI-Standardized SDS Format

Section	Title	Description
1	Chemical Product and Company Identification	Product name, manufacturer/supplier name, address, and emergency contact.
2	Composition/Information on Ingredients	Chemical identity, CAS numbers, impurities, and concentration ranges.
3	Hazards Identification	Emergency overview, potential health effects, signs, and symptoms of exposure.

(continued)

Section	Title	Description
4	First-Aid Measures	Actions to take in case of exposure (inhalation, ingestion, skin, and eye contact).
5	Fire-Fighting Measures	Suitable extinguishing media, fire hazards, and special protective equipment.
6	Accidental Release Measures	Cleanup procedures, spill containment, and environmental precautions.
7	Handling and Storage	Safe handling and storage guidelines, including incompatibilities
8	Exposure Controls/Personal Protection	OSHA/ACGIH limits, PPE, and ventilation recommendations.
9	Physical and Chemical Properties	Appearance, odor, pH, melting point, boiling point, solubility, etc.
10	Stability and Reactivity	Chemical stability, hazardous reactions, and conditions to avoid.
11	Toxicological Information	Routes of exposure, acute/chronic effects, and LD_{50}/LC_{50} data.
12	Ecological Information	Environmental impact, persistence, and bioaccumulation potential.
13	Disposal Considerations	Waste treatment methods and proper disposal guidelines.
14	Transport Information	DOT/IATA/IMDG classification, UN number, packing group, and hazards.
15	Regulatory Information	Safety, health, and environmental regulations (e.g., OSHA, TSCA, WHMIS).
16	Other Information	Date of preparation or revision, references, and disclaimer.

Appendix F: (Good Laboratory Practices)

General Laboratory Checklist

Item	Description	Yes/no	Comments
1	Lab personnel wear appropriate PPE (gloves, lab coats, and goggles)	□ / □	
2	Emergency exits and safety showers are accessible	□ / □	
3	First aid kit and spill kit are available and stocked	□ / □	
4	Food and drinks are prohibited in lab areas	□ / □	

(continued)

Item	Description	Yes/no	Comments
5	Lab is clean, orderly, and free of obstructions	□ / □	
6	All chemicals are labeled and stored appropriately	□ / □	
7	SOPs are up-to-date and accessible	□ / □	
8	Hazardous waste is stored in proper containers	□ / □	
9	Equipment is calibrated and in good working condition	□ / □	
10	Records are legible, complete, and properly dated	□ / □	

GLP Personnel Training Checklist

Item	Description	Completed	Comments
1	Laboratory safety orientation completed	□ / □	
2	Proper PPE use demonstrated	□ / □	
3	Trained on all SOPs relevant to duties	□ / □	
4	Spill and emergency procedures understood	□ / □	
5	Waste handling procedures explained	□ / □	
6	Equipment use and maintenance training done	□ / □	

Sample GLP Audit Form

Audit Title: Laboratory GLP Compliance Audit

Date: _____

Auditor Name: _____

Department: _____

Location: _____

Section A: Personnel and Training

Audit item	Compliance (Y/N)	Notes
Staff are trained and training records are current	□ / □	
Job descriptions and responsibilities are defined	□ / □	
Unauthorized personnel are restricted from lab access	□ / □	

Section B: Laboratory Operations

Audit item	Compliance (Y/N)	Notes
SOPs are followed for all routine operations	□ / □	
Deviations from SOPs are documented and justified	□ / □	
Experiments are properly documented (dates, initials, and results)	□ / □	

Section C: Equipment and Calibration

Audit item	Compliance (Y/N)	Notes
Equipment is labeled and uniquely identified	□ / □	
Calibration records are available and current	□ / □	
Maintenance is documented and performed regularly	□ / □	

Section D: Sample and Reagent Handling

Audit item	Compliance (Y/N)	Notes
Samples are properly labeled and traceable	□ / □	
Reagents are within expiry and correctly stored	□ / □	
Chain of custody is maintained for samples	□ / □	

Section E: Waste and Safety Management

Audit item	Compliance (Y/N)	Notes
Hazardous waste is properly segregated and labeled	□ / □	
Fire extinguishers, eyewash stations, and first aid are accessible	□ / □	
Safety data sheets (SDSs) are available and up to date	□ / □	

Final Comments:

Auditor Signature: _____

Date: _____

Appendix G: (General Questions and Answers)

Short Questions
1. What is the primary purpose of chemical safety in laboratories?
Answer: To minimize the risks associated with handling hazardous chemicals.

2. Name three routes through which chemicals can enter the human body.
Answer: Inhalation, ingestion, and skin contact.

3. What does PPE stand for in a laboratory context?
Answer: Personal protective equipment.

4. Why is proper labeling of chemical containers important?
Answer: To help identify chemicals and provide safe handling instructions.

5. What should you do immediately if a chemical splashes into your eye?
Answer: Rinse the eye for at least 15 min and seek medical attention.

6. Define a corrosive substance.
Answer: A chemical that can cause the destruction of living tissue or severe corrosion of materials upon contact.

7. What is the purpose of a fume hood in a laboratory?
Answer: To provide ventilation and protect users from harmful fumes.

8. Why should acids and bases be stored separately?
Answer: To prevent dangerous reactions that can occur if they come into contact.

9. What is the significance of the skull and crossbones symbol on a chemical label?
Answer: It indicates acute toxicity.

10. What is the first step in responding to a chemical spill?
Answer: Contain the spill using appropriate materials, and follow safety guidelines.

11. What are pyrophoric chemicals?
Answer: Chemicals that can ignite spontaneously upon exposure to air.

12. Name two examples of water-reactive chemicals.
Answer: Sodium metal and calcium carbide.

13. What does the term "oxidizer" mean in chemical safety?
Answer: A chemical that can cause or enhance the combustion of other materials.

14. What is the recommended storage method for flammable liquids?
Answer: In a flammable liquids cabinet or an approved container.

15. Why should the chemical inventory be kept up to date?
Answer: To ensure proper storage, compliance, and emergency preparedness.

16. What is the purpose of an SDS?
Answer: It provides information on chemical properties, hazards, and safe handling procedures.

17. How should compressed gas cylinders be stored?
Answer: Securely upright, with caps in place, and away from ignition sources.

18. Define an explosive chemical.
Answer: A substance that can undergo rapid chemical changes, producing heat and gas.

19. What is good laboratory practice (GLP)?
Answer: A set of principles to ensure safe, reliable, and reproducible results.

20. What is the purpose of separation techniques in chemical analysis?
Answer: To isolate and purify compounds for further study.

21. Why is it important to segregate incompatible chemicals?
Answer: To prevent accidental reactions that can be hazardous.

22. What is the difference between acute and chronic exposure?
Answer: Acute is short-term, high-level exposure and chronic is long-term, lower-level exposure.

23. Name a type of local exhaust ventilation.
Answer: Fume hood.

24. What is the hazard of storing oxidizers near flammable materials?
Answer: They can react and cause fires or explosions.

25. Why should concentrated acids be handled with caution?
Answer: They can cause severe burns and dangerous reactions.

26. Name two common flammable solvents.
Answer: Acetone and ethanol.

27. What is the main hazard of hydrofluoric acid?
Answer: It is highly corrosive and can penetrate the skin, causing severe damage.

28. When handling toxic chemicals, what is essential?
Answer: Use of PPE and fume hoods.

29. What is the importance of emergency showers?
Answer: They allow immediate decontamination of large chemical spills on the body.

30. What is the recommended way to neutralize an acid spill?
Answer: Slowly add a suitable base, while stirring and monitoring.

31. Why is proper labeling of waste containers important?
Answer: To prevent the mixing of incompatible wastes and to ensure safe disposal.

32. What does LD_{50} stand for?
Answer: The lethal dose that kills 50% of a test population.

33. How should empty chemical containers be treated?
Answer: They should be triple-rinsed and disposed of according to regulations.

34. Why should gloves be worn when handling chemicals?
Answer: To prevent skin contact with hazardous substances.

35. What is the purpose of a safety shower?
Answer: To rapidly wash chemicals off the body in case of a spill.

36. What is a mutagen?
Answer: A chemical that can cause genetic mutations.

37. Why is proper housekeeping important in the lab?
Answer: To reduce hazards and maintain a safe work environment.

38. What is the recommended action in case of a chemical fire?
Answer: Activate the fire alarm, evacuate, and call for help.

39. Name one reason to use plastic over glass containers for some chemicals.
Answer: Plastic may resist breakage and corrosion better than glass.

40. What should be done before starting work with a new chemical?
Answer: Review the SDS and understand its hazards.

41. Why are water-reactive chemicals dangerous?
Answer: They can react violently, releasing heat and flammable gases.

42. What does a corrosive pictogram look like?
Answer: A test tube pouring liquid on a hand and a metal surface.

43. Why is it important to know a chemical's reactivity?
Answer: To prevent dangerous combinations or reactions.

44. How should chemicals be transported within a facility?
Answer: In appropriate containers, using carts with containment features.

45. What is the best practice for storing chemicals?
Answer: Group by hazard class and separate incompatibles.

46. Why should food and drink be prohibited in the lab?
Answer: To prevent the accidental ingestion of chemicals.

47. Name one common laboratory separation technique.
Answer: Filtration.

48. What is the difference between a hazard and a risk?
Answer: A hazard is a potential source of harm and risk is the likelihood of harm occurring.

49. What is GLP?
Answers: GLP stands for good laboratory practice, a system of management controls for laboratories to ensure the quality, integrity, and reliability of non-clinical research data.

50. Why is GLP important in laboratories?
Answer: It ensures consistency, traceability, and accuracy in data generation and reporting.

51. Who is responsible for implementing GLP?
Answer: All laboratory personnel, including supervisors, analysts, and quality assurance staff.

52. Name one document required for GLP compliance.
Answer: Standard operating procedures (SOPs).

53. What is the purpose of maintaining laboratory notebooks?
Answer: To record experimental procedures, results, and observations in a traceable and auditable manner.

54. Define calibration in the context of GLP.
Answer: Calibration is the process of adjusting and verifying instruments to ensure accurate measurements.

55. What is the role of quality assurance (QA) in GLP?
Answer: To independently monitor studies and ensure compliance with GLP principles.

56. Mention one common GLP violation.
Answer: Failure to maintain proper documentation.

57. How often should balances be calibrated?
Answer: Regularly, as per SOP or the manufacturer's guidelines.

58. Why is reagent labeling critical in GLP?
Answer: To avoid misuse and ensure traceability.

59. What information should a reagent label contain?
Answer: Name, concentration, preparation date, expiry date, and initials of the preparer.

60. What is a chain of custody?
Answer: A documented history of sample handling to ensure integrity and traceability.

61. What is spectroscopy?
Answer: A technique that measures the interaction of light with matter.

62. Name two types of spectroscopy.
Answer: UV-Vis spectroscopy and infrared (IR) spectroscopy.

63. What does IR spectroscopy detect?
Answer: Functional groups based on molecular vibrations.

64. What is the principle of UV-Vis spectroscopy?
Answer: It is based on the absorption of ultraviolet or visible light by electrons in molecules.

65. What is chromatography?
Answer: A technique used to separate components of a mixture based on their movement through a stationary phase.

66. Name one type of detector used in gas chromatography.
Answer: Flame ionization detector (FID).

67. What is the main principle of NMR spectroscopy?
Answer: It relies on the magnetic properties of atomic nuclei.

68. What is the retention time in chromatography?
Answer: The time a compound takes to travel through the column to the detector.

69. What is a calibration curve?
Answer: A graph used to determine the concentration of an analyte based on the instrument response.

70. What is the function of a blank sample?
Answer: To correct for background signal or interference.

71. What is titrimetry?
Answer: A quantitative analytical method based on measuring the volume of a reagent needed to react with an analyte.

72. Give one advantage of atomic absorption spectroscopy (AAS).
Answer: High sensitivity for detecting metal ions.

73. What is filtration?
Answer: A method to separate solids from liquids using a porous barrier.

74. Define centrifugation.
Answer: A process that uses centrifugal force to separate particles based on their density.

75. What is distillation used for?
Answer: To separate liquids based on differences in boiling points.

76. What is the principle of chromatography as a separation technique?
Answer: Separation is based on differential adsorption between stationary and mobile phases.

77. What is the solvent extraction?
Answer: The separation of compounds is based on their solubility in two immiscible liquids.

78. Give an example of a separation technique used in biological samples.
Answer: Gel electrophoresis.

79. What is crystallization?
Answer: A technique used to purify solids by forming crystals from a solution.

80. How does paper chromatography work?
Answer: Based on capillary action and the differential solubility of components.

81. What separation technique is best for volatile compounds?
Answer: Gas chromatography (GC).

82. What is a separating funnel used for?
Answer: To separate immiscible liquid phases.

83. What is a molar solution?
Answer: A solution that contains one mole of solute per liter of solution.

84. How do you prepare a 1 M NaCl solution?
Answer: Dissolve 58.44 g of NaCl in water and dilute it to 1 liter.

85. What is a standard solution?
Answer: A solution of known and precise concentration.

86. What is the difference between a solute and a solvent?
Answer: Solute is the substance dissolved and a solvent is the medium doing the dissolving.

87. Why is volumetric glassware used for solution preparation?
Answer: For high-accuracy measurement of liquid volumes.

88. How is percent solution (w/v) calculated?
Answer: Grams of solute per 100 mL of solution.

89. What is serial dilution?
Answer: A stepwise dilution of a substance to achieve a range of concentrations.

90. Why is pH adjustment important in solution preparation?
Answer: To ensure optimal chemical or biological activity.

91. What tool is commonly used to measure pH?
Answer: A pH meter.

92. What is a buffer solution?
Answer: A solution that resists changes in pH upon the addition of an acid or a base.

93. What is a beaker used for?
Answer: To hold, mix, or heat liquids; not for precise volume measurement.

94. What is a volumetric flask used for?
Answer: To prepare solutions of precise volumes.

95. What is the main use of a burette?
Answer: To deliver known volumes of a solution during titration.

96. How is a pipette different from a burette?
Answer: A pipette delivers a fixed volume, whereas a burette delivers a variable volume.

97. What is a conical flask (Erlenmeyer flask) used for?
Answer: Mixing and heating solutions, especially during titrations.

98. Why is glassware cleaned before use?
Answer: To prevent contamination and ensure accurate results.

99. What is the purpose of a graduated cylinder?
Answer: To measure liquid volume more accurately than with a beaker.

100. How is a watch glass used in the lab?
Answer: To hold small quantities of solids or to cover beakers.

Multiple-Choice Questions
1. What is the primary purpose of chemical safety?
    A. To increase chemical production
    B. To reduce costs
    C. To minimize risks from hazardous chemicals
    D. To promote chemical experimentation
    Answer: C

2. Which of the following is NOT a route of chemical exposure?
    A. Inhalation
    B. Skin contact
    C. Digestion
    D. Injection
    Answer: C

3. What does PPE stand for?
    A. Personal preparation equipment
    B. Personal protective equipment
    C. Public protection engineering
    D. Primary protective engineering
    Answer: B

4. Why is proper labeling of chemical containers important?
    A. To make them look organized
    B. To prevent mixing chemicals
    C. To identify chemicals and provide safe handling instructions
    D. To track purchase dates
    Answer: C

5. What should you do if a chemical splashes into your eyes?
    A. Rub your eyes
    B. Rinse with water for at least 15 min
    C. Use a cloth to wipe it
    D. Ignore it
    Answer: B

6. What is a corrosive substance?
    A. A chemical that catches fire
    B. A chemical that reacts with water
    C. A chemical that can destroy living tissue or materials
    D. A chemical that evaporates quickly
    Answer: C

7. What is the purpose of a fume hood?
    A. To store chemicals
    B. To mix chemicals safely
    C. To provide ventilation and protection from harmful fumes
    D. To heat chemicals
    Answer: C

8. Why should acids and bases be stored separately?
    A. They react violently with each other
    B. They look similar
    C. They evaporate quickly
    D. They are both toxic
    Answer: A

9. Which symbol indicates acute toxicity?
    A. Skull and crossbones
    B. Flame
    C. Exclamation mark
    D. Corrosion
    Answer: A

10. What is the first step in responding to a chemical spill?
    A. Evacuating the lab
    B. Contain the spill safely
    C. Call 911
    D. Add water to the spill
    Answer: B

11. What is the first action to take in the case of a chemical spill on the skin?
    A. Report to supervisor
    B. Wipe it off with a paper towel
    C. Wash with copious water
    D. Apply lotion
    Answer: C

12. What should you never do in a chemistry lab?
    A. Wear goggles
    B. Eat or drink
    C. Use fume hood
    D. Label reagents
    Answer: B

13. What is the most appropriate footwear for a lab?
    A. Sandals
    B. Sneakers
    C. Open-toed shoes
    D. Flip-flops
    Answer: B

14. Which type of hazard symbol indicates a health hazard?
    A. Flame
    B. Skull and crossbones
    C. Exclamation mark
    D. Silhouette of a person with a star on their chest
    Answer: D

15. What is the primary purpose of a fume hood?
    A. Store chemicals
    B. Heat samples
    C. Vent hazardous vapors
    D. Dry glassware
    Answer: C

16. Safety data sheets (SDSs) provide information about:
    A. Experiment results
    B. Proper waste disposal only
    C. Chemical properties, hazards, and handling
    D. Inventory records
    Answer: C

17. Which of the following is a biological hazard?
    A. Mercury
    B. Asbestos
    C. Bacteria
    D. Acetone
    Answer: C

18. Which personal protective equipment (PPE) is essential for handling corrosives?
    A. Safety goggles
    B. Ear plugs
    C. Face shield
    D. A and C
    Answer: D

19. Lab coats are primarily used to protect:
    A. Eyes
    B. Feet
    C. Skin and clothing
    D. Chemicals
    Answer: C

20. How should broken glass be disposed of?
    A. In general trash
    B. Down the sink
    C. In a puncture-resistant container
    D. In a recycling bin
    Answer: C

21. Which GHS symbol represents explosive substances?
    A. Flame over circle
    B. Exploding bomb
    C. Gas cylinder
    D. Corrosion
    Answer: B

22. The exclamation mark symbol indicates:
    A. Serious chronic hazard
    B. Toxic by inhalation
    C. Irritant or skin sensitizer
    D. Flammable gas
    Answer: C

23. Which chemical has both physical and health hazards?
    A. Water
    B. Sodium chloride
    C. Hydrogen peroxide
    D. Ethanol
    Answer: D

24. The "skull and crossbones" GHS symbol means:
    A. Environmental hazard
    B. Gas under pressure
    C. Acutely toxic
    D. Irritant
    Answer: C

25. Which of the following is NOT a GHS pictogram?
    A. Biohazard
    B. Flame
    C. Corrosion
    D. Health hazard
    Answer: A

26. Which hazard class does sulfuric acid fall under?
    A. Explosive
    B. Flammable
    C. Corrosive
    D. Oxidizer
    Answer: C

27. Flammable chemicals should be stored in:
    A. Regular wooden cabinets
    B. Glass shelves
    C. Flammable storage cabinets
    D. Cardboard boxes
    Answer: C

28. Which color code in NFPA diamond indicates health hazard?
    A. Red
    B. Yellow
    C. Blue
    D. White
    Answer: C

29. Oxidizing agents should never be stored with:
    A. Bases
    B. Acids
    C. Flammables
    D. Inert gases
    Answer: C

30. Which of the following is a chronic health hazard?
    A. Sodium bicarbonate
    B. Carbon monoxide
    C. Ethyl alcohol
    D. Benzene
    Answer: D

31. Before using any equipment, you should first:
    A. Plug it in
    B. Ask someone
    C. Read the SOP or manual
    D. Test randomly
    Answer: C

32. Cracked glassware should be:
    A. Repaired
    B. Reused
    C. Discarded
    D. Ignored
    Answer: C

33. Which glassware is most likely to explode under pressure if misused?
    A. Test tube
    B. Volumetric flask
    C. Round-bottomed flask
    D. Sealed Erlenmeyer flask
    Answer: D

34. Which of the following is safest to use for heating liquids?
    A. Volumetric flask
    B. Erlenmeyer flask
    C. Graduated cylinder
    D. Beaker
    Answer: B

35. You should always inspect glassware for:
    A. Weight
    B. Color
    C. Cracks and chips
    D. Stamps
    Answer: C

36. What is the purpose of a boiling chip?
    A. Raise boiling point
    B. Prevent bumping
    C. Clean the liquid
    D. Absorb water
    Answer: B

37. Glassware should be cleaned with:
    A. Concentrated acid
    B. Soap and water
    C. Bleach only
    D. Any solvent
    Answer: B

38. What is the best method to handle hot glassware?
    A. With bare hands
    B. Using tongs or heat-resistant gloves
    C. With cotton pads
    D. Wait until it's cool
    Answer: B

39. If you notice a leaking gas line on equipment:
    A. Use tape to seal it
    B. Inform your instructor immediately
    C. Ignore it
    D. Light a flame to check
    Answer: B

40. The best material for flame-resistant laboratory glassware is:
    A. Soda-lime glass
    B. Borosilicate glass
    C. Plastic
    D. Lead glass
    Answer: B

41. OSHA stands for:
    A. Office for Scientific Hazard Awareness
    B. Occupational Safety and Health Administration
    C. Organization of Safety and Hazard Analysis
    D. Occupational Safety for Health Academics
    Answer: B

42. What is the main purpose of OSHA?
    A. Regulate chemical production
    B. Promote workplace safety and health
    C. Train chemists
    D. Certify equipment
    Answer: B

43. Which document must be accessible for every hazardous chemical?
    A. Risk assessment form
    B. SDS
    C. Lab notebook
    D. Fire inspection log
    Answer: B

44. Chemical hygiene plan (CHP) is required under:
    A. FDA regulations
    B. EPA policies
    C. OSHA Laboratory Standard
    D. GLP guidelines
    Answer: C

45. The Laboratory Standard (29 CFR 1910.1450) applies to:
    A. Factories
    B. Hospitals
    C. Academic research labs
    D. All government offices
    Answer: C

46. GHS stands for:
    A. General hazard symbol
    B. Globally harmonized system
    C. General handling standard
    D. Global health and safety
    Answer: B

47. The GHS aims to:
    A. Limit chemical sales
    B. Restrict international transport
    C. Standardize chemical labeling and classification
    D. Increase costs of hazardous materials
    Answer: C

48. Which section of the SDS describes first-aid measures?
    A. Section 1
    B. Section 4
    C. Section 9
    D. Section 16
    Answer: B

49. Section 2 of SDS includes:
    A. Storage guidelines
    B. Manufacturer contact
    C. Hazard identification
    D. Firefighting methods
    Answer: C

50. Which of the following is NOT a required element of a GHS label?
    A. Pictogram
    B. Signal word
    C. Manufacturer's logo
    D. Hazard statement
    Answer: C

51. Which signal word indicates higher hazard under GHS?
    1. Attention
    2. DANGER
    3. CAUTION
    4. LOW RISK
    Answer: B

52. Under OSHA, employers must:
    A. Issue fines for spills
    B. Provide SDS and training
    C. Reduce research time
    D. Perform experiments
    Answer: B

53. An eyewash station should be accessible within:
    A. 10 s of exposure
    B. 30 min
    C. 5 min
    D. 2 h
    Answer: A

54. Which is NOT a responsibility of lab workers under chemical hygiene?
    A. Wearing PPE
    B. Reading SDS
    C. Modifying OSHA regulations
    D. Following SOPs
    Answer: C

55. Which type of hazard does a compressed gas pose?
    A. Only chemical
    B. Only fire
    C. Both physical and chemical
    D. Only environmental
    Answer: C

56. Which of the following is the primary objective of GLP?
    A. Maximizing profit
    B. Ensuring data integrity
    C. Speeding up experiments
    D. Minimizing paperwork
    Answer: B

57. GLP primarily applies to:
    A. Clinical trials
    B. Manufacturing processes
    C. Nonclinical laboratory studies
    D. Marketing campaigns
    Answer: C

58. Which document provides detailed instructions for laboratory procedures?
    A. Research proposal
    B. SOP
    C. Annual report
    D. Certificate of analysis
    Answer: B

59. Which department ensures GLP compliance?
    A. Accounting
    B. Logistics
    C. Quality assurance
    D. Marketing
    Answer: C

60. The term "traceability" in GLP refers to:
    A. Finding lost chemicals
    B. Tracking the source and history of data
    C. Auditing financial records
    D. Locating staff
    Answer: B

61. Which of the following is NOT a GLP principle?
    A. Proper documentation
    B. Staff training
    C. Product marketing
    D. Equipment calibration
    Answer: C

62. In GLP, raw data must be:
    A. Erased after processing
    B. Digitized only
    C. Retained and archived
    D. Ignored after reporting
    Answer: C

63. Which information is essential on a reagent label?
    A. Formula and color
    B. Brand and volume
    C. Name, date, and concentration
    D. Storage room number
    Answer: C

64. Equipment used in GLP laboratories must be:
    A. Brand new
    B. The most expensive
    C. Calibrated and maintained
    D. Returned after use
    Answer: C

65. Who is ultimately responsible for GLP compliance?
    A. Only the laboratory manager
    B. Only quality assurance
    C. All laboratory personnel
    D. External auditors
    Answer: C

66. Which technique identifies compounds based on light absorption?
    A. Chromatography
    B. Spectroscopy
    C. Centrifugation
    D. Electrophoresis
    Answer: B

67. In IR spectroscopy, what is detected?
    A. Color changes
    B. Molecular vibrations
    C. DNA sequences
    D. Light refraction
    Answer: B

68. Which spectroscopy method uses magnetic fields?
    A. UV-Vis
    B. IR
    C. NMR
    D. AAS
    Answer: C

69. What is the main detector in flame atomic absorption spectroscopy?
    A. Mass analyzer
    B. Photodiode array
    C. Flame photometer
    D. Photomultiplier tube
    Answer: D

70. What does the peak in a chromatogram represent?
    A. Equipment noise
    B. Detector malfunction
    C. Presence of a compound
    D. Air bubble
    Answer: C

71. Titration is based on:
    A. Measuring current
    B. Volume of a reaction solution
    C. Melting point differences
    D. Particle size
    Answer: B

72. The purpose of a blank sample is to:
    A. Waste chemicals
    B. Calibrate the balance
    C. Eliminate background noise
    D. Identify impurities
    Answer: C

73. Gas chromatography is best used for:
    A. Nonvolatile compounds
    B. Heavy metals
    C. Volatile organic compounds
    D. Solid residues
    Answer: C

74. In UV-Vis spectroscopy, absorbance is plotted against:
    A. Temperature
    B. Retention time
    C. Wavelength
    D. Pressure
    Answer: C

75. A calibration curve is used to:
    A. Determine sample temperature
    B. Calibrate the pH meter
    C. Identify solvents
    D. Quantify unknown concentrations
    Answer: D

76. What is the principle behind chromatography?
    A. Boiling point
    B. Differential solubility
    C. Density
    D. pH differences
    Answer: B

77. Which method separates components using centrifugal force?
    A. Distillation
    B. Extraction
    C. Filtration
    D. Centrifugation
    Answer: D

78. Paper chromatography uses what as the stationary phase?
    A. Glass
    B. Paper
    C. Silica gel
    D. Gas
    Answer: B

79. Which of the following is best for separating two immiscible liquids?
    A. Distillation
    B. Chromatography
    C. Separating funnel
    D. Centrifuge
    Answer: C

80. Which separation technique uses boiling points?
    A. Extraction
    B. Filtration
    C. Distillation
    D. Electrophoresis
    Answer: C

81. Gel electrophoresis separates substances based on:
    A. Density
    B. Magnetic properties
    C. pH
    D. Size and charge
    Answer: D

82. The mobile phase in gas chromatography is:
    A. Liquid solvent
    B. Paper
    C. Inert gas
    D. Water
    Answer: C

83. Which of the following is a physical separation method?
    A. Centrifugation
    B. Precipitation
    C. Combustion
    D. Neutralization
    Answer: A

84. Which technique involves crystal formation from a saturated solution?
    A. Chromatography
    B. Crystallization
    C. Filtration
    D. Evaporation
    Answer: B

85. Column chromatography is an example of:
    A. Electrical separation
    B. Chromogenic detection
    C. Solid–liquid separation
    D. Adsorption-based separation
    Answer: D

86. Which unit represents molarity?
    A. g/mL
    B. mol/kg
    C. mol/L
    D. g/L
    Answer: C

87. To prepare 1 M NaCl, how many grams of NaCl are needed for 1 L of solution?
    A. 35.5 g
    B. 58.44 g
    C. 100 g
    D. 22.4 g
    Answer: B

88. Which of the following is a standard solution?
    A. Tap water
    B. Unknown acid
    C. Calibrated NaOH solution
    D. Milk
    Answer: C

89. Which equipment gives the highest volume accuracy?
    A. Beaker
    B. Measuring cylinder
    C. Volumetric flask
    D. Test tube
    Answer: C

90. A buffer solution is used to:
    A. Increase solubility
    B. Maintain constant pH
    C. Promote evaporation
    D. Enhance boiling point
    Answer: B

91. Which tool is used to measure pH accurately?
    A. Refractometer
    B. Colorimeter
    C. pH meter
    D. Voltmeter
    Answer: C

92. Serial dilution is used to:
    A. Increase solution concentration
    B. Decompose solutes
    C. Obtain a range of concentrations
    D. Distill a solution
    Answer: C

93. Which glassware is most suitable for titration?
    A. Beaker
    B. Pipette and burette
    C. Test tube
    D. Crucible
    Answer: B

94. Which glassware is used to accurately measure a fixed volume?
    A. Erlenmeyer flask
    B. Graduated cylinder
    C. Pipette
    D. Watch glass
    Answer: C

95. A beaker is primarily used for:
    A. Precise volume measurement
    B. Storage only
    C. Mixing and heating liquids
    D. Titration endpoint
    Answer: C

96. The volumetric flask is calibrated to:
    A. Any volume
    B. One specific volume
    C. 100 mL intervals
    D. 10 mL increments
    Answer: B

97. Which equipment is ideal for measuring ~25 mL with high precision?
    A. Beaker
    B. Pipette
    C. Erlenmeyer flask
    D. Test tube
    Answer: B

98. A conical flask is preferred during titration because:
    A. It's transparent
    B. It's wide
    C. It allows easy swirling
    D. It has a flat bottom
    Answer: C

99. Before use, laboratory glassware should be:
    A. Colored
    B. Broken
    C. Clean and dry
    D. Wrapped in foil
    Answer: C

100. Which of the following is used for evaporating solutions?
    A. Test tube
    B. Petri dish
    C. Evaporating dish
    D. Beaker
    Answer: C

Index

https://doi.org/10.1515/9783112218105-028